T0205945

Sequence Spaces

Mathematics and its Applications
Modelling, Engineering, and Social Sciences
Series Editor: Hemen Dutta, Department of Mathematics, Gauhati University

Tensor Calculus and Applications
Simplified Tools and Techniques
Bhaben Kalita

Discrete Mathematical Structures
A Succinct Foundation
Beri Venkatachalapathy Senthil Kumar and Hemen Dutta

Methods of Mathematical Modelling
Fractional Differential Equations
Edited by Harendra Singh, Devendra Kumar, and Dumitru Baleanu

Mathematical Methods in Engineering and Applied Sciences
Edited by Hemen Dutta

Sequence Spaces
Topics in Modern Summability Theory
Mohammad Mursaleen and Feyzi Başar

For more information about this series, please visit: www.crcpress.com/Mathematics-and-its-applications/book-series/MES

ISSN (online): 2689-0224
ISSN (print): 2689-0232

Sequence Spaces
Topics in Modern Summability Theory

Mohammad Mursaleen
Department of Mathematics
Aligarh Muslim University
Aligarh, India

Feyzi Başar
Department of Mathematical Education
İnönü University
Central Campus
Malatya, Turkey

CRC Press
Taylor & Francis Group
Boca Raton London New York

CRC Press is an imprint of the
Taylor & Francis Group, an **informa** business

CRC Press
Taylor & Francis Group
6000 Broken Sound Parkway NW, Suite 300
Boca Raton, FL 33487-2742

First issued in paperback 2021

© 2020 by Taylor & Francis Group, LLC
CRC Press is an imprint of Taylor & Francis Group, an Informa business

No claim to original U.S. Government works

ISBN-13: 978-0-367-81917-0 (hbk)
ISBN-13: 978-1-03-217319-1 (pbk)
DOI: 10.1201/9781003015116

Publisher's Note

The publisher has gone to great lengths to ensure the quality of this reprint but points out that some imperfections inthe original copies may be apparent

**Visit the Taylor & Francis Web site at
http://www.taylorandfrancis.com**

**and the CRC Press Web site at
http://www.crcpress.com**

Contents

Foreword

Summability theory is more than a century old. It began with a paper in 1890 by E. Cesàro dealing with multiplication of series. The main aim of summability in its early days was the development of summability methods for divergent series and divergent integrals. The topic then developed its own identity far beyond its beginnings. An important and central theme in summability was the introduction of matrix methods such as Cesàro, Abel, Hölder, Riesz, Hausdorff, Nörlund and others. Summability theory relied initially on classical analysis, and as such it was considered a branch of Classical Analysis. The book by Hardy [97] marks the highlight of that era. The use of functional analysis methods began with the seminal research by Karl Zeller and his colleagues (see [239]) and continued with the fundamental contributions of A. Willansky and others (see [228]). It is gratifying to note that the topic has found its way into introductory textbooks on functional analysis (see [147] and [227]).

Over the past century there have been many landmarks in the theory and applications of summability theory, both in the contexts of classical analysis and functional analysis. For example, Tauberian theory, one of the classical topics in the theory, compares summability methods for series and integrals with the aim of deciding which of these methods converge and providing asymptotic estimates. There are profound and celebrated results in this area, such as the Hardy-Littlewood theorems and Norbert Wiener's breakthroughs and his simple proof of one of those theorems based on Fourier analysis (see the charming book by Korevaar [131], which traces a century of developments on Tauberian theorems). There are also applications of various Tauberian methods to prime number theory. Closer to the content of the present monograph, there have been remarkable applications of functional analysis methods in summability to iterative methods of linear and nonlinear operator equations in Hilbert and Banach spaces, in addition to the applications covered in this monograph. Summability theory in return has led to introduction of new classes of matrices and many interesting spaces of summable sequences and double sequences.

Professors M. Mursaleen and F. Başar are two of the renowned researchers in the field of summability in the last two decades. They have cultivated a research school on summability in their respective countries, India and Turkey. They have mentored two generations of students and researchers on various aspects of summability theory, sequence spaces, different notions of convergence

Foreword

and other topics. They have also collaborated on many joint research papers. This monograph reflects their achievements in these endeavors. The book is written for graduate students and researchers with an interest in sequence spaces, matrix transformations in the context of summability, various spaces of summable sequences and other topics mentioned in the preface. The book is a welcome addition to the literature. I look forward to adding it to my bookshelf as a companion to the other books [32, 52, 97, 131, 147, 168, 227, 228] and [239].

M. Zuhair Nashed
University of Central Florida,
Orlando, Florida

Preface

This book is intended for graduate students and researchers with a (special) interest in sequence spaces, matrix transformations and related topics. Besides a preface and index, the book consists of six chapters with abstracts and is organized as follows:

In Chapter 1; we present some basic definitions, notations and various basic ideas that will be required throughout the book. In this chapter, we state and prove Hahn-Banach, Baire's category, Banach-Steinhaus, bounded inverse, closed graph and open mapping theorems together with uniform boundedness principle, which are basic for functional analysis.

In Chapter 2, we investigate the geometric properties of normed Euler sequence spaces and Cesàro sequence space $ces(p)$, and some sequence spaces involving lacunary sequence space equipped with the Luxemburg norm besides topological, and some other usual properties.

Chapter 3 is devoted to some classes of matrix transformations with establishing the necessary and sufficient conditions on the elements of a matrix to map a sequence space X into a sequence space Y. This is a natural generalization of the problem to characterize all summability methods given by infinite matrices that preserve convergence.

In Chapter 4, we study the notion of almost convergence and the related matrix transformations with their some applications.

In Chapter 5, after giving some elementary examples following Yeşilkayagil and Başar [234], Dündar and Başar [75], Başar and Karaisa [38], and Srivastava and Kumar [205], we determine the spectrum and the fine spectrum of the lambda matrix Λ, the upper triangle double band matrix Δ^+, the generalized difference operator defined by a double sequential band matrix $B(\tilde{r}, \tilde{s})$ and the generalized difference operator Δ_{uv} acting on the sequence spaces c_0, c; ℓ_p and ℓ_1 with respect to Goldberg's classification, where $1 < p < \infty$.

In Chapter 6, we summarize the literature on some sets of fuzzy valued sequences and series. Talo and Başar [213] have extended the main results of Başar and Altay [35] to fuzzy numbers and defined the α-, β- and γ-duals of a set of fuzzy valued sequences, and gave the duals of the classical sets of fuzzy valued sequences together with the characterization of the classes of infinite matrices of fuzzy numbers transforming one of the classical sets into another one. Also, Kadak and Başar [104, 105] have recently studied the power series of fuzzy numbers and examined the alternating and binomial series of fuzzy numbers and some sets of fuzzy-valued functions with the level sets, and gave some properties of the level sets together with some inclusion relations, in [103, 108]. Finally, following Talo and Başar [215]; we introduce the classes $\ell_\infty(F)$, $c(F)$, $c_0(F)$ and $\ell_p(F)$ consisting of all bounded, convergent, null and absolutely p-summable fuzzy valued sequences with the level sets and the sets $\ell_\infty(F; f)$, $c(F; f)$, $c_0(F; f)$ and $\ell(F; f)$ of fuzzy valued sequences defined by a modulus function.

Mohammad Mursaleen & Feyzi Başar
Aligarh & İstanbul
February 2019

Acknowledgements

Professor Bilâl Altay, Department of Mathematical Education, Faculty of Education, İnönü University, 44280–Malatya, Turkey, provided all kinds of technical support in the preparation of this book. He has corrected the errors in the TEX files and drawing figures with Latex commands. Additionally, he has prepared the cover composition during the revision of the study. For all of these, the authors express their sincere gratitude to Professor Altay.

The second named author is very grateful to Associate Professor Özer Talo, Department of Mathematics, Faculty of Art and Sciences, Celal Bayar University, 45040–Manisa, Turkey, for valuable suggestions, corrections and remarks concerning results from Chapter 6. We also thank Assistant Professor Uğur Kadak, Bozok University, Faculty of Art and Sciences, Department of Mathematics, 66100–Yozgat, Turkey, for his relevant remarks on Chapter 6.

We would like to thank Associate Professor Medine Yeşilkayagil, School of Applied Sciences, Uşak University, 1 Eylül Campus, 64200 – Uşak, Turkey, for reading in detail the whole of the manuscript and added required lines, pointed out several typos and suggested numerous improvements.

Also, our sincerest thanks due to Professor Mohamed Bakari, Department of American Culture and Literature, Faculty of Art and Sciences, Fatih University, 34500–Büyükçekmece/İstanbul, Turkey, for kindly correcting some grammatical errors in the manuscript.

Finally, we are very grateful to the three reviewers for their praised reports on the work.

The authors are thankful for Ann Chapman from the CRC/Taylor & Francis Group and Narayani Govindarajan from Nova Techset for her careful reading and valuable corrections on the earlier version of the book which improved the presentation and readability.

Mohammad Mursaleen & Feyzi Başar
Aligarh & İstanbul
February 2019

Authors

Mohammad Mursaleen is currently a Principal Investigator for a SERB Core Research Grant at the Department of Mathematics, Aligarh Muslim University. He has published more than 330 research papers in the field of summability, sequence spaces, approximation theory, fixed point theory, measures of noncompactness. He has also published eight books and completed several national and international projects, in several countries. Besides several master's students, he has guided twenty Ph.D. students, and served as a reviewer for various international scientific journals. He is also member of editorial boards, for many international scientific journals. Recently, he has been recognized as Highly Cited Researcher 2019 by Web of Science.

Feyzi Başar is a Professor Emeritus since July 2016, at İnönü University, Turkey. He has published an e-book for graduate students and researchers and more than 150 scientific papers in the field of summability theory, sequence spaces, FK-spaces, Schauder bases, dual spaces, matrix transformations, spectrums of certain linear operators represented by a triangle matrix over some sequence space, the alpha-, beta- and gamma-duals and some topological properties of the domains of some double and four-dimensional triangles in the certain spaces of single and double sequences, sets of the sequences of fuzzy numbers, multiplicative calculus. He has guided 17 master's and 10 Ph.D. students and served as a referee for 121 international scientific journals. He is a member of an editorial board of 21 scientific journals. Feyzi Başar is also a member of scientific committees of 17 mathematics conferences, gave talks at 14 different universities as invited speaker and participated in more than 70 mathematics symposiums with a paper.

List of Abbreviations and Symbols

$\overline{A}$: the closure of a set A

A^0 : the interior of a set A

A^C : the complement of a set A

$P(A)$: the collection of all subsets of a set A

$\mathbb{N}_0$: set of natural numbers, i.e., $\mathbb{N}_0 = \{0, 1, 2, \ldots\}$

$\mathbb{N}_k$: set of integers which are greater than or equal to $k \in \mathbb{N}_0$

$\mathbb{Z}$: set of integers, i.e., $\mathbb{Z} = \{\ldots, -2, -1, 0, 1, 2, \ldots\}$

$\mathbb{Q}$: set of rational numbers

$\mathbb{R}$: set of real numbers, the real field

$\mathbb{R}^+$: set of non-negative real numbers

$\mathbb{C}$: set of complex numbers, the complex field

$\mathbb{K}$: either of the fields of $\mathbb{R}$ or $\mathbb{C}$

$Re[z]$: real part of $z \in \mathbb{C}$

$Im[z]$: imaginer part of $z \in \mathbb{C}$

$\mathbb{R}^2$: set of all pairs of real numbers

$\mathbb{R}^n$: n-dimensional Euclidean space

$\mathbb{C}^n$: n-dimensional complex Euclidean space

$[a]$: integer part of a number a

$\mathcal{F}$: collection of all finite subsets of $\mathbb{N}_0$

$P[0, 1]$: set of all polynomials defined on the interval $[0, 1]$

$C[0, 1]$: space of all continuous real or complex valued functions on the interval $[0, 1]$

$C[a, b]$: space of all continuous real or complex valued functions on the interval $[a, b]$

$C_B[a, b]$: space of all continuous and bounded functions on $[a, b]$

$C_F[a, b]$: set of all continuous fuzzy-valued functions on the interval $[a, b]$

$B_F[a, b]$: set of all bounded fuzzy-valued functions on the interval $[a, b]$

$L_p(X)$: collection of all complex measurable functions on X with $1 \leq p < \infty$

e : sequence whose elements are equal to 1

$e^{(k)}$: sequences whose only non-zero term is a 1 in k^{th} place for each $k \in \mathbb{N}_0$

δ_{ij} : Kronecker delta which is $= 1$ if $i = j$ and $= 0$ if $i \neq j$

ω : space of all sequences over the complex field

$\omega(F)$: set of all sequences of fuzzy numbers

$C(F)$: set of all fuzzy valued Cauchy sequences

ϕ	:	set of all finitely non-zero sequences
ℓ_∞	:	space of bounded sequences over the complex field
$\ell_\infty(F)$	:	set of bounded sequences of fuzzy numbers
$\ell_\infty(F;f)$	:	set of bounded sequences of fuzzy numbers defined by a modulus function
f	:	space of almost convergent sequences over the complex field
$f(X)$	:	set of all almost convergent sequences in X
$wf(X)$	:	set of all weakly almost convergent sequences in X
$[f]$	:	space of all strongly almost convergent sequences over the complex field
f_0	:	space of almost null sequences over the complex field
c	:	space of convergent sequences over the complex field
$c(F)$	:	set of convergent sequences of fuzzy numbers
$c(F;f)$	:	set of convergent sequences of fuzzy numbers defined by a modulus function
c_0	:	space of null sequences over the complex field
$c_0(F)$	:	set of null sequences of fuzzy numbers
$c_0(F;f)$	:	set of null sequences of fuzzy numbers defined by a modulus function
ℓ_1	:	space of absolutely summable sequences over the complex field
$\widehat{\ell}$	:	set of all absolutely almost convergent sequences
ℓ_p	:	space of absolutely p-summable sequences over the complex field
$\widehat{\ell}_p$	:	set of all absolutely p-almost convergent sequences
$\ell_p(F)$	:	set of absolutely p-summable sequences of fuzzy numbers
$\ell_p(F;f)$	:	set of absolutely p-summable sequences of fuzzy numbers defined by a modulus function
ℓ_Φ	:	Orlicz sequence space
fs	:	space of almost convergent series over the complex field
bs	:	space of bounded series over the complex field
$bs(F)$	:	set of bounded series of fuzzy numbers
fs_0	:	space of series almost converging to zero over the complex field
cs	:	space of convergent series over the complex field
$cs(F)$	:	set of convergent series of fuzzy numbers
cs_0	:	space of series converging to zero over the complex field
$cs_0(F)$	:	set of series of fuzzy numbers converging to zero
bv	:	space of sequences of bounded variation over the complex field
bv_σ	:	space of sequences of σ-bounded variation over the complex field
$bv(F)$	:	set of sequences of bounded variation of fuzzy numbers
bv_0	:	space of sequences of both bounded variation and null over the complex field
bv_p	:	space of sequences of p-bounded variation over the complex field
$bv_p(F)$	:	set of p-bounded variation sequences of fuzzy numbers
$\ell_\infty(p)$	:	space of all sequences (x_k) such that $\sup_{k\in\mathbb{N}_0}|x_k|^{p_k} < \infty$
$c(p)$	:	space of all sequences (x_k) such that $|x_k - l|^{p_k} \to 0$, as $k \to \infty$
$c_0(p)$	:	space of all sequences (x_k) such that $|x_k|^{p_k} \to 0$, as $k \to \infty$

$\ell(p)$	:	space of all sequences (x_k) such that $\sum_k |x_k|^{p_k} < \infty$
X^α	:	alpha dual of a sequence space X
X^β	:	beta dual of a sequence space X
X^γ	:	gamma dual of a sequence space X
λ^*	:	continuous dual of a sequence space λ
λ^f	:	f-dual of a sequence space λ
$x^{[n]}$	:	n^{th} section of a sequence $x = (x_k)$
$\{(Ax)_n\}$	:	A-transform of a sequence x
$\mathcal{A}x$	:	$\{(Ax)_n^i\}_{i,n=0}^\infty$
c_A	:	convergence domain of a matrix A
$\chi(A)$	:	characteristic of a matrix A
L	:	Banach limit
S	:	shift operator
C_1	:	Cesàro mean of order one
$\triangle$	:	forward difference operator, i.e., $(\triangle x)_k = x_k - x_{k+1}$ and $(\triangle^2 x)_k = \triangle(x_k - x_{k+1})$
Δ	:	backward difference operator, i.e., $(\Delta x)_k = x_k - x_{k-1}$
$|C_1|$	:	absolute summability of Cesàro mean of order one
C_r	:	Cesàro mean of order r
E_1	:	original Euler matrix
E_q	:	Euler mean of order q
E^r	:	Euler-Knopp matrix of order r
T^r	:	Taylor matrix
R^t	:	Riesz mean with respect to the sequence $t = (t_k)$
N^t	:	Nörlund mean with respect to the sequence $t = (t_k)$
θ	:	zero vector in a linear space X
$L(\mathbb{R})$	:	set of all fuzzy numbers on $\mathbb{R}$
$L(\mathbb{R})^+$	:	set of all non-negative fuzzy numbers on $\mathbb{R}$
$L(\mathbb{R})^-$	:	set of all non-positive fuzzy numbers on $\mathbb{R}$
W	:	set of all closed bounded intervals A with endpoints $\underline{A}$ and $\overline{A}$
$[u]_\alpha$	:	α-level set of $u \in L(\mathbb{R})$
$supp(u)$	:	set of real numbers t such that $u(t) > 0$
$u \not\sim v$	:	neither $u \preceq v$ nor $v \preceq u$
BSFN	:	a bounded sequence of fuzzy numbers
$x_k \sim \infty$	:	$x = (x_k)$ is definitely divergent
$\mathcal{A}$	:	sequence of infinite matrices $A^i = \{a_{nk}(i)\}$
$\mathcal{B}x$	:	$\{(Bx)_m^i\}_{i,m=0}^\infty$
$|K|$	:	cardinality of K
D_∞	:	Hausdorff metric on the set $\ell_\infty(F)$
D_p	:	Hausdorff metric on the set $\ell_p(F)$
$(\lambda : \mu)$	:	class of all matrices A such that $A : \lambda \to \mu$
$(c : c)$	:	class of conservative matrices
$(c : c; p)$	:	class of Teoplitz matrices
$(c : c)_{reg}$	:	class of regular matrices
$(cs : c; p)$	:	class of series to sequence regular matrices

$(c : v_\sigma)$	:	class of sequence to sequence sigma-conservative matrices
$(c : f)$	:	class of almost conservative matrices
$(c : f)_{reg}$	:	class of almost regular matrices
$(f : c)$	:	class of strongly conservative matrices
$(f : c; p)$	:	class of strongly regular matrices
$(\ell_\infty : c)$	:	class of Schur (coercive) matrices
$(\ell_\infty : f)$	:	class of sequence to sequence almost coercive matrices
$(\ell_\infty : fs)$	:	class of sequence to series almost coercive matrices
$(bs : f)$	:	class of series to sequence almost coercive matrices
$(bs : fs)$	:	class of series to series almost coercive matrices
$\emptyset$	:	empty set
$(AB)_{ij}$	:	i^{th} row and j^{th} column entry of the matrix product AB
I	:	unit matrix
$G(A)$	:	graph of a continuous operator A
$D(T)$	:	domain of a linear operator T
$R(T)$	:	range of a linear operator T
$Ker(T)$	:	kernel or null space of a linear operator T
$r_\sigma(T)$	:	spectral radius of an operator $T \in B(X)$
T^*	:	adjoint of a bounded linear operator T
T_α	:	resolvent operator of T with each $\alpha \in \mathbb{C}$
$B(x_0; r)$	:	open ball of radius r with center x_0
$S(x_0; r)$	:	sphere of radius r with center x_0
$S[\theta, \delta]$	:	closed sphere of radius δ with center origin $\theta = (0, 0, 0, \ldots)$
S_X	:	the unit sphere in X
$\mathcal{L}(X)$	:	set of all linear and continuous operators on a space X into itself
$\mathcal{L}(X : Y)$	:	set of all linear and continuous operators $T : X \to Y$
$\mathcal{B}(X)$	:	set of all bounded linear operators on a space X into itself
$\mathcal{B}(X : Y)$	:	set of all bounded linear operators $T : X \to Y$
$\mathcal{C}(X : Y)$	:	set of all compact operators $T : X \to Y$
$\mathcal{F}(X : Y)$	:	set of all finite rank operators $T : X \to Y$
X'	:	set of bounded linear functionals on a space X
X^*	:	continuous dual of a space X
$\sigma(T, X)$	:	spectrum of a linear operator T on a space X
$\rho(T, X)$	:	resolvent set of a linear operator T on a space X
$\sigma_e(T, X)$	:	eigenspace of a linear operator T corresponding to the eigenvalue α
$\sigma_a(T, X)$	:	approximate spectrum of a linear operator T on a space X
$\sigma_p(T, X)$	:	point (discrete) spectrum of a linear operator T on a space X
$\sigma_c(T, X)$	:	continuous spectrum of a linear operator T on a space X
$\sigma_r(T, X)$	:	residual spectrum of a linear operator T on a space X
$\sigma_{ap}(T, X)$	:	approximate point spectrum of a linear operator T on a space X
$\sigma_\delta(T, X)$	:	defect spectrum of a linear operator T on a space X
$\sigma_{co}(T, X)$	:	compression spectrum of a linear operator T on a space X

Chapter 1

Basic Functional Analysis

Keywords. Metric sequence spaces, normed linear spaces, bounded linear operators, Köthe-Toeplitz duals, Hahn-Banach theorem, Baire category theorem, uniform boundedness principle, Banach-Steinhaus theorem, bounded inverse theorem, closed graph theorem, open mapping theorem, compact operators, Schauder basis, separability, reflexivity, weak convergence, Hilbert spaces, topological vector spaces, FK-spaces.

1.1   Metric Spaces

In $\mathbb{R}$, the set of all real numbers or in $\mathbb{C}$, the set of all complex numbers, the concept of absolute value plays an important role in defining two basic concepts, i.e., the concepts of convergence and continuity, on which the whole theory of real (or complex) variables depends. Metric space is a generalization of $\mathbb{R}$ (or $\mathbb{C}$), insofar as it is a space with a metric or a distance function. In the theory of metric spaces, the concept of distance is generalized by replacing $\mathbb{R}$ (or $\mathbb{C}$) with an arbitrary non-empty set X in such a way that one can have a notion of convergence and continuity in a more general setting.

Definition 1.1.1. *A metric space is a set X together with a function d, called a metric or distance function, which assigns a real number $d(x, y)$ to every pair x, y belonging to X satisfying the following axioms:*

(M1) (positive): $d(x, y) \geq 0$ for all x, y in X.

(M2) (strictly positive): $d(x, y) = 0$ iff $x = y$ for all x, y in X.

(M3) (symmetry): $d(x, y) = d(y, x)$ for all x, y in X.

(M4) (triangle inequality): $d(x, z) \leq d(x, y) + d(y, z)$ for all x, y, z in X.

Definition 1.1.2. *Let X be a non-empty set. Define d for $x, y \in X$ by*

$$d(x, y) = \left\{ \begin{array}{ll} 0 & , \quad x = y, \\ 1 & , \quad x \neq y. \end{array} \right. \tag{1.1.1}$$

The metric d given by (1.1.1) is called the trivial metric or discrete metric on X. The metric space (X, d) is called discrete metric space and is denoted by X_d.

Examples 1.1.3. *We have the following:*

(a) *The usual distance $d(x, y) = |x - y|$ is a metric for the set $\mathbb{R}$ of all real numbers.*

(b) *On the plane $\mathbb{R}^2$, the metric d_1 is defined by $d_1[(x_1, y_1), (x_2, y_2)] = |x_1 - x_2| + |y_1 - y_2|$. Another metric d_2 on $\mathbb{R}^2$ is the "usual distance" (measured using Pythagoras's theorem):*

$$d_2[(x_1, y_1), (x_2, y_2)] = \sqrt{(x_1 - x_2)^2 + (y_1 - y_2)^2}.$$

Note that a non-empty set X may have more than one metric.

(c) *On the set $\mathbb{C}$ of all complex numbers, the metric d is defined by $d(z, w) = |z - w|$, where $|\cdot|$ represents the modulus of the complex number rather than the absolute value of a real number.*

(d) *On the plane $\mathbb{R}^2$, another metric d_∞ is defined with the supremum or maximum as*

$$d_\infty[(x_1, y_1), (x_2, y_2)] = \max\{|x_1 - x_2|, |y_1 - y_2|\}.$$

(e) *Let $C[0, 1]$ be the set of all continuous real-valued functions on the interval $[0, 1]$. We define the metrics d_1, d_2 and d_∞ on $C[0, 1]$ by analogy to the above examples:*

$d_1(f, g) = \int_0^1 |f(x) - g(x)| dx.$

$d_2(f, g) = \sqrt{\int_0^1 [f(x) - g(x)]^2 dx}.$

$d_\infty(f, g) = \max\limits_{0 \leq x \leq 1} |f(x) - g(x)|.$

Definition 1.1.4. *A sequence (x_n) in a metric space (X, d) is said to be convergent to x in X if for every $\varepsilon > 0$ there is $N > 0$ such that $d(x, x_n) < \varepsilon$ whenever $n \geq N$; it is said to be Cauchy if $d(x_m, x_n) < \varepsilon$ whenever $n, m \geq N$. A metric space (X, d) is said to be complete if every Cauchy sequence in X is convergent in X.*

Now, we may give the definition of closure and the interior of a set.

Definition 1.1.5. *Let (X, d) be a metric space and let $S \subset X$. A point $x_0 \in X$ is a closure point of S if, for every $\varepsilon > 0$, there is a point $x \in S$ with $d(x_0, x) < \varepsilon$. The closure $\overline{S}$ of S is the set of all closure points of S. We call x_0 an interior point of a set $S \subset X$ if S is a neighborhood of x_0. The interior S° of S is the set of all interior points of S. S° is open and is the largest open set in S.*

Definition 1.1.6. *A subset S of a metric space (X, d) is said to be dense in X iff $\overline{S} = X$. S is said to be nowhere dense in X if $(\overline{S})^0 = \emptyset$.*

A metric space (X, d) is said to be separable if it contains a countable dense subset.

Examples 1.1.7. *We give the following examples for separable/non-separable spaces:*

(i) *The set of rational numbers $\mathbb{Q}$ dense in $\mathbb{R}$, hence $\mathbb{R}$ is separable.*

(ii) *The set of all rational polynomials $P[0, 1]$ is dense in $C[0, 1]$ with sup-norm $\|\cdot\|_\infty$ as well with integral norm $\|\cdot\|_p$, $(1 \le p < \infty)$, hence $C[0, 1]$ is separable.*

(iii) *ϕ is dense in the spaces c_0 and ℓ_p with the norms $\|\cdot\|_\infty$ and $\|\cdot\|_p$, respectively, i.e., c_0 and ℓ_p are separable, where $1 \le p < \infty$ and ϕ denotes the set of all finetely non-zero sequences.*

(iv) *Finite sets, $\mathbb{N}_0$ and $\mathbb{Z}$ are nowhere dense in $\mathbb{R}$.*

(v) *ℓ_∞ is not separable.*

Proof. We prove here only (v). It is easy to see that the set $E := \{x = (x_j) \in \ell_\infty : x_j \in \{0, 1\}, j \in \mathbb{N}_0\} \subset \ell_\infty$ is uncountable, and for every distinct $x, y \in E$, $\|x - y\|_\infty = 1$. We have to show that E is not dense in ℓ_∞. Let if possible, E be dense in ℓ_∞. Then, there exists $z \in \ell_\infty$ such that $\|x - z\|_\infty < 1/4 (= \epsilon)$ for $x \in E$. Now,

$$1 = \|x - y\|_\infty \le \|x - z\|_\infty + \|z - y\|_\infty < \frac{1}{4} + \|z - y\|_\infty$$

for all $y \in E$. This implies that $\|z - y\|_\infty > 3/4$, i.e., E is not dense in ℓ_∞. Hence, ℓ_∞ cannot be separable. $\qquad \square$

Definition 1.1.8. *Let M and S be two subsets of a metric space (X, d) and $\epsilon > 0$. Then, the set S is called ϵ-net of M if for any $x \in M$ there exists $s \in S$ such that $d(x, s) < \epsilon$. If the set S is finite, then the ϵ-net S of M is called finite ϵ-net.*

Definition 1.1.9. *The set M is said to be totally bounded if it has a finite ϵ-net for every $\epsilon > 0$.*

Definition 1.1.10. *A subset M of a metric space X is compact if every sequence (x_n) in M has a convergent subsequence, and in this case the limit of that subsequence is in M.*

Definition 1.1.11. *The set M is said to be relatively compact if the closure $\overline{M}$ of M is a compact set.*

If the set M is relatively compact, then M is totally bounded. If the metric space (X, d) is complete, then the set M is relatively compact if and only if it is totally bounded. It is easy to prove that a subset M of a metric space X is relatively compact if and only if every sequence (x_n) in M has a convergent subsequence; in that case, the limit of that subsequence need not be in M.

1.2   Metric Sequence Spaces

The space bv is the space of all sequences of bounded variation, that is, consisting of all sequences (x_k) such that $(x_k - x_{k+1})$ in ℓ_1, and $bv_0 = bv \cap c_0$. Let $e = (1, 1, \ldots)$ and $e^{(k)} = (0, 0, \ldots, 0, 1(k\text{th place}), 0, \ldots)$.

Examples 1.2.1. *We give the following examples for metric sequence spaces:*

(i) *The most popular metric d_ω which is known as the Fréchet metric on the space ω of all real or complex valued sequences is defined by*

$$d_\omega(x, y) = \sum_k \frac{|x_k - y_k|}{2^k(1 + |x_k - y_k|)}; \quad x = (x_k),\ y = (y_k) \in \omega.$$

For simplicity in notation, here and in what follows, the summation without limits runs from 0 to ∞, and use the convention that any term with negative subscript is equal to zero.

(ii) *The space of bounded sequences is denoted by ℓ_∞, i.e.,*

$$\ell_\infty := \left\{ x = (x_k) \in \omega : \sup_{k \in \mathbb{N}_0} |x_k| < \infty \right\}.$$

The natural metric on the space ℓ_∞ known as the sup-metric is defined by

$$d_\infty(x, y) = \sup_{k \in \mathbb{N}_0} |x_k - y_k|; \quad x = (x_k),\ y = (y_k) \in \ell_\infty.$$

(iii) *The spaces of convergent and null sequences are denoted by c and c_0, that is,*

$$c := \left\{ x = (x_k) \in \omega : \exists l \in \mathbb{C}\ \text{ such that }\ \lim_{k \to \infty} |x_k - l| = 0 \right\},$$

$$c_0 := \left\{ x = (x_k) \in \omega : \lim_{k \to \infty} x_k = 0 \right\}.$$

The metric d_∞ is also a metric for the spaces c and c_0.

(iv) The space of absolutely convergent series is denoted by ℓ_1, i.e.,

$$\ell_1 := \left\{ x = (x_k) \in \omega : \sum_k |x_k| < \infty \right\}.$$

The natural metric on the space ℓ_1 is defined by

$$d_1(x,y) = \sum_k |x_k - y_k|; \quad x = (x_k), \ y = (y_k) \in \ell_1.$$

(v) The space of absolutely p-summable sequences is denoted by ℓ_p, that is,

$$\ell_p := \left\{ x = (x_k) \in \omega : \sum_k |x_k|^p < \infty \right\}, \quad (0 < p < \infty).$$

In the case $1 < p < \infty$, the metric d_p on the space ℓ_p is given by

$$d_p(x,y) = \left(\sum_k |x_k - y_k|^p \right)^{1/p}; \quad x = (x_k), \ y = (y_k) \in \ell_p.$$

Also in the case $0 < p < 1$, the metric $\widetilde{d}_p$ on the space ℓ_p is given by

$$\widetilde{d}_p(x,y) = \sum_k |x_k - y_k|^p; \quad x = (x_k), \ y = (y_k) \in \ell_p.$$

(vi) The space of bounded series is denoted by bs, i.e.,

$$bs := \left\{ x = (x_k) \in \omega : \sup_{n \in \mathbb{N}_0} \left| \sum_{k=0}^{n} x_k \right| < \infty \right\}.$$

The natural metric on the space bs is defined by

$$d(x,y) = \sup_{n \in \mathbb{N}_0} \left| \sum_{k=0}^{n} (x_k - y_k) \right|; \quad x = (x_k), \ y = (y_k) \in bs. \quad (1.2.1)$$

(vii) The space of convergent series and the space of the series converging to zero are denoted by cs and cs_0, respectively, that is,

$$cs := \left\{ x = (x_k) \in \omega : \exists l \in \mathbb{C} \text{ such that } \lim_{n \to \infty} \left| \sum_{k=0}^{n} x_k - l \right| = 0 \right\},$$

$$cs_0 := \left\{ x = (x_k) \in \omega : \lim_{n \to \infty} \left| \sum_{k=0}^{n} x_k \right| = 0 \right\}.$$

The relation d defined by (1.2.1) is the natural metric on the spaces cs and cs_0.

(viii) The space of sequences of bounded variation is denoted by bv, i.e.,

$$bv := \left\{ x = (x_k) \in \omega : \sum_k |x_k - x_{k+1}| < \infty \right\}.$$

Define the forward difference sequence $\triangle u = \{(\triangle u)_k\}$ by $(\triangle u)_k = u_k - u_{k+1}$ for all $k \in \mathbb{N}_0$. The natural metric on the space bv is defined by

$$d(x,y) = \sum_k \left| (\triangle(x-y))_k \right|; \quad x = (x_k),\ y = (y_k) \in bv.$$

Let $p = (p_k)_{k \in \mathbb{N}_0}$ be a bounded sequence of positive real numbers with $\sup_{k \in \mathbb{N}_0} p_k = H$ and $M = \max\{1, H\}$. The following spaces were introduced and studied by Lascarides and Maddox [137], and Simons [197]:

$$\ell_\infty(p) := \left\{ x = (x_k) \in \omega : \sup_{k \in \mathbb{N}_0} |x_k|^{p_k} < \infty \right\},$$

$$c(p) := \left\{ x = (x_k) \in \omega : \exists l \in \mathbb{C} \text{ such that } \lim_{k \to \infty} |x_k - l|^{p_k} = 0 \right\},$$

$$c_0(p) := \left\{ x = (x_k) \in \omega : \lim_{k \to \infty} |x_k|^{p_k} = 0 \right\},$$

$$\ell(p) := \left\{ x = (x_k) \in \omega : \sum_k |x_k|^{p_k} < \infty \right\}.$$

If $p_k = p$ for all $k \in \mathbb{N}_0$ for some constant $p > 0$, then these sets are reduced to ℓ_∞, c, c_0 and ℓ_p, respectively. The metrics d_∞ and d_p on the spaces $\ell_\infty(p)$, $c(p)$, $c_0(p)$ and $\ell(p)$ are defined by

$$d_\infty(x,y) = \sup_{k \in \mathbb{N}_0} |x_k - y_k|^{p_k},$$

$$d_p(x,y) = \left(\sum_k |x_k - y_k|^{p_k} \right)^{1/M};$$

respectively, where $0 \leq p_k < \sup_{k \in \mathbb{N}_0} p_k < \infty$.

1.3   Normed Linear Spaces

The Euclidean distance between two points $x = (x_1, x_2)$ and $y = (y_1, y_2)$ belonging to two-dimensional Euclidean space $\mathbb{R}^2$ is given by

$$\|x - y\| = \sqrt{(x_1 - y_1)^2 + (x_2 - y_2)^2}.$$

In this way, we can view $\| \cdot \|$ as a real valued-function defined on the real Euclidean plane, we desire to extend this concept to a linear space, in general, which leads us to seek a conception of "norm."

Definition 1.3.1. *Let X be a linear space over the field $\mathbb{K}$ of real or complex numbers. A function $\| \cdot \| : X \to \mathbb{R}$ is said to be a norm on X if the following axioms hold for arbitrary points $x, y \in X$ and any scalar α:*

(N1) (positive definiteness): $\|x\| = 0$, if and only if $x = \theta$, where θ denotes the zero vector.

(N2) (absolute homogeneity): $\|\alpha x\| = |\alpha| \|x\|$.

(N3) (triangle inequality): $\|x + y\| \leq \|x\| + \|y\|$.

A normed linear space is a pair $(X, \| \cdot \|)$, where X is a linear space and $\| \cdot \|$ is a norm defined on X. When no confusion is likely we denote $(X, \| \cdot \|)$ by X.

Note that $\| \cdot \|$ is always non-negative: By (N2) and (N3), we have $0 = \|\theta\| = \|x - x\| \leq \|x\| + \| - x\| = 2\|x\|$, i.e., $\|x\| \geq 0$.

We have the following important relation between a metric space and a normed linear space:

Remark 1.3.2. *Each norm $\| \cdot \|$ of X defines a metric d on X given by $d(x,y) = \|x - y\|$ for all $x, y \in X$ and is called as induced metric. But it is known that not every metric on a linear space can be obtained from a norm.*

It is easy to check the first part. For the second part, let us consider the linear space ω; the metric d_ω cannot be obtained from a norm. Indeed, if $d_\omega(x,y) = \|x - y\|$ then we have

$$
\begin{aligned}
d_\omega(\alpha x, \alpha y) &= \|\alpha x - \alpha y\| \\
&= \sum_k \frac{|\alpha| |x_k - y_k|}{2^k (1 + |\alpha| |x_k - y_k|)} \neq |\alpha| d_\omega(x,y),
\end{aligned}
$$

that is, $\|\alpha(x - y)\| \neq |\alpha| \|x - y\|$. This means that the space ω is not a normed linear space.

Definition 1.3.3. *A seminorm ν on a linear space X is a function $\nu : X \to \mathbb{R}$ such that*

(i) $\nu(\alpha x) = |\alpha| \nu(x)$ for all $\alpha \in \mathbb{K}$ ($\mathbb{R}$ or $\mathbb{C}$) and all $x \in X$ (absolute homogeneity).

(ii) $\nu(x + y) \leq \nu(x) + \nu(y)$ for all $x, y \in X$ (subadditivity).

Note that by (i), we have $\nu(0x) = 0 \cdot \nu(x) = 0$.

Note that every norm is a seminorm but not conversely. For converse, define $\nu(x) = |\lim_{n \to \infty} x_n|$ on c. Take $x_n = 1/(n+1)$ for all $n \in \mathbb{N}_0$. Then, $\nu(x) = 0$ while $x \neq \theta$. Hence, ν is not a norm while it is a seminorm on c.

Definition 1.3.4. *A normed linear space X is complete if every Cauchy sequence in X converges in X, that is, if $\|x_m - x_n\| \to 0$, as $m, n \to \infty$; where $x_n \in X$, then there exists $x \in X$ such that $\|x_n - x\| \to 0$, as $n \to \infty$. A complete normed linear space is said to be a Banach space.*

Definition 1.3.5. *Let X be a normed linear space. We say that the series $\sum_k x_k$ with $x_k \in X$, converges to $s \in X$ if and only if the sequence of partial sums $(s_n) = \left(\sum_{k=0}^{n} x_k\right)_{n \in \mathbb{N}_0}$, converges to s, that is, $\|s_n - s\| \to 0$, as $n \to \infty$, and we write $\sum_k x_k = s$. A series $\sum_k x_k$ in X is said to be absolutely convergent if $\sum_k \|x_k\| < \infty$.*

Remark 1.3.6. *In $\mathbb{R}$ or $\mathbb{C}$, every absolutely convergent series is convergent. This is a direct consequence of the completeness of $\mathbb{R}$ or $\mathbb{C}$. But, in general, an absolutely convergent series need not be convergent in a normed space. For example, consider the space $X = P[0,1]$ with respect to $\|f\|_\infty = \sup_{t \in [0,1]} |f(t)|$. Then, the series $\sum_n x^n / n!$ is not convergent in X, since*

$$\sum_n \frac{x^n}{n!} = 1 + x + \frac{x^2}{2!} + \frac{x^3}{3!} + \cdots = e^x \notin P[0,1].$$

On the other hand, it is absolutely convergent. Since

$$\sum_n \left\| \frac{x^n}{n!} \right\| = \sum_n \frac{1}{n!} \quad for \ \ x = 1,$$

which is convergent by the ratio test and for $x = 0$, $\sum_n \|x^n / n!\| = 0$ for $|x| < 1$, $\sum_n \|x^n\|$ is convergent and $(1/n!)$ is a positive monotone decreasing sequence, then the series $\sum_n \|x^n / n!\|$ is also convergent by Abel's test.

Theorem 1.3.7. *If a Cauchy sequence has a convergent subsequence, then the whole sequence is convergent.*

Proof. Let (x_n) be a Cauchy in a normed linear space X and (x_{n_k}) be a subsequence of (x_n) converging to $x \in X$, say. Then, (x_{n_k}) is also Cauchy. Therefore, for every $\epsilon > 0$ there exists a $n_0 \in \mathbb{N}_0$ such that

$$\|x_n - x\| \leq \|x_n - x_{n_k}\| + \|x_{n_k} - x\| < \frac{\epsilon}{2} + \frac{\epsilon}{2} = \epsilon$$

for all $n_k \geq n_0$. Hence, (x_n) converges to x. $\square$

Remark 1.3.8. *If a subsequence of a sequence in a normed linear space X is convergent then the sequence itself need not be convergent. For example, consider the sequence $(x_n) = \{(-1)^n\}$ in the usual normed linear space $\mathbb{R}$. It is trivial that (x_n) is not convergent, but its subsequence $(x_{2n}) = (1, 1, 1, \ldots)$ converges to 1.*

Theorem 1.3.9. *A normed linear space X is complete if and only if every absolutely convergent series is convergent.*

Proof. Let X be complete and $\sum_n x_n$ be an absolutely convergent series. Then, since $\sum_k \|x_k\| < \infty$ it is immediate that

$$\lim_{m \to \infty} \|s_n - s_m\| = \lim_{m \to \infty} \left\| \sum_{k=m+1}^{n} x_k \right\| \leq \lim_{m \to \infty} \sum_{k=m+1}^{n} \|x_k\| = 0$$

for $n > m$. Hence, (s_n) is a Cauchy sequence in X and is convergent since X is complete, that is, $\sum_n x_n$ is convergent.

Conversely, let every absolutely convergent series be convergent and (x_n) be a Cauchy sequence in X. Then, we can find an increasing sequence $(n_k)_{k \in \mathbb{N}_0}$ of natural numbers such that

$$\|x_{n_{k+1}} - x_{n_k}\| < \frac{1}{2^k} \quad \text{for all} \quad k \in \mathbb{N}_0.$$

Therefore, $\sum_k \|x_{n_{k+1}} - x_{n_k}\| < \infty$. It follows that $\sum_k (x_{n_{k+1}} - x_{n_k})$ converges. Therefore, there is $x \in X$ such that $\sum_{k=0}^{m} (x_{n_{k+1}} - x_{n_k}) \to x$, say, as $m \to \infty$, that is, $x_{n_{m+1}} - x_{n_1} \to x$ implies $x_{n_{m+1}} \to x + x_{n_1}$, as $m \to \infty$. Hence, (x_{n_k}) converges. That is, the Cauchy sequence (x_n) has a convergent subsequence (x_{n_k}) and so, by Theorem 1.3.7, the whole sequence (x_n) is convergent. Therefore, X is complete. $\qquad \square$

Examples 1.3.10. *We have the following examples:*

(i) c_0, c *and* ℓ_∞ *are Banach spaces with the sup-norm* $\|x\|_\infty = \sup_{k \in \mathbb{N}_0} |x_k|$.

We consider only the space c. *Let* $\{x^{(m)}\}$ *be a Cauchy sequence in* c, *we have*

$$\lim_{m,n \to \infty} \left\| x^{(n)} - x^{(m)} \right\|_\infty = 0.$$

Now, for each $\epsilon > 0$, *there exists* N *such that* $\|x^{(n)} - x^{(m)}\|_\infty < \epsilon$ *for all* $n, m \geq N$, *i.e.,*

$$\sup_{i \in \mathbb{N}_0} \left| x_i^{(n)} - x_i^{(m)} \right| < \frac{\epsilon}{3}; \quad m, n \geq N$$

and hence,

$$\left| x_i^{(n)} - x_i^{(m)} \right| < \frac{\epsilon}{3} \quad \text{for} \quad i \in \mathbb{N}_0 \quad \text{and for all} \quad n, m \geq N. \qquad (1.3.1)$$

Hence, for each i *the sequence of real numbers* $\{x_i^{(n)}\} = \{x_i^{(0)}, x_i^{(1)}, \ldots\}$ *is a Cauchy sequence in* $\mathbb{R}$ *and hence convergent, say, to* x_i, *i.e.,* $\left| x_i^{(m)} - x_i \right| \to 0$, *as* $m \to \infty$, *for each* $i \in \mathbb{N}_0$.

Now, fix $n \geq N$ *and letting* $m \to \infty$ *in (1.3.1), we get*

$$\left| x_i^{(n)} - x_i \right| < \frac{\epsilon}{3} \quad \text{for each} \quad i \in \mathbb{N}_0. \qquad (1.3.2)$$

So that

$$\sup_{i \in \mathbb{N}_0} \left| x_i^{(n)} - x_i \right| < \frac{\epsilon}{3} \quad \text{for all } n \geq N,$$

that is, $\|x^{(n)} - x\|_\infty \to 0$, *as* $n \to \infty$; *where* $x = (x_i)$. *Hence,* $x^{(n)} \to x$, *as* $n \to \infty$, *i.e., a Cauchy sequence* $\{x^{(n)}\}$ *converges to* x. *Now, we have to show that* $x \in c$.

Now, the sequence $\{x_i^{(N)}\} \in c$ *and is a Cauchy sequence, hence*

$$\left| x_i^{(N)} - x_j^{(N)} \right| < \frac{\epsilon}{3} \quad \text{for all } i, j \geq M. \tag{1.3.3}$$

Consequently by (1.3.2) and (1.3.3), we have

$$\begin{aligned}
|x_i - x_j| &= \left| x_i - x_i^{(N)} + x_i^{(N)} - x_j^{(N)} + x_j^{(N)} - x_j \right| \\
&\leq \left| x_i - x_i^{(N)} \right| + \left| x_i^{(N)} - x_j^{(N)} \right| + \left| x_j^{(N)} - x_j \right| \\
&< \frac{\epsilon}{3} + \frac{\epsilon}{3} + \frac{\epsilon}{3} = \epsilon.
\end{aligned}$$

Therefore, $x = (x_i)$ *is a Cauchy sequence in* $\mathbb{R}$ *and hence convergent, i.e.,* $x \in c$. *That is,* c *is a Banach space.*

(ii) Let $1 \leq p < \infty$. *Then,* ℓ_p *is complete with* $\|x\|_p = \left(\sum_k |x_k|^p \right)^{1/p}$.

Let $\{x_k^{(m)}\}_{m \in \mathbb{N}_0}$ *be a Cauchy sequence in* ℓ_p. *Then, there is* $N \in \mathbb{N}_0$ *such that for all* $r, s \geq N$, $\|x^{(r)} - x^{(s)}\|_p < \epsilon$. *Hence,*

$$\sum_k \left| x_k^{(r)} - x_k^{(s)} \right|^p < \epsilon^p, \tag{1.3.4}$$

which implies that

$$\left| x_k^{(r)} - x_k^{(s)} \right| < \epsilon \quad \text{for each } k \text{ and for all } r, s \geq N.$$

Hence, $\{x_k^{(m)}\}$ *is a Cauchy sequence in* $\mathbb{R}$ *and so is convergent to* x_k *in* $\mathbb{R}$. *Define* $x = (x_k)_{k \in \mathbb{N}_0}$. *We show that* $x \in \ell_p$. *From (1.3.4), we have*

$$\sum_k \left| x_k^{(r)} - x_k^{(s)} \right|^p < \epsilon^p \quad \text{for all } r, s \geq N. \tag{1.3.5}$$

Therefore, we get by letting $s \to \infty$ *in (1.3.5) that*

$$\sum_k \left| x_k^{(r)} - x_k \right|^p < \epsilon^p \quad \text{for all } r \geq N. \tag{1.3.6}$$

This implies that the sequence $\left\{ x_k^{(r)} - x_k \right\}_k \in \ell_p$ *for each r. Also,* $x^{(r)} \in \ell_p$ *by hypothesis. Hence, by Minkowski's inequality*

$$\sum_k |x_k|^p = \sum_k \left| x_k^{(r)} - x_k^{(r)} + x_k \right|^p \leq \sum_k \left| x_k^{(r)} \right|^p + \sum_k \left| x_k^{(r)} - x_k \right|^p.$$

That is, $x = (x_k) \in \ell_p$. *Finally, by (1.3.6), we have*

$$\left\| x^{(m)} - x \right\|_p = \left[\sum_k \left| x_k^{(m)} - x_k \right|^p \right]^{1/p} < \epsilon$$

for all $m \geq N$, *i.e.,* $x^{(m)} \to x$, *as* $m \to \infty$, *in* ℓ_p. *Hence,* ℓ_p *is complete.*

(iii) *The space* ϕ *is a normed linear space but not a Banach space with respect to any norm.*

Examples 1.3.11. *We have the following examples:*

(i) *The space* $C[a, b]$ *of all continuous functions on* $[a, b]$ *is complete normed linear space with* $\|f\|_\infty = \sup_{t \in [a,b]} |f(t)|$.

Let (f_n) *be a Cauchy sequence in* $C[a, b]$ *for each* $t \in [a, b]$. *Then, there is* $N \in \mathbb{N}_0$ *such that*

$$\|f_n - f_m\| = \max_{t \in [a,b]} |f_n(t) - f_m(t)| < \epsilon \text{ for all } n, m \geq N$$

$$\Rightarrow |f_n(t) - f_m(t)| < \epsilon \text{ for all } n, m \geq N.$$

Hence, for a fixed $t_0 \in [a, b]$, $|f_n(t_0) - f_m(t_0)| < \epsilon$ *for all* $n, m \geq N$.

$\Rightarrow$: $\{f_n(t_0)\}$ *is a Cauchy sequence in* $\mathbb{R}$, *hence convergent, say to* $f(t_0)$, *i.e.,* $f_n(t_0) \to f(t_0)$, *as* $n \to \infty$, *which is the pointwise convergent to* f. *We have to show that it is uniformly convergent. For given* $\epsilon > 0$, *choose* N *such that* $|f_n(t) - f(t)|$ *for all* $n, m \geq N$. *Then,*

$$\begin{aligned} |f_n(t) - f(t)| &= |f_n(t) - f(t) + f_m(t) - f_m(t)| \\ &\leq |f_n(t) - f_m(t)| + |f_m(t) - f(t)| \\ &\leq \sup_{t \in [a,b]} |f_n(t) - f_m(t)| + |f_m(t) - f(t)| \\ &= \|f_n - f_m\| + |f_m(t) - f(t)|. \end{aligned}$$

By choosing m *sufficiently large, we can make each term on the right-hand side less than* $\epsilon/2$. *Hence,*

$$M_n = \sup_{t \in [a,b]} |f_n(t) - f(t)| < \epsilon \text{ for all } n \geq N,$$

i.e., $M_n \to 0$, *as* $n \to \infty$. *Therefore,* $f_n(t) \to f(t)$, *as* $n \to \infty$, *uniformly on* $[a, b]$. *Since* (f_n) *is a sequence of continuous functions which converge to* f *uniformly on* $[a, b]$, *we have that* $f \in C[a, b]$.

(ii) *The space $C[a,b]$ of all continuous functions on $[a,b]$ is not complete with $\|f\| = \int_a^b |f(t)|dt,\ f \in C[a,b]$. Let us take $[a,b] = [-1,1]$ and define $\{f_n(t)\}$ by*

$$f_n(t) := \begin{cases} 0 & , \quad -1 \le t < 0, \\[2mm] nt & , \quad 0 \le t < \dfrac{1}{n}, \\[2mm] 1 & , \quad \dfrac{1}{n} \le t \le 1; \end{cases}$$

see Figure 1.1. It is clear that (f_n) is a sequence of continuous functions

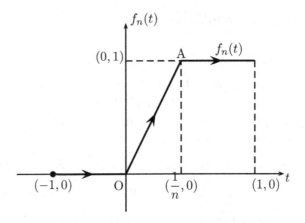

FIGURE 1.1: The sequence (f_n) of continuous functions.

on $[-1,1]$. We will show that (f_n) is a Cauchy sequence. Let $m > n$. Then, $\dfrac{1}{m} < \dfrac{1}{n}$.

Note that $\int_{-1}^1 |f_n(t) - f_m(t)|dt$ represents the area between the graphs of $f_n(t)$ and $f_m(t)$; see Figure 1.2. Therefore, this area is equal to

$$\frac{1}{2}|AB| \times |OC| = \frac{1}{2}\left|\frac{1}{n} - \frac{1}{m}\right|.$$

Hence,

$$\lim_{m,n\to\infty} \|f_n - f_m\| =$$

$$= \lim_{m,n\to\infty} \int_{-1}^1 |f_n(t) - f_m(t)|dt$$

$$= \frac{1}{2} \lim_{m,n\to\infty} \left|\frac{1}{n} - \frac{1}{m}\right| = 0.$$

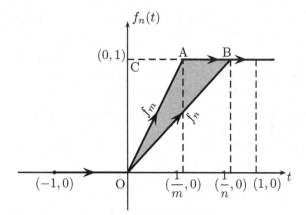

FIGURE 1.2: The graphs of $f_n(t)$ and $f_m(t)$.

Hence, $(f_n)_{n=1}^{\infty}$ *is Cauchy in* $C[-1,1]$.

Next, we show that (f_n) *converges to an element, say* f*, not in* $C[-1,1]$*. Since the interval* $0 \le t \le 1/n$ *reduces to* $t = 0$ *whenever* $n \to \infty$ *and the interval* $1/n < t \le 1$ *to* $0 < t \le 1$*, we therefore consider the function* $f : [-1,1] \to [0,1]$ *defined by*

$$f(t) = \begin{cases} 0 & , \quad -1 \le t \le 0, \\ 1 & , \quad 0 < t \le 1; \end{cases}$$

see Figure 1.3.

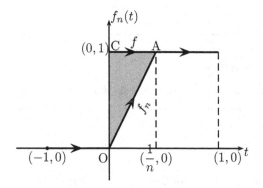

FIGURE 1.3: Graph of the function f.

We will show that $\|f_n - f\| \to 0$*, as* $n \to \infty$*.*

$\|f_n - f\| = \int_{-1}^{1} |f_n(t) - f(t)| dt =$ *area of the triangle* $O\overset{\triangle}{C}A$ *shaded* $= 1/(2n) \to 0$*, as* $n \to \infty$*. Thus, the Cauchy sequence* (f_n) *in* $C[-1,1]$

converges to f. However, f is not continuous at 0. Hence, $f \notin C[-1,1]$. Therefore, $C[-1,1]$ is not complete.

(iii) *The Lebesgue spaces L_p play a central role in many branches of analysis and in particular in functional analysis, Fourier analysis and wavelets. We present here some basic theorems and inequalities for L_p.*

Given a measure space $(X, \mathcal{A}, \mu)$, where $\mathcal{A}$ is a σ-algebra in X, μ a positive measure on $(X, \mathcal{A})$. We define $L_p(X)$ with $1 \le p < \infty$ as the collection of all complex measurable functions f on X such that

$$\int_X |f(x)|^p d\mu < \infty,$$

i.e., $f \in L_p(X) = L_p(X, \mathcal{A}, \mu)$ if and only if f is $\mathcal{A}$-measurable and $\int_X |f(x)|^p d\mu < \infty$. If $1 \le p < \infty$ and $f \in L_p(X)$, then we define

$$\|f\|_p = \left(\int_X |f(x)|^p d\mu \right)^{1/p}. \tag{1.3.7}$$

For $1 \le p < \infty$, $L_p(X)$ is a complex normed linear space with the norm $\| \cdot \|_p$ defined by (1.3.7).

Triangle inequality follows from Minkowski's inequality. Absolute homogeneity also follows easily. For positive definiteness, we define the relation "$\sim$" as $f \sim g$ iff $f = g$ almost everywhere. Then, "$\sim$" defines an equivalence relation and $L_p(X)$ is regarded as the set of equivalence classes of the functions. Then, we easily obtain that $\|f\|_p = 0$ iff $f = \theta$.

1.4 Bounded Linear Operators

Definition 1.4.1. *Let X and Y be linear spaces over the field of real numbers $\mathbb{R}$ or complex numbers $\mathbb{C}$. Then, a function $T : X \to Y$ is called a linear operator if*

$$T(\alpha x_1 + \beta x_2) = \alpha T x_1 + \beta T x_2$$

for all $x_1, x_2 \in X$ and for all scalars $\alpha, \beta \in \mathbb{C}$. By $\mathcal{L}(X : Y)$, we denote the set of all linear operators $T : X \to Y$. A complex valued linear operator is called a linear functional.

Definition 1.4.2. *Let X and Y be normed linear spaces. Then an operator $T : X \to Y$ is said to be bounded if there exists a constant $M > 0$ such that*

$$\|Tx\| \le M \|x\| \quad \text{for all} \ \ x \in X.$$

Note that a bounded functional f on X satisfies

$$|f(x)| \leq M\|x\| \quad \text{for all} \quad x \in X.$$

We denote the set of all bounded linear operators by $\mathcal{B}(X : Y)$. Let $T \in \mathcal{B}(X : Y)$. Then, the operator norm of T is defined by

$$\|T\| = \sup_{x \in X \setminus \{\theta\}} \frac{\|Tx\|}{\|x\|}.$$

Note that if Y is complete then $\mathcal{B}(X : Y)$ is also complete with the operator norm. By $X' = \mathcal{B}(X, \mathbb{C})$, we denote the continuous (bounded) dual of X, that is, the set of all continuous linear functionals on X. If X is a Banach space, then we write X^* for X' with its norm given by

$$\|f\| = \sup_{\substack{x \in X \\ \|x\|=1}} |f(x)| \quad \text{for all} \quad f \in X'.$$

The dual space is an important concept in functional analysis. There are two types of duals: The algebraic dual and the continuous dual. When dealing with topological vector spaces, one is typically only interested in the continuous linear functionals. This gives rise to the notion of the "continuous dual" or "topological dual." For any finite-dimensional normed linear space or topological vector space, such as Euclidean n-space, the continuous dual and the algebraic dual coincide. This is however false for any infinite-dimensional normed linear space.

Definition 1.4.3. *Let X and Y be normed linear spaces. Then an operator $T : X \to Y$ is said to be continuous at a point $x_0 \in X$ if for every $\varepsilon > 0$ there exists $\delta = \delta(\varepsilon, x_0) > 0$ such that*

$$\|Tx - Tx_0\| < \varepsilon \quad \text{whenever} \quad \|x - x_0\| < \delta.$$

If we can find such a δ, which depends upon ε only, then T is said to be uniformly continuous. An operator $T : X \to Y$ is said to be continuous on X if it is continuous at all points of X.

Theorem 1.4.4. *Let X and Y be normed spaces, and $T : X \to Y$ be a linear operator. Then, the following statements hold:*

(i) If T is continuous at the origin, then it is uniformly continuous on X.

(ii) T is continuous on X if and only if it is bounded.

Proof. (i) If T is continuous at the origin, then for every $\epsilon > 0$ there is a $\delta > 0$ such that

$$\|Tx\| < \epsilon \text{ whenever } \|x\| < \delta.$$

Hence, by linearity of T

$$\|T(x-y)\| = \|Tx - Ty\| < \epsilon$$

whenever $\|x - y\| < \delta$, that is, T is uniformly continuous on X.

(ii) Let T be bounded. It follows from the linearity of T and the definition of boundedness that

$$\|Tx - Tx_0\| = \|T(x - x_0)\| \le M\|x - x_0\|.$$

By setting $\frac{\delta \epsilon}{M}$, we see that T is continuous at x_0. Hence, T is uniformly continuous on X which implies continuity of T.

Conversely, assume that T is continuous. Then, T is continuous at an arbitrary point $x_0 \in X$. Thus, given any $\epsilon > 0$, there is a $\delta > 0$ such that

$$\|Tx - Tx_0\| < \epsilon \text{ for all } x \in X \text{ satisfying } \|x - x_0\| < \delta. \qquad (1.4.1)$$

Take a point $y \in X \setminus \{\theta\}$ and write $x = x_0 + \frac{\delta}{\|y\|}y$. Then $x - x_0 = \frac{\delta}{\|y\|}y$. Hence $\|x - x_0\| = \delta$, so that by (1.4.1) and by the linearity of T, we see that

$$\|Tx - Tx_0\| = \|T(x - x_0)\| = \|T(\frac{\delta}{\|y\|}y)\| = \frac{\delta}{\|y\|}\|Ty\| < \epsilon$$

Therefore

$$\|Ty\| < \frac{\epsilon}{\delta}\|y\| = C\|y\|,$$

where $C = \epsilon/\delta$ and hence T is bounded.

This completes the proof. $\square$

Remark 1.4.5. *For any normed spaces X and Y, $\mathcal{B}(X:Y) \subset \mathcal{L}(X:Y)$ and the inclusion is proper.*

Since every bounded linear operator is obviously also linear, the inclusion is trivial. To show it proper, let $Y = \mathbb{C}$ and $X = \ell_1$ with the norm of ℓ_∞.

Consider $f(x) = \sum_k x_k$ with $x = (x_k) \in \ell_1$. It is obvious that $f : \ell_1 \to \mathbb{C}$ and that f is linear on ℓ_1. We have to show that f is not bounded. On the contrary, suppose that f is bounded. Then, there is an integer $N > 0$ such that

$$|f(x)| \le N\|x\| \text{ for all } x \in \ell_1. \qquad (1.4.2)$$

Let $y = (y_k) \in \ell_1$ be defined by

$$y_k := \begin{cases} 1 & , \quad 1 \le k \le N+1, \\ 0 & , \quad k > N+1 \end{cases}$$

for all $k \in \mathbb{N}_0$. Then, $\|y\|_\infty = \sup_{k \in \mathbb{N}_0} |y_k| = 1$, $f(y) = N+1$ and by (1.4.2) we get $N + 1 \le N$, a contradiction. Hence, f is not bounded.

Theorem 1.4.6. *The following statements hold:*

(i) The space $\mathcal{B}(X:Y)$ is a normed linear space with

$$\|T\| = \sup_{x \in X \setminus \{\theta\}} \frac{\|Tx\|}{\|x\|} \quad \text{for } T \in \mathcal{B}(X:Y).$$

(ii) If Y is a Banach space, then $\mathcal{B}(X:Y)$ is also a Banach space with the norm given in Part (i).

Proof. (i) Let $T_1, T_2 \in \mathcal{B}(X:Y)$. Then, for all $x \in X$,

$$
\begin{aligned}
\|(T_1 + T_2)x\| &= \|T_1 x + T_2 x\| \\
&\leq \|T_1 x\| + \|T_2 x\| \leq (\|T_1\| + \|T_2\|)\|x\|,
\end{aligned}
$$

that is,

$$\|T_1 + T_2\| \leq \|T_1\| + \|T_2\|.$$

It is trivial that $\|\lambda T\| \leq |\lambda| \|T\|$, $\lambda \in \mathbb{K}$. Now, if $\|T\| = 0$ then from $\|Tx\| \leq \|T\|\|x\|$, we get $Tx = \theta$ on X, i.e., $T = \theta$. Hence, $\mathcal{B}(X:Y)$ is a normed linear space.

(ii) To show $\mathcal{B}(X:Y)$ is complete while Y is complete, let (T_n) be a Cauchy sequence in $\mathcal{B}(X:Y)$. Then, for each $x \in X$,

$$\lim_{m,n \to \infty} \|T_n x - T_m x\| \leq \lim_{m,n \to \infty} \|T_n - T_m\|\|x\| = 0,$$

so that $(T_n x)$ is a Cauchy sequence in Y. Since Y is complete, $T_n x \to Tx$, as $n \to \infty$, say. We have to show that $T \in \mathcal{B}(X:Y)$ and $\|T_n - T\| \to 0$, as $n \to \infty$. It is easy to check that T is linear. Now

$$\|Tx\| = \lim_{n \to \infty} \|T_n x\| \leq \|x\| + \|T_N\|\|x\|,$$

where $N = N(1)$ comes from the Cauchy condition on (T_n). Therefore, T is bounded. Finally, we have for every $\epsilon > 0$

$$
\begin{aligned}
\|T_n - T\| &= \sup_{x \in X \setminus \{\theta\}} \frac{\|T_n x - Tx\|}{\|x\|} \\
&< \frac{\|T_n x - Tx\|}{\|x\|} + \frac{\epsilon}{2} \\
&= \left\| T_n \left(\frac{x}{\|x\|} \right) - T \left(\frac{x}{\|x\|} \right) \right\| + \frac{\epsilon}{2} < \epsilon, \quad \text{if } n > N(\epsilon).
\end{aligned}
$$

Since $T_n(x/\|x\|) \to T(x/\|x\|)$, as $n \to \infty$, in the norm of Y, $T_n \to T$, as $n \to \infty$, in the norm of $\mathcal{B}(X:Y)$. Hence, $\mathcal{B}(X:Y)$ is a Banach space. $\quad\square$

Examples 1.4.7. ([228, p. 91], [111, 29]) *We shall write $X^* \cong Y$ if Y is the continuous dual of X. We have the following:*

(i) $\ell_p^* \cong \ell_q$, *where* $1 < p < \infty$ *with* $p^{-1} + q^{-1} = 1$.

(ii) $c^* \cong \ell_1$.

(iii) We should note that dual of ℓ_∞ *is not a sequence spaces.*

1.5 Köthe-Toeplitz Duals

Besides continuous duals, there are some other types of duals, namely α-, β-, γ-, δ- and f-duals (cf. [29], [32, Theorem 2.5.9], [228, Theorem 7.3.5]), which are very useful in the study of sequence spaces and matrix transformations.

For any two sequences $x = (x_k)_{k \in \mathbb{N}_0}$ and $y = (y_k)_{k \in \mathbb{N}_0}$, let $xy = (x_k y_k)_{k \in \mathbb{N}_0}$. If $z = (z_k)_{k \in \mathbb{N}_0}$ is any sequence and Y is any subset of ω, then we shall write

$$z^{-1} * Y := \{x = (x_k) \in \omega : zx = (z_k x_k) \in Y\}.$$

Definition 1.5.1. *Let* X *be a sequence space. Then, the Köthe-Toeplitz dual or* α-*dual* X^α*, generalized Köthe-Toeplitz-dual or* β-*dual* X^β *and bounded-dual or* γ-*dual* X^γ *of* X *are defined by*

$$X^\alpha := \left\{ a = (a_k) \in \omega : \sum_k |a_k x_k| < \infty \text{ for all } x \in X \right\} = \bigcap_{x \in X} (x^{-1} * \ell_1),$$

$$X^\beta := \left\{ a = (a_k) \in \omega : \sum_k a_k x_k \text{ converges for all } x \in X \right\} = \bigcap_{x \in X} (x^{-1} * cs),$$

$$X^\gamma := \left\{ a = (a_k) \in \omega : \sup_{n \in \mathbb{N}_0} \left| \sum_{k=0}^{n} a_k x_k \right| < \infty \text{ for all } x \in X \right\} = \bigcap_{x \in X} (x^{-1} * bs),$$

respectively.

Examples 1.5.2. *Let* † *stands for* α, β *or* γ. *Then, the following statements hold:*

(i) $\ell_1^\dagger = \ell_\infty$

(ii) $\ell_\infty^\dagger = c_0^\dagger = c^\dagger = \ell_1$

(iii) $\ell_p^\dagger = \ell_q$, *where* $1 < p, q < \infty$ *with* $p^{-1} + q^{-1} = 1$.

Remark 1.5.3. *Let* † *denote any of the symbols* α, β *or* γ. *Then, the following statements hold:*

(i) $\phi \subset X^\alpha \subset X^\beta \subset X^\gamma$ *for any* $X \subset \omega$.

(ii) *Let* X *and* Y *be any two subsets in* ω. *If* $X \subset Y$, *then* $Y^\dagger \subset X^\dagger$.

(iii) $X^{\dagger\dagger\dagger} = X^\dagger$.

1.6 Basic Theorems

We state now well-known theorems of functional analysis, which will be frequently used in this book.

The Hahn-Banach theorem is a central tool in functional analysis. It allows the extension of bounded linear functionals defined on a subspace of some linear space to the whole space. We begin with giving the definition of sublinear functional.

Definition 1.6.1. *Let* X *be a real vector space. A sublinear functional on* X *is a function* $f : X \to \mathbb{R}$ *such that*

(i) $f(x + y) \le f(x) + f(y)$ *for all* $x, y \in X$.

(ii) $f(\alpha x) = \alpha f(x)$ *for all* $x \in X$ *and* $\alpha \ge 0$.

Hahn-Banach Theorem. *Let* p *be a sublinear functional on the real linear space* X *and suppose that* Y *is a subspace of* X, *and* f *be a linear functional on* Y *such that* $f(x) \le p(x)$ *for all* $x \in Y$. *Then, there exists a linear functional* F *on* X *such that* $F(x) \le p(x)$ *for all* $x \in X$ *and* $F|_Y = f$, *i.e.,* $F(x) = f(x)$ *for all* $x \in Y$.

Proof. First we show the existence of F. Let $\mathcal{K}$ be the set of all linear extensions g_α of f, i.e., the set of all pairs (Y_α, g_α) in which Y_α is a linear subspace of X containing Y and g_α is a real linear functional on Y_α with the properties that $g_\alpha(y) = f(y)$ for all $y \in Y$, and $g_\alpha(x) \le p(x)$ for all $x \in Y_\alpha$. Clearly, $\mathcal{K} \ne \emptyset$ since $f \in \mathcal{K}$. On $\mathcal{K}$, we can define a partial ordering by defining the relation $(Y_\alpha, g_\alpha) \le (Y_\beta, g_\beta)$ if $Y_\alpha \subseteq Y_\beta$ and $g_\alpha = g_\beta|_{Y_\alpha}$, meaning that g_β is an extension of g_α. It is clear that any totally ordered subset $\{(Y_\lambda, g_\lambda) : \lambda \in I\}$ has an upper bound given by the subspace $Y' = \cup_\lambda Y_\lambda$ and the functional defined by $g'(y) = g_\lambda(y)$ for $y \in Y_\lambda$ and $\lambda \in I$. That Y' is a subspace and that g' is well-defined both follow since $\{(Y_\lambda, g_\lambda) : \lambda \in I\}$ is linearly ordered. By Zorn's lemma, $\mathcal{K}$ has a maximal element (Y_0, g_0). By the definition of $\mathcal{K}$, g_0 is a linear extension of f, which satisfies $g_0(x) \le p(x)$, $x \in Y_0$.

Now, we have to show that F is defined on the whole of X. Assume that $Y_0 \ne X$. Let $x \in X \backslash Y_0$ and Y_1 be the linear space spanned by Y_0 and x. Note that $x \ne \theta$ since $\theta \in Y_0$. Each element $z \in Y_1$ can be expressed uniquely in the form $z = y + \lambda x$ with $y \in Y_0$ and $\lambda \in \mathbb{R}$, because x is assumed not to be in the vector space Y_0. Define a linear functional g_1 on Y_1 by setting $g_1(y + \lambda x) = g_0(y) + \lambda c$. Clearly g_1 is linear. Furthermore, for $\lambda = 0$ we have $g_1(y) = g_0(y)$. Hence, g_1 is a proper extension of g_0. Consequently, if we can prove that $g_1 \in \mathcal{K}$ by showing that $g_1(z) \leq p(z)$ for every $z \in Y_1$, this will contradict the maximality of g_0, so that $Y_0 \neq X$ is false. Note that if $y_1, y_2 \in Y_0$, then

$$g_0(y_1) - g_0(y_2) = g_0(y_1 - y_2) \leq p(y_1 - y_2) \leq p(y_1 + x) + p(-x - y_2)$$

and hence

$$-p(-x - y_2) - g_0(y_2) \leq p(y_1 + x) - g_0(y_1).$$

Consequently, we have

$$A = \sup_{y \in Y_0} \{-p(-x - y) - g_0(y)\} \leq \inf_{y \in Y_0} \{p(y + x) - g_0(y)\} = B.$$

Choose any number $c \in [A, B]$. Then, by construction of A and B,

$$c \leq p(y + x) - g_0(y) \tag{1.6.1}$$

for all $y \in Y_0$ and

$$-p(-x - y) - g_0(y) \leq c \tag{1.6.2}$$

for all $y \in Y_0$. Multiplying (1.6.1) by $\lambda > 0$ and substitute y/λ for y, we obtain

$$\lambda c \leq p(y + \lambda x) - g_0(y). \tag{1.6.3}$$

Multiplying (1.6.2) by $\lambda < 0$ and substitute y/λ for y, we again obtain (1.6.3). Hence, we obtain $g_1(z) = g_1(y + \lambda x) = g_0(y) + \lambda c \leq p(y + \lambda x) = p(z)$ for all $\lambda \in \mathbb{R}$ and $y \in Y_0$. That is, $(Y_1, g_1) \in \mathcal{K}$ and $(Y_0, g_0) \leq (Y_1, g_1)$ with $Y_0 \neq Y_1$. This contradicts the maximality of (Y_0, g_0) and hence $F = g_0$ is defined on all of X.

This completes the proof of the theorem.   □

Definition 1.6.2. *Let X, Y be normed linear spaces, and $E \subseteq X$ and $\mathcal{A} = (T_n)$ be a family of linear operators from X to Y. We say that $\mathcal{A}$ is pointwise bounded on E if for each $x \in E$, there exists $c_x > 0$ such that $\|T_n x\| \leq c_x$ for all $n \in \mathbb{N}_0$. The family $\mathcal{A}$ is said to be uniformly bounded on E if there exists $c > 0$ such that $\|T_n\| \leq c$ for all $n \in \mathbb{N}_0$ and $x \in E$.*

A family $\mathcal{A}$ of linear operators from X to Y is pointwise bounded if and only if $\mathcal{A}$ is pointwise bounded on X and $\mathcal{A}$ is uniformly bounded if $\mathcal{A} \subseteq \mathcal{B}(X : Y)$, i.e., $T_n \in \mathcal{B}(X : Y)$ and the sequence of norm $(\|T_n\|)$ is bounded.

Clearly, uniform boundedness implies pointwise boundedness.

The uniform boundedness principle asserts that the converse is also true if $\mathcal{A} \subseteq \mathcal{B}(X : Y)$ and X is Banach space.

Definition 1.6.3. *A subset M of a metric space X is said to be*

(i) *of the first category in X if M can be expressed as a countable union of nowhere dense sets.*

(ii) *of the second category in X if M is not of the first category in X.*

Now, we may state and prove the Baire category theorem, which will be used in proving our next theorem.

Baire Category Theorem. [134, Theorem 4.7, p. 246] *If a metric space $X \neq \emptyset$ is complete, it is second category.*

Hence, if $X \neq \emptyset$ is complete and $X = \bigcup_{k=0}^{\infty} A_k$, where A_k's are closed subsets of X, then at least one A_k contains a nonempty open subset, i.e., A_k has a nonempty interior.

Proof. The idea of the proof is simple. Suppose the complete metric space $X \neq \emptyset$ were of the first category in itself. Then,

$$X = \bigcup_{k=1}^{\infty} M_k \qquad (1.6.4)$$

with each M_k nowhere dense in X. We construct a Cauchy sequence (p_k) whose limit p (which exists by completeness) is in no M_k , thereby contradicting the representation (1.6.4).

By assumption, M_1 is nowhere dense in X, so that by definition $\overline{M}_1$ does not contain a nonempty open set. But X does (for instance, X itself). This implies $\overline{M}_1 \neq X$. Hence, the complement $\overline{M}_1^C = X - \overline{M}_1$ of $\overline{M}_1$ is not empty and is open. We may thus choose a point p_1 in $\overline{M}_1^C$ and an open ball about it, say,

$$B_1 = B(p_1; \varepsilon_1) \subset \overline{M}_1^C; \quad \varepsilon_1 < \frac{1}{2}.$$

By assumption, M_2 is nowhere dense in X, so that $\overline{M}_2$ does not contain a nonempty open set. Hence, it does not contain the open ball $B(p_1; \frac{1}{2}\varepsilon_1)$. This implies that $\overline{M}_2^C \cap B(p_1; \varepsilon_1/2)$ is not empty and is open so that we may choose an open ball in this set, say,

$$B_2 = B(p_2; \varepsilon_2) \subset \overline{M}_2^C \cap B(p_1; \varepsilon_1/2); \quad \varepsilon_2 < \frac{1}{2}\varepsilon_1.$$

By induction we thus obtain a sequence of balls $(B_k) = \{B(p_k; \varepsilon_k)\}; \varepsilon_k < 2^{-k}$ such that $B_k \cap M_k = \varnothing$ and

$$B_{k+1} \subset B(p_k; \varepsilon_k/2) \subset B_k; \quad k \in \mathbb{N}_1.$$

Since $\varepsilon_k < 2^{-k}$, the sequence (p_k) of the centers is Cauchy and converges, say, $p_k \to p \in X$ because X is complete by assumption. Also, for every $m \in \mathbb{N}_1$ and $n > m$ we have $B_n \subset B(p_m; \varepsilon_m/2)$, so that

$$d(p_m, p) \leq d(p_m, p_n) + d(p_n, p) < \frac{1}{2}\varepsilon_m + d(p_n, p) \to \frac{1}{2}\varepsilon_m,$$

as $n \to \infty$. Hence, $p \in B_m$ for every $m \in \mathbb{N}_1$. Since $B_m \subset \overline{M}_m^C$, we now see that $p \notin M_m$ for every $m \in \mathbb{N}_1$ so that $p \notin \bigcup M_m = X$. This contradicts $p \in X$.

This completes the proof. $\qquad\qquad\qquad\qquad\qquad\qquad\qquad\qquad\qquad\square$

Uniform Boundedness Principle. *Let $T_n \in \mathcal{B}(X : Y)$, X be a Banach space and Y be a normed linear space such that $(\|T_n x\|)$ is bounded for every $x \in X$, say*

$$\|T_n x\| \leq C_x \quad \text{for all} \ \ n \in \mathbb{N}_0, \qquad\qquad (1.6.5)$$

where $C_x \in \mathbb{R}^+$. Then, $(\|T_n\|)$ is bounded, i.e., there exists a $C > 0$ such that

$$\|T_n\| \leq C \quad \text{for all} \ \ n \in \mathbb{N}_0.$$

Proof. For every $n \in \mathbb{N}_0$, let $A_k = \{x \in X : \|T_n x\| \leq k\}$. Suppose that $\mathcal{A} = (T_n)$ is pointwise bounded. Then, for each $x \in X$ there exists $C_x > 0$ such that $\|T_n x\| \leq C_x$ for all $n \in \mathbb{N}_0$. Then, each $x \in X$ belongs to some A_k by (1.6.5).

Hence, $X = \bigcup_{k=0}^{\infty} A_k$, where each A_k is closed. Let (x_j) be a sequence in A_k converging to $x \in A_k$. Since (T_n) is pointwise bounded, we have $\|T_n x_j\| \leq k$ for all fixed $n \in \mathbb{N}_0$ and hence $\|T_n x\| \leq k$, (Since T_n is continuous and $\|\cdot\|$ is also continuous).

Hence, by the definition of A_k, $x \in A_k$. That is, A_k is closed.

Since X is complete, Baire's category theorem implies that some A_k has nonempty interior, i.e., some A_k contains an open ball, say

$$B_0 = B(x_0; r) \subset A_{k_0}. \qquad\qquad\qquad\qquad (1.6.6)$$

Let $x \in X$ be arbitrary, not zero, we set

$$z = x_0 + \alpha x, \quad \text{where} \ \ \alpha = \frac{r}{2\|x\|}. \qquad\qquad (1.6.7)$$

Then, $\|z - x_0\| = r/2 < r$ implies $z \in B_0$. By (1.6.6) and from the definition of A_{k_0}, we thus have $\|T_n z\| \leq k_0$ for all $n \in \mathbb{N}_0$. Also, $\|T_n x_0\| \leq k_0$, since $x_0 \in B_0$. From (1.6.7), we get $x = (z - x_0)/\alpha$ which implies that

$$\|T_n x\| = \frac{1}{\alpha}\|T_n(z - x_0)\| \leq \frac{1}{\alpha}\left(\|T_n z\| + \|T_n x_0\|\right) \leq \frac{4}{r}\|x\|k_0 \ \text{for all} \ n \in \mathbb{N}_0$$

because $\dfrac{1}{\alpha} = \dfrac{2}{r}\|x\|$. Hence,

$$\|T_n\| = \sup_{\substack{x \in X \\ \|x\|=1}} \|T_n x\| \leq \frac{4}{r}k_0 \quad \text{for all} \ \ n \in \mathbb{N}_0.$$

That is, $\|T_n\| \leq C$, where $C = 4k_0/r$ (independent of x) for all $n \in \mathbb{N}_0$. Hence, (T_n) is uniformly bounded.

This completes the proof of the theorem. $\qquad\square$

The following theorem is an immediate consequence of uniform boundedness principle:

Banach-Steinhaus Theorem. *Let X be a Banach space and Y be a normed linear space, and (T_n) be a sequence of bounded linear operators from X into Y such that $(T_n x)$ converges for every $x \in X$. Let $T : X \to Y$ be defined for $x \in X$ by $Tx = \lim_{n\to\infty} T_n x$. Then, $(\|T_n\|)$ is bounded and $T \in \mathcal{B}(X : Y)$.*

Proof. T is clearly linear (because $\lim$ and T_n are linear). Also, as we have seen earlier, for $x \in X$,

$$\|Tx\| = \lim_{n\to\infty} \|T_n x\| \leq \limsup_{n\to\infty} \|T_n\|\|x\| = M\|x\| \text{ (because } (\|T_n\|) \text{ is bounded).}$$

Hence, $T \in \mathcal{B}(X : Y)$. $\qquad\square$

Remark 1.6.4. *In case $Y = \mathbb{C}$ or $\mathbb{R}$, then $\|T_n x\|$ is replaced by $|A_n(x)|$.*

Bounded Inverse Theorem. *Let X, Y be Banach spaces and $A \in \mathcal{B}(X : Y)$ be surjective. Then, $A^{-1} \in \mathcal{B}(Y : X)$.*

Proof. Let X, Y be Banach spaces and $A \in \mathcal{B}(X : Y)$ be surjective. It follows from bounded inverse theorem that A is bicontinuous. Thus, A is a linear homeomorphism. Since A is onto, A^{-1} exists. Since A is an open map, A^{-1} is continuous. Hence, $A^{-1} \in \mathcal{B}(Y : X)$. $\qquad\square$

Closed Graph Theorem. *Let X and Y be Banach spaces, and $A \in \mathcal{L}(X : Y)$. Then, A is continuous if and only if its graph is closed.*

Proof. Suppose that A is continuous and $G(A)$ is its graph. Let $(x, y) \in \overline{G(A)}$. Then, $\exists\, x_n \in X$ such that $x_n \to x$ and $Ax_n \to y$, as $n \to \infty$. But A is continuous, so $Ax_n \to Ax$ and hence $y = Ax$. Therefore, $(x, y) = (x, Ax) \in G(A)$, i.e., $\overline{G(A)} \subset G(A)$ and so $G(A)$ is closed.

Conversely, let $G(A)$ be closed. Since Y is complete, $G(A)$ is also a complete subspace of $X \times Y$. Now, consider the mapping $f : G(A) \to X$ given by $f((x, Ax)) = x$. This mapping is clearly linear and bijective. Also, f is continuous, since $\|f(x, Ax)\| = \|x\| \leq \|(x, Ax)\|$. Hence, $f^{-1} : X \to G(A)$ is continuous by bounded inverse theorem. Finally,

$$\|Ax\| \leq \|(x, Ax)\| = \|f^{-1}(x)\| \leq \|f^{-1}\|\|x\|,$$

so that A is bounded and hence continuous.

This completes the proof of the theorem. $\qquad\square$

Open Mapping Theorem. *Let X and Y be Banach spaces, and suppose that $A \in \mathcal{B}(X : Y)$ is surjective. Then, A is an open mapping (that is, for every open set in X the image is an open set in Y).*

Proof. Let G be open in X and let $y \in A(G)$ so that $y = Ax$ for some $x \in G$. Now, there exists $S(x, \delta) \subset G$, where $A(S(x, \delta)) \subset A(G)$. Provided we can show that there is a sphere $S(y) \subset A(S(x, \delta))$ we will have $S(y) \subset A(G)$, so that $A(G)$ will be open. To ensure the provision it is enough to prove that there exists a sphere $S(\theta) \subset A(S_0)$, where S_0 is the unit sphere in X and $S(\theta)$ is a sphere about the origin in Y. Now, we prove that there exists $S(\theta) \subset A(S_0)$. Let

$$S_k = S\left(\theta, 2^{-k}\right) = \left\{ x : \|x\| < \frac{1}{2^k} \right\} \quad \text{for all } k \in \mathbb{N}_0.$$

Now, if $x \in X$, we may write $x = k(x/k)$, where $k = [2\|x\|] + 1$, then $(x/k) \in S_1$ and so $x \in kS_1$. Hence, $X = \bigcup_{k=1}^{\infty} kS_1$. Since A is onto, we get

$$Y = A(X) = A(\cup kS_1) = \cup kA(S_1) = \overline{\cup kA(S_1)}$$

(because the union is already the whole space Y.)

Since Y is complete, it is of second category. Hence, there exists k such that $\overline{kA(S_1)}$ is not nowhere dense set. Thus, $\overline{kA(S_1)}$ and hence, $\overline{A(S_1)}$ contain some sphere $S(a, r)$, say. Hence,

$$
\begin{aligned}
S(\theta, r) &= S(a, r) - a \\
&\subset \overline{A(S_1)} - a \subset \overline{A(S_1)} - \overline{A(S_1)} \\
&\subset 2\overline{A(S_1)} = \overline{A(S_0)},
\end{aligned}
$$

which implies that

$$S(\theta, r) \subset \overline{A(S_0)} \tag{1.6.8}$$

(we have used the fact that $a \in \overline{A(S_1)}$, $\overline{A(S_1)}$ is convex and that $-y \in \overline{A(S_1)}$ whenever $y \in \overline{A(S_1)}$).

From (1.6.8), it follows that

$$S(\theta, r_2^n) \subset \overline{A(S_n)}. \tag{1.6.9}$$

Finally, we show that $S(\theta, r/2) \subset A(S_0)$. Take $y \in S(\theta, r/2)$. Then, $y \in \overline{A(S_1)}$ by (1.6.9). Hence, $\|y - y_1\| < r/4$ for some $y_1 \in A(S_1)$. Also $y - y_1 \in S(\theta, r/4) \subset \overline{A(S_2)}$ and so $\|y - y_1 - y_2\| < r/8$ for some $y_2 \in A(S_2)$. Continuing in this way, we find that

$$\left\| y - \sum_{k=0}^{n} y_k \right\| < \frac{r}{2^{n-1}}, \tag{1.6.10}$$

where $y_k = Ax_k \in A(S_k)$ for some $x_k \in S_k$. From (1.6.10), we get $y = \sum_k Ax_k$ and since $\|x_k\| < 2^{-k}$, $\sum_k \|x_k\| < 1$, we see that $\sum_k x_k = x$, say. Also $\|x\| \leq \sum_k \|x_k\| < 1$, $x \in S_0$ and by continuity of A,

$$Ax = \sum_k Ax_k = y.$$

Thus, $y \in S(\theta, r/2)$ implies $y = Ax$ for some $x \in S_0$, i.e., $y \in A(S_0)$, which shows that $S(\theta, r/2) \subset A(S_0)$. Hence, $A(S_0)$ contains an open ball but $y = Ax$. Since $y \in A(S_0)$ was arbitrary, $A(S_0)$ is open. Hence, the mapping is open. $\quad \square$

1.7 Compact Operators

Definition 1.7.1. *Let X and Y be normed spaces. Then, an operator T from X to Y is said to be a compact linear if T is linear and if for every bounded subset M of X, the closure $\overline{T(M)}$ of the image $T(M)$ is compact.*

Further, we write $C(X : Y)$ for the class of all compact operators from X to Y. Let us remark that every operator in $C(X : Y)$ is bounded, that is $C(X : Y) \subset B(X : Y)$. More precisely, the class $C(X : Y)$ is a closed subspace of the Banach space $B(X : Y)$ with the operator norm.

On the other hand, an operator $L \in B(X : Y)$ is said to be *of finite rank* if $\dim R(L) < \infty$, where $R(L)$ denotes the range space of L. An operator of finite rank is clearly compact. Thus, if we write $\mathcal{F}(X : Y)$ for the *class of all finite rank operators* from X to Y, then we have $\mathcal{F}(X : Y) \subset C(X : Y) \subset B(X : Y)$.

Note that, if $A : X \to Y$ is a bounded operator between the normed linear spaces X and Y, then $\overline{\{Ax : \|x\| \leq 1\}}$ is a closed and bounded subset of Y. Since every closed and bounded subset of a finite dimensional normed linear space is compact, it follows that, if $A : X \to Y$ is a bounded operator of finite rank, then the above set is compact as well. This is not necessarily true if rank $A = \infty$. For example, suppose A is identity operator on an infinite dimensional normed linear space X, then the above set is the same as the closed unit ball of X, which is not compact.

In other words, we say that a linear operator $A : X \to Y$ between normed linear spaces X and Y is said to be a compact operator if the set $\overline{\{Ax : \|x\| \leq 1\}}$ is compact in Y.

The following results are easy to verify:

(i) Every bounded operator of finite rank is a compact operator.

(ii) The identity operator on a normed linear space is a compact operator if and only if the space is of finite dimension.

Theorem 1.7.2. *Let X and Y be two normed linear spaces, and $A : X \to Y$ be a linear operator. Then, the following statements are equivalent:*

(a) A is a compact operator.

(b) $\overline{\{Ax : \|x\| \leq 1\}}$ is compact in Y.

(c) For every bounded subset E of X, $\overline{A(E)}$ is compact in Y.

(d) For every bounded sequences (x_n) in X, the sequence (Ax_n) has a convergent subsequence in Y.

Proof. Clearly, (c) implies (a) and (b). Now, assume that (a) holds, i.e., $\overline{\{Ax : \|x\| \leq 1\}}$ is compact. Let E be a bounded subset of X. Then, it is known that there exists a $r > 0$ such that $E \subseteq \{x \in X : \|x\| < r\}$. Now, from the relations

$$\overline{A(E)} \subseteq \overline{\{Ax : x \in X, \|x\| < r\}} \subseteq \overline{\{Ax : x \in X, \|x\| \leq r\}},$$

and the fact that a closed subset of a compact set is compact, it follows that (a) implies (b) and (c), and (b) implies (c).

Now, we prove the equivalence of (c) and (d). Assume that (c) holds, and let (x_n) be a bounded sequence in X. Suppose that $\|x\| \leq c$ and let $E = \{x \in X : \|x\| \leq c\}$. Then, (Ax_n) is a sequence in the compact set $\overline{A(E)}$, so that it has a convergent subsequence. Thus, (d) holds.

Conversely, assume that (d) holds, and let E be a bounded subset of X. To show that $\overline{A(E)}$ is compact, it is enough to prove that every sequence in it has a convergent subsequence. To show this, suppose that (y_n) is a sequence in $\overline{A(E)}$. Then, there exists (x_n) in E such that $\|y_n - Ax_n\| \leq 1/n$ for all $n \in \mathbb{N}_1$. Now, by the hypothesis (d), (Ax_n) has a convergent subsequence, say (Ax_{n_j}). Then, it follows that (y_{n_j}) is a convergent subsequence of (y_n).

This completes the proof. $\qquad\square$

Theorem 1.7.3. *Let X be a normed linear space, Y be a Banach space and $A \in \mathcal{B}(X : Y)$. If (A_n) is a sequence in $\mathcal{C}(X : Y)$ such that $\|A_n - A\| \to 0$, as $n \to \infty$, then $A \in \mathcal{C}(X : Y)$.*

Proof. Let (A_n) be in $\mathcal{C}(X : Y)$ such that $\|A_n - A\| \to 0$, as $n \to \infty$. In order to show that $A \in \mathcal{C}(X : Y)$, it is enough to show that for every bounded sequence (x_n) in X, the sequence (Ax_n) has a convergent subsequence in Y. So, let (x_n) be a bounded sequence in X and $\epsilon > 0$ be given. By the assumption on (A_n), there exists $N \in \mathbb{N}_0$ such that $\|A - A_n\| < \epsilon$ for all $n \geq N$. Since $A_N \in \mathcal{C}(X : Y)$, there exists a subsequence $(\tilde{x}_n)$ of (x_n) such that $(A_N \tilde{x}_n)$ converges. In particular, there exists $n_0 \in \mathbb{N}_0$ such that $\|A_N \tilde{x}_n - A_N \tilde{x}_m\| < \epsilon$ for all $n, m \geq n_0$. Now, for every $n, m \geq n_0$,

$$
\begin{aligned}
\|A\tilde{x}_n - A\tilde{x}_m\| &\leq \|A\tilde{x}_n - A_N\tilde{x}_n\| + \|A_N\tilde{x}_n - A_N\tilde{x}_m\| + \|A_N\tilde{x}_m - A\tilde{x}_m\| \\
&\leq \|A - A_N\|\|\tilde{x}_n\| + \|A_N\tilde{x}_n - A_N\tilde{x}_m\| + \|A_N - A\|\|\tilde{x}_m\| \\
&\leq (2c + 1)\epsilon,
\end{aligned}
$$

where $c > 0$ is such that $\|x_n\| \leq c$ for all $n \in \mathbb{N}_0$. Thus, $(A\tilde{x}_n)$ is a Cauchy sequence. Since Y is complete, $(A\tilde{x}_n)$ converges. Hence, $A \in \mathcal{C}(X : Y)$.

This completes the proof. $\qquad\square$

Examples 1.7.4. *Let $1 \leq p \leq \infty$. Then, the following statements hold:*

(i) Let $A : \ell_p \to \ell_p$ be the right shift operator on ℓ_p, i.e., A is defined by

$$
(Ax)_i := \begin{cases} 0 & , \quad i = 1, \\ x_{i-1} & , \quad i > 1 \end{cases}
$$

Since $Ae^{(n)} = e^{(n+1)}$ and

$$\|e^{(n)} - e^{(m)}\|_p = \begin{cases} 2^{1/p} & , & 1 \leq p < \infty, \\ 1 & , & p = \infty \end{cases}$$

for all $n, m \in \mathbb{N}_0$, it follows that, corresponding to the bounded sequence $(e^{(n)})$, the sequence $(Ae^{(n)})$ does not have a convergent subsequence. Hence, the operator A is not compact.

(ii) *Using similar arguments as in (i) above, it can be seen that the left shift operator on ℓ_p is also not a compact operator on ℓ_p for any p with $1 \leq p \leq \infty$.*

(iii) *Let (λ_n) be a bounded sequence of non-negative scalars such that $\lambda_n \to \lambda \neq 0$, as $n \to \infty$, and let $A : \ell_p \to \ell_p$ be the diagonal operator associated with this sequence, i.e., A is defined by $(Ax)_i = \lambda_i x_i$ for all $i \in \mathbb{N}_0$; $x = (x_i) \in \ell_p$. Then, we know that $A \in \mathcal{B}(\ell_p)$ and $\|A\| = \sup_{n \in \mathbb{N}_0} |\lambda_n|$. Since*

$$Ae^{(n)} = \lambda_n e^{(n)}, \quad \|e^{(n)} - e^{(m)}\|_p = 1 \quad \text{for all} \quad n, m \in \mathbb{N}_0,$$

it follows that

$$\begin{aligned} \|Ae^{(n)} - Ae^{(m)}\|_p &= \|\lambda_n e^{(n)} - \lambda_m e^{(m)}\|_p \\ &\geq \|\lambda_n(e^{(n)} - e^{(m)})\|_p - \|(\lambda_n - \lambda_m)e^{(m)}\|_p \\ &= |\lambda_n| - |\lambda_n - \lambda_m|. \end{aligned}$$

Since $\lambda_n \to \lambda \neq 0$, as $n \to \infty$, it follows that $(Ae^{(n)})$ does not have any convergent subsequence. Hence, A is not a compact operator.

1.8   Schauder Basis and Separability

Now, we introduce the concept of a Schauder basis. For finite dimensional spaces, the concepts of Schauder and algebraic bases coincide. In most cases of interest, however, the concepts differ. Every linear space has an algebraic basis. But, there are linear metric spaces without a Schauder basis, as we shall see later in this section.

A Schauder basis or countable basis is similar to the *usual (Hamel) basis* of a vector space, the difference is that Hamel bases use linear combinations that are finite sums, while for Schauder bases they may be infinite sums. This makes Schauder bases more suitable for the analysis of infinite-dimensional topological vector spaces including Banach spaces. A Hamel basis is free from topology while a Schauder basis depends on the metric in question since it involves the notion of "convergence" in its definition and hence topology.

Definition 1.8.1. *A sequence $(b_k)_{k\in\mathbb{N}_0}$ in a linear metric space (X, d) is called a Schauder basis or briefly basis for X if for every $x \in X$ there exists a unique sequence $(\alpha_k)_{k\in\mathbb{N}_0}$ of scalars such that $x = \sum_k \alpha_k b_k$, that is $d(x, x^{[n]}) \to 0$, as $n \to \infty$; where $x^{[n]} = \sum_{k=0}^{n} \alpha_k b_k$ is known as the n-section of x. The series $\sum_k \alpha_k b_k$ which has the sum x is called the expansion of x, and (α_k) is called the sequence of coefficients of x with respect to the basis (b_k).*

Theorem 1.8.2. *If a normed space X has Schauder basis then, it is separable.*

Proof. Let X be a Banach space having a Schauder basis $(b_k)_{k\in\mathbb{N}_0}$, such that $\|b_k\| = 1$ for all $k \in \mathbb{N}_0$, i.e., every $x \in X$ has a unique representation $x = \sum_k \alpha_k b_k$, where $\alpha_k \in \mathbb{R}$ for all $k \in \mathbb{N}_0$. We fix such an $x \in X$ and show how to approximate it by elements from a countable set. Given $\varepsilon > 0$, there exists $N \in \mathbb{N}_0$ such that $\left\|x - \sum_{k=0}^{N} \alpha_k b_k\right\| < \dfrac{\varepsilon}{2}$. For each $\alpha_k \in \mathbb{R}$, we can find $\beta_k \in \mathbb{Q}$ such that $|\alpha_k - \beta_k| < \varepsilon/2^{k+1}$. Then, by the triangle inequality

$$\left\|x - \sum_{k=0}^{N} \beta_k b_k\right\| \leq \left\|x - \sum_{k=0}^{N} \alpha_k b_k\right\| + \left\|\sum_{k=0}^{N} \alpha_k b_k - \sum_{k=0}^{N} \beta_k b_k\right\| \frac{\varepsilon}{2} + \sum_{k=0}^{N} \frac{\varepsilon}{2^{k+1}} < \varepsilon.$$

Thus, every element in X can be approximated by finite linear combination of the elements of the Schauder basis. This proves that $\mathbb{Q}$ is dense in X and hence X is separable. □

Examples 1.8.3. *We have the following:*

    (i) *In ϕ if we get, say, the metric of c_0, $\{e^{(k)}\}_{k\in\mathbb{N}_0}$ is both a Schauder basis as well as a Hamel basis.*

    (ii) *In c_0, $\{e^{(k)}\}_{k\in\mathbb{N}_0}$ is a Schauder basis but not a Hamel basis. Since each $x = (x_k) \in c_0$ has the representation $\sum_k x_k e^{(k)}$ which is unique. For if there is any other representation, say, $\sum_k b_k e^{(k)}$ which of course diverges for $b = (b_k) \notin c_0$, and converges for $b \in c_0$, to b, not x, if $(b_k) \neq (x_k)$. But this is not a Hamel basis, since its span is ϕ, a proper subset of c_0. On the other hand, any Hamel basis of c_0 is uncountable and hence is automatically not a Schauder basis.*

    (iii) *$\{e^{(k)}\}_{k\in\mathbb{N}_0}$ is a Schauder basis for $\ell(p)$ under the paranorm $p(x) = \left(\sum_k |x_k|^{p_k}\right)^{1/M}$ on $\ell(p)$.*

    (iv) *The space ℓ_∞ has no Schauder basis, since it is not separable.*

    (v) *The spaces ω, c_0 and ℓ_p with $1 \leq p < \infty$ have $\{e^{(k)}\}_{k\in\mathbb{N}_0}$ as their Schauder bases.*

    (vi) *We define $b^{(k)}$ by $b^{(k)} := \begin{cases} e & , \quad k = 0, \\ e^{(k)} & , \quad k \in \mathbb{N}_1 \end{cases}$ for all $k \in \mathbb{N}_0$. Then, the sequence $\{b^{(k)}\}_{k\in\mathbb{N}_0}$ is a Schauder basis for c. More precisely, every sequence $x = (x_k) \in c$ has a unique representation $x = le + \sum_{k\in\mathbb{N}_1}(x_k - l)e^{(k)}$, where $x_k \to l$, as $k \to \infty$.*

Note that the separability implies the existence of Schauder basis, is not true (see P. Enflo [77]).

1.9 Reflexivity

Definition 1.9.1. *Let (X_1, d_1) and (X_2, d_2) be metric linear spaces over $\mathbb{K}$. Then a mapping T of X_1 into X_2 is said to be isometric or an isometry if $d_2(Tx, Ty) = d_1(x, y)$ for all $x, y \in X_1$. The space X_1 is said to be isometric with the space X_2 if there exists a bijective isometry on X_1 onto X_2. The spaces X_1 and X_2 are called isometric spaces.*

A normed linear space X is said to be reflexive if X and X^{**} are linearly isometric under the linear isometry $J : X \to X^{**}$, where X^{**} is the bidual (or second dual) of X.

Theorem 1.9.2. *If X is finite dimensional normed linear space, then it is reflexive.*

Proof. Let $\dim(X) = n < \infty$. Then, X^* and X^{**} also have dimension n. Let $J : X \to X^{**}$ be a linear isometry. Since J is linear and one to one, $J(X)$ is an n dimensional subspace of the n dimensional linear space X^{**}. Therefore, $J(X) = X^{**}$, i.e., J is onto. Hence, J is an onto isometry, i.e., X and X^{**} are linearly isometric under J, and consequently X becomes reflexive. □

Theorem 1.9.3. *Let X be a reflexive normed linear space. Then, the following statements hold:*

(a) Every closed subspace of X is reflexive.

(b) X^* is reflexive.

(c) X is separable iff X^* is separable.

Proof. (a) Let Y be a closed subspace of X and $J_y : Y \to Y^{**}$ be the linear isometry of Y into Y^{**}. We know that $J : X \to X^{**}$ is onto. Let $g^{**} \in Y^{**}$. Define $f^{**} \in X^{**}$ by letting $f^{**}(f^*) = g^{**}(f^*|_Y)$ for $f^* \in X^*$. Since J is onto, there exists $x \in X$ such that $J(x) = f^{**}$. We claim that, in fact $x \in Y$. For, if $x \in Y$, then since Y is closed, there would exists $f_0^* \in X$ such that $f_0^* = 1$, but $f_0^* = 0$ on Y. Then $1 = f_0^*(x) = f^{**}(f_0^*) = g^{**}(0) = 0$, a contradiction. Hence, $x \in Y$ and we have $J_y(x) = g^{**}$.

(b) Let $J'^* : X^* \to X^{***}$ be a linear isometry of X^* into $(X^*)^{**}$. We must show that J' is onto. Let $f^{***} \in X^{***}$. Define $f^* \in X^*$ by $f^* = f^{***} \circ J$. We must show that $J'^* = f^{***}$. If $f^{**} \in X^{**}$, then since J is onto, there exists $x \in X$ with $J(x) = f^{**}$. Now,

$$J'^*(f^{**}) = f^{**}(f^*) = J(x)(f^*) = f^*(x) = f^{***} \circ J(x) = f^{***}(f^{**}).$$

Thus, $J'^* = f^{***}$. This shows that J' is onto.

(c) It can be easily seen that if X^* is separable then so is X.

Conversely, let X be separable. Since X is reflexive, X^{**} is isometric to X, and hence is itself separable. Therefore, X^* is separable since $X^{**} = (X^*)^*$. $\square$

Examples 1.9.4. *We have the following:*

(i) *c_0, c, ℓ_1, ℓ_∞ and $C[a,b]$ are not reflexive. For example, note that the duals of these normed linear spaces are linearly isometric to ℓ_1, while the dual of ℓ_1 is linearly isometric to ℓ_∞. Since c_0 and c are separable, they cannot be isometric to the non-separable space ℓ_∞.*

(ii) *Let $1 < p < \infty$. Then, ℓ_p and $L^p[a,b]$ are reflexive. For example, let $X = L^p[a,b]$ with $1 < p < \infty$ and $Y = L^q[a,b]$ with $p^{-1} + q^{-1} = 1$. The linear isometries $F : Y \to X^*$ and $G : X \to Y^*$ given by*

$$F(y)(x) = \int_a^b xy\,dm = G(x)(y); \ x, y \in X$$

are onto. Let now, $f^{**} \in X^{**}$. Since $f^{**} \circ F \in Y^*$, there exists $x \in X$ with $G(x) = F^{**} \circ F$. For any $f^* \in X^*$, there is unique $y \in Y$ such that $F(y) = f^*$. Thus, for every $f^* \in X^*$ with $F(y) = f^*$, we have

$$f^{**}(f^*) = f^{**}F(y) = G(x)(y) = F(y)(x) = f^*(x).$$

This shows that $X = L^p[a,b]$ is reflexive, where $1 < p < \infty$.

(iii) *$L^1[a,b]$ and $L^\infty[a,b]$ are not reflexive. For example, since*

$$(L^1[a,b])^{**} = (L^\infty[a,b])^* \neq L^1[a,b];$$

$L^1[a,b]$ is not reflexive. Similarly, $L^\infty[a,b]$ is not reflexive, since otherwise its closed subspace $C[a,b]$ would also be reflexive which is also not the case.

As a direct consequence of Part (i) of Example 1.9.4, we have:

Remark 1.9.5. *A reflexive normed linear space is a Banach space but the converse need not be true.*

1.10   Weak Convergence

Weak convergence has various applications in analysis. This is more suitable for considerations of the duality between a normed linear space X and its dual X^*.

Definition 1.10.1. *A sequence* (x_n) *is said to be weakly convergent to* x *in a normed linear space* X, *written as* $x_n \overset{w}{\to} x$, *if* $f(x_n) \to f(x)$ *for every* $f \in X^*$. *We call* x *the weak limit of* (x_n). *We say that* (x_n) *converges to* x *in norm if* $\|x_n - x\| \to 0$, *as* $n \to \infty$, *and we simply write it as* $x_n \to x$, *as* $n \to \infty$. *We also call convergence in norm as strong convergence.*

Theorem 1.10.2. *Let* X *be a normed linear space and* $x_n \overset{w}{\to} x$ *in* X. *Then, the following statements hold:*

(i) *The weak limit of* (x_n) *is unique.*

(ii) *Every subsequence of* (x_n) *converges weakly to* x.

(iii) *The sequence* $(\|x_n\|)$ *is bounded.*

Proof. (i) Suppose that $x_n \overset{w}{\to} x$ and $x_n \overset{w}{\to} y$. Then, $f(x_n) \to f(x)$ and $f(x_n) \to f(y)$. Therefore,

$$f(x - y) = f(x - x_n + x_n - y) = f(x - x_n) + f(x_n - y) \to 0.$$

Hence, $f(x - y) = 0$ for every $f \in X^*$. Since f is linear and continuous on X, we have $x - y = 0$, i.e., $x = y$.

(ii) This follows from the fact that $\{f(x_n)\}$ is a convergent sequence of numbers so that every subsequence converges to the same limit.

(iii) $x_n \overset{w}{\to} x$ implies $f(x - x_n) \to 0$, as $n \to \infty$, for every $f \in X^*$. Therefore, by Hahn-Banach theorem there exists $f_n \in X^*$ such that $f_n(x_n - x) = \|x_n - x\|$ and $\|f_n\| = 1$. For each $f \in X^*$, define $F_n(f) = f(x_n - x)$. Then, (F_n) is a sequence of continuous linear functionals on the Banach space X^*. Since $f(x_n - x) \to 0$, as $n \to \infty$, it follows that $\limsup_{n \to \infty} |F_n(f)| < \infty$ on X^*. The Banach-Steinhaus theorem does not yield $M = \sup_{n \in \mathbb{N}_0} \|F_n\| < \infty$. Hence,

$$\|x_n - x\| = |f_n(x_n - x)| = |F_n(f_n)| \leq \|F_n\| \|f_n\| \leq M,$$

so that $\|x_n\| \leq M + \|x\|$ for every $n \in \mathbb{N}_0$. This implies that $\sup_{n \in \mathbb{N}_0} \|x_n\| < \infty$. This completes the proof. $\qquad \square$

In the following theorem, we give the relation between the weak convergence and the convergence in norm.

Theorem 1.10.3. *Let* (x_n) *be a sequence in a normed linear space* X. *Then, the following statements hold:*

(i) *Strong convergence implies weak convergence with the same limit but the converse need not be true.*

(ii) *In general, weak convergence is not equivalent to strong convergence.*

(iii) *If* $\dim(X) < \infty$, *then weak convergence implies strong convergence.*

(iv) *In* ℓ_1, *weak convergence is equivalent to strong convergence.*

Proof. (i) Since $|f(x_n) - f(x)| \leq \|f\| \|x_n - x\|$ for every $f \in X^*$, if $\|x_n - x\| \to 0$, as $n \to \infty$, then $f(x_n) \to f(x)$. Hence, $x_n \overset{w}{\to} x$.

(ii) Let $X = \ell_p$ with $1 < p < \infty$. Then, every $f \in \ell_p^*$ can be written as $f(x) = \sum_k a_k x_k$ for all $x = (x_k) \in \ell_p$, where $a = (a_k) \in \ell_q$ and $p^{-1} + q^{-1} = 1$. If we write $e^{(1)} = (1, 0, 0, \ldots)$, $e^{(2)} = (0, 1, 0, \ldots)$, ..., then with the ℓ_p-norm we have $\|e^{(n)} - e^{(m)}\| = 2^{1/p}$ for all $n \neq m$ so that $\{e^{(n)}\}$ does not converge in norm. However, $\{e^{(n)}\}$ is weakly convergent to θ, since if $f \in \ell_p^*$ we have $f(e^{(n)}) = a_n$, and $a = (a_n) \in \ell_q$ implies $a_n \to 0$, i.e., $f(e^{(n)}) \to 0$. Hence, $\{e^{(n)}\}$ is weakly convergent to θ.

(iii) Suppose that $x_n \overset{w}{\to} x$ and $\dim(X) = r < \infty$. Let $\{e^{(1)}, e^{(2)}, \ldots, e^{(r)}\}$ be any basis for X and, say, $x_k = \sum_{i=1}^r \alpha_i^{(n)} e^{(i)}$ and $x = \sum_{i=1}^r \alpha_i e^{(i)}$. Since $f(x_n) \to f(x)$ for every $f \in X^*$, take in particular $f_1, f_2, \ldots, f_k$ defined by

$$f_i(e^{(m)}) = \begin{cases} 1 & , \quad m = i, \\ 0 & , \quad m \neq i. \end{cases}$$

Then, $f_i(x_n) = \alpha_i^{(n)}$ and $f_i(x) = \alpha_i$, as $i \to \infty$. Hence, $f_i(x_n) \to f_i(x)$ implies $\alpha_i^{(n)} \to \alpha_i$ and so

$$\begin{aligned}
\lim_{n \to \infty} \|x_n - x\| &= \lim_{n \to \infty} \left\| \sum_{i=1}^r \left[\alpha_i^{(n)} - \alpha_i \right] e^{(i)} \right\| \\
&\leq \lim_{n \to \infty} \sum_{i=1}^r \left| \alpha_i^{(n)} - \alpha_i \right| \left\| e^{(i)} \right\| = 0,
\end{aligned}$$

that is, $x_n \to x$, as $n \to \infty$, in norm.

(iv) See to the proof of Theorem 3.4.6 of Chapter 3.

This completes the proof. $\qquad \square$

1.11 Hilbert Spaces

Definition 1.11.1. *Let H be a complex vector space. An inner product on H is a function $\langle \cdot, \cdot \rangle : H \times H \to \mathbb{C}$ if the following axioms hold for arbitrary points $x, y, z \in H$ and arbitrary scalars $\alpha, \beta \in \mathbb{C}$:*

(H1) $\langle x, x \rangle \in \mathbb{R}$ *and* $\langle x, x \rangle \geq 0$. $\|x\|^2 = \langle x, x \rangle \geq 0$ *with the equality* $\|x\|^2 = 0$ *if and only if* $x = \theta$.

(H2) $\langle x, x \rangle = 0$ *if and only if* $x = \theta$.

(H3) $\langle \alpha x + \beta y, z \rangle = \alpha \langle x, z \rangle + \beta \langle y, z \rangle$, *i.e.,* $x \longmapsto \langle x, z \rangle$ *is linear.*

(H4) $\langle x, y \rangle = \overline{\langle y, x \rangle}$.

In this case $(H, \langle \cdot, \cdot \rangle)$ *is called an inner product space. A complete inner product space is called a Hilbert space.*

Note that combining properties (H3) and (H4) that $x \mapsto \langle z, x \rangle$ is antilinear for fixed $z \in H$, i.e.,

$$\langle z, \alpha x + \beta y \rangle = \bar{\alpha} \langle z, x \rangle + \bar{\beta} \langle z, y \rangle.$$

Remark 1.11.2. *An inner product on X defines a norm by $\|x\| = \langle x, x \rangle^{1/2}$ and a metric by $d(x, y) = \|x - y\| = \langle x - y, x - y \rangle^{1/2}$. Therefore, an inner product space is a normed space. Note that all normed spaces are not inner product spaces.*

Theorem 1.11.3. *(Schwarz Inequality) . Let $(H, \langle \cdot, \cdot \rangle)$ be an inner product space. Then, $|\langle x, y \rangle| \leq \|x\| \|y\|$ for all $x, y \in H$ and equality holds iff x and y are linearly dependent.*

Remark 1.11.4. *All Hilbert spaces are Banach spaces but not conversely (see Part (ii) of Example 1.11.5, below).*

Examples 1.11.5. *We have the following:*

 (i) *The space ℓ_2 is a Hilbert space with the inner product defined by $|\langle x, y \rangle| = \sum_k x_k \bar{y}_k$.*

 (ii) *The space ℓ_p $(p \neq 2)$ is not an inner product space and hence is not a Hilbert space.*

 (iii) *The space $L^2[a, b]$ is a Hilbert space with the inner product defined by $|\langle x, y \rangle| = \int_a^b x(t) \bar{y}(t) dt$.*

 (iv) *The spaces $L^p[a, b]$ with $p \neq 2$ and $C[a, b]$ are not inner product spaces and hence are not Hilbert spaces.*

1.12   Topological Vector Spaces

Definition 1.12.1. *Let X be a nonempty set and suppose that τ is a collection of subsets of X. τ is called a topology on X provided that the following axioms hold:*

 (i) *$\emptyset \in \tau$ and $X \in \tau$.*

 (ii) *Arbitrary union of elements of τ is in τ.*

 (iii) *Finite intersection of elements of τ is in τ.*

The elements of τ are called open sets. A topological space is a pair (X, τ), where X is a non-empty set and τ is a topology on X.

Examples 1.12.2. *We can give the following:*

(i) *For any X, let τ be the set of all subsets of X. Then, τ is called the discrete topology on X.*

(ii) *For any X, let $\tau = \{\emptyset, X\}$. τ is called the indiscrete topology on X.*

(iii) *Let $X = \{0, 1, 2\}$ and let $\tau = \{\emptyset, X, \{0\}, \{1, 2\}\}$. Then, τ is a topology on X.*

(iv) *Let X be any metric space and let τ be the set of open sets in the usual metric space sense. Then, τ is a topology on X. Thus, the notion of topological space is more general than the notion of metric space.*

A topological space (X, τ) is said to be *metrizable* if τ is induced by a metric d on X. Not every topology is metrizable.

We wish to endow a vector space X with a topological structure that is compatible with the algebraic structure of X. We thus need to choose a topology on X which goes well with the linear structure of X. That is, we need such a topological structure to be imposed on a vector space which can be "compatible" with the inherent algebraic structure of that space. This leads us to the notion of topological vector space, in short, TVS.

Definition 1.12.3. *A topological vector space (or linear topological space) X is a vector space over $\mathbb{K}$ ($\mathbb{C}$ or $\mathbb{R}$) which is endowed with a topology such that vector addition $+ : X \times X \to X$ and scalar multiplication $\cdot : \mathbb{K} \times X \to X$ are continuous functions.*

Every normed linear space is a topological vector space because it has a natural topological structure: The norm induces a metric and the metric induces a topology. Therefore, all Banach spaces and Hilbert spaces, are examples of topological vector spaces.

Definition 1.12.4. *Let (X, T) be a topological space and $x \in X$. Then, a set U is called a neighborhood of x if there is an open set G with $x \in G \subset U$. Thus, any open set G containing x is a neighborhood of x.*

Example 1.12.5. *We have the following:*

(i) *A TVS whose topology is defined by a norm is a lcTVS. The balls centered at the origin are convex. All normed linear spaces, and therefore all Banach spaces and Hilbert spaces, are examples of topological vector spaces.*

(ii) *Any vector space X with the indiscrete topology $\tau_I = \{X, \emptyset\}$ is a TVS.*

(iii) A vector space X with the discrete topology $\tau_D = P(E)$ is not a TVS unless $X = \{\theta\}$, since the scalar multiplication need not be continuous, where $P(E)$ denotes the collection of all subsets of E.

Definition 1.12.6. *Let E be a subset of a linear space X. Then, E is called*

(i) convex if for each x and y in E, $\alpha x + (1 - \alpha)y$ is in E for all α in the unit interval, that is, whenever $0 \leq \alpha \leq 1$. In other words, E contains all line segments between points in E.

(ii) a cone (when the underlying field is ordered) if for every $x \in E$ and $0 \leq \lambda$, $\lambda x \in E$. That is, E is cone if $\lambda E \subset E$ for every $0 \leq \lambda$.

(iii) balanced if for all x in E, λx is in E if $|\lambda| \leq 1$. If the underlying field $\mathbb{K}$ is the real numbers, this means that if x is in E, E contains the line segment between x and $-x$. For a complex linear space X, it means for any $x \in E$, E contains the disk with x on its boundary, centered on the origin, in the one-dimensional complex subspace generated by x.

(iv) absorbent or absorbing if for every $x \in X$, there exists $\varepsilon > 0$ such that $\lambda x \in E$, whenever λ is a scalar satisfying $|\lambda| \leq \varepsilon$. The set E can be scaled out to absorb every point in the space.

(v) absolutely convex if it is both balanced and convex. That is, if $x, y \in E$, $|\lambda| + |\mu| \leq 1$ imply $\lambda x + \mu y \in E$.

A locally convex space is defined either in terms of convex sets, or equivalently in terms of seminorms.

Example 1.12.7. *The following statements hold:*

(i) Any vector subspace E of X is balanced and convex, but it need not be absorbing.

(ii) Let $(X, \| \cdot \|)$ be a normed linear space. Then, both the open and closed balls are absorbing, balanced and convex.

Definition 1.12.8. *A linear topological space is called locally convex if and only if every neighbourhood U contains an absolutely convex set V. That is, a locally convex space is a linear topological space in which the origin has a local base of absolutely convex absorbent sets. Because translation is (by definition of "linear topological space") continuous, all translations are homeomorphisms, so every base for the neighborhoods of the origin can be translated to a base for the neighborhoods of any given vector.*

Definition 1.12.9. *A subset A of a TVS E is called:*

(a) bounded if for any neighborhood W of θ in E, there exists $r > 0$ such that $A \subseteq \lambda W$ for all $\lambda \in \mathbb{K}$ with $|\lambda| \geq r$, or, equivalently, $A \subseteq rW$ if W is balanced.

(b) precompact (or totally bounded) if for each neighborhood W of θ, there exists a finite subset $D = \{x_0, x_1, \ldots, x_n\}$ of E such that

$$A \subseteq D + W = \bigcup_{i=0}^{n} (x_i + W).$$

Clearly, compactness $\Rightarrow$ precompactness $\Rightarrow$ boundedness; the reverse implications need not be true.

Definition 1.12.10. *A net $\{x_\alpha : \alpha \in I\}$ in a TVS E is said to be convergent to $x \in E$ if given any neighborhood U of θ in E, there exists an $\alpha_0 \in I$ such that $x_\alpha - x \in U$ for all $\alpha \geq \alpha_0$. In this case, we write $x_\alpha \to x$. $(x_{\alpha \in I})$ is said to be a Cauchy net if given any neighborhood U of θ in E, there exists an $\alpha_0 \in I$ such that $x_\alpha - x_\beta \in U$ for all $\alpha, \beta \geq \alpha_0$.*

Definition 1.12.11. *Let E be a TVS, and let $A \subseteq E$. Then, A is called:*

(i) complete if every Cauchy net in A converges to a point in A.

(ii) sequentially complete if every Cauchy sequence in A converges to a point in A.

(iii) quasi-complete (or boundedly complete) if every bounded Cauchy net in A converges to a point of E (i.e. if every bounded closed subset of E is complete). Clearly, completeness $\Rightarrow$ quasi-completeness $\Rightarrow$ sequential completeness; the reverse implications need not hold.

Theorem 1.12.12. *A subset A of a TVS E is compact iff it is complete and precompact.*

1.13   Linear Metric Spaces

Of course, linear metric space is a special case of topological vector space. First, we define such a metric which makes the vector addition operation continuous.

Definition 1.13.1. *Let X be a linear space and d be a metric on X. Then, d is called* translation invariant *if $d(x + z, y + z) = d(x, y)$ for all $x, y, z \in X$.*

Remark 1.13.2. *(cf. [29, p. 141]) It is easy to see that if a linear space is endowed with a translation invariant metric, then it makes the vector addition operation continuous. But, the translation invariance alone does not guarantee the continuity of the scalar multiplication operation.*

Thus, we need the scalar multiplication operation on X to be continuous and then, we arrive at a class of spaces which are both metric and linear, and in which the metric and linear structures of the space are naturally compatible.

We define it, as follows:

Definition 1.13.3. *Let X be a linear space which is also a metric space with the translation invariant metric d on X. Then, (X, d) or X for short, is said to be a linear metric space, if the algebraic operations on X are continuous functions. That is, X is a linear metric space if and only if it is both a linear and a metric space such that*

(i) the vector addition map $(x, y) \mapsto x + y$ is a continuous function from $X \times X$ into X, and

(ii) the scalar multiplication map $(\lambda, x) \mapsto \lambda x$ is a continuous function from $\mathbb{R} \times X$ into X.

Moreover, if a linear metric space is complete, then it is called Fréchet space. Some authors call a complete linear metric space as an F-space and a locally convex F-space as a Fréchet space.

Examples 1.13.4. *We have the followings:*

(i) Let the function d be defined by

$$d \ : \ \mathbb{R} \times \mathbb{R} \ \longrightarrow \ \mathbb{R}^+$$
$$(a, b) \ \longmapsto \ d(a, b) = |a^3 - b^3|.$$

Then, we see that the scalar multiplication operation on $\mathbb{R}$ is continuous, but do not make $\mathbb{R}$ a linear metric space because the metric d is not translation invariant.

(ii) Let us take the discrete metric on $\mathbb{R}$. Then, we do not obtain a linear metric space even though this metric is translation invariant.

(iii) ℓ_p with $1 \leq p < \infty$ and ℓ_∞ are Fréchet spaces with respect to their usual metrics.

(iv) $\mathbb{R}^n$ is a Fréchet space for any $n \in \mathbb{N}_0$.

The concept of paranorm is closely related to linear metric space and, in fact, total paranormed space, and linear metric space both are same. Paranorm is another version of linear metric space (cf. [147]).

Definition 1.13.5. *A paranorm is a function $p : X \to \mathbb{R}$ defined on a linear space X if the following axioms hold:*

(P.1) $p(x) = 0$ if $x = \theta$ (zero element of X).

(P.2) $p(x) \geq 0$ for all $x \in X$.

(P.3) $p(-x) = p(x)$ for all $x \in X$.

(P.4) $p(x + y) \leq p(x) + p(y)$ for all $x, y \in X$; (triangle inequality).

(P.5) *If (λ_n) is a sequence of scalars with $\lambda_n \to \lambda$, as $n \to \infty$ and (x_n) is a sequence of vectors with $p(x_n - x) \to 0$, as $n \to \infty$, then $p(\lambda_n x_n - \lambda x) \to 0$, as $n \to \infty$; (continuity of scalar multiplication).*

In this case (X, p) or X for short, is called a paranormed space. If $p(x) = 0$ implies $x = \theta$, then a paranorm p is called total paranorm, and (X, p) is called a total paranormed space.

If we suppose that (X, d) is a linear metric space and if we define $p(x) = d(x, 0)$ for each $x \in X$, then it is straightforward to see that the properties of d imply all the properties of p. If (X, p) is paranormed (total paranormed) space then (X, d) is a semilinear metric (respectively, linear metric) space whenever d is defined by $d(x, y) = p(x - y)$ for all $x, y \in X$. The converse is also true. In fact, a linear metric space and a total paranormed space both are same.

Example 1.13.6. [29, Example 1.8] *The space $\ell(p)$ is a linear metric space with the metric $d(x, y) = \sum_k |x_k - y_k|^{p_k}$, where $0 < p_k \le 1$ for all $k \in \mathbb{N}_0$. Moreover, it is a Fréchet space.*

Example 1.13.7. [29, Example 1.9] *The space $\ell_\infty(p)$ is not a linear metric space with the paranorm $g(x) = \sup_{k \in \mathbb{N}_0} |x_k|^{p_k/M}$, since scalar multiplication is not continuous. Moreover, it turns out to be a linear metric space if and only if $\inf_{k \in \mathbb{N}_0} p_k > 0$.*
Let $p_k = 1/k$ and $x_k = 1$ for all $k \in \mathbb{N}_1$. Then, $x \in \ell_\infty(p)$. Let $0 < |\lambda| < 1$. Then, $|\lambda|^{p_k} = |\lambda|^{1/k} < 1$ for each k and $|\lambda|^{1/k} \to 1$, as $k \to \infty$, so that

$$g(\lambda x) = \sup_{k \in \mathbb{N}_1} |\lambda x_k|^{p_k/M} = \sup_{k \in \mathbb{N}_1} |\lambda|^{1/k} = 1.$$

Hence, $g(\lambda x) \to 0$ does not hold, as $\lambda \to 0$, and thus the scalar multiplication is not continuous, i.e., $\ell_\infty(p)$ is not a linear metric space.
Now, suppose that $\inf_{k \in \mathbb{N}_0} p_k = \beta > 0$ and $x = (x_k) \in \ell_\infty(p)$. Then,

$$g(\lambda x) = \sup_{k \in \mathbb{N}_0} |\lambda x_k|^{p_k/M} \le \max\{|\lambda|, |\lambda|^{\beta/M}\} g(x).$$

Thus, $g(\lambda x) \to 0$, as $\lambda \to 0$, which means that $\ell_\infty(p)$ is a linear metric space.
Conversely, if $\inf_{k \in \mathbb{N}_0} p_k = 0$, then the above example shows that there is an $x \in \ell_\infty(p)$ for which $g(\lambda x) \to 0$ does not hold, as $\lambda \to 0$ which contradicts that g is a paranorm.

We have the following important relation:

Theorem 1.13.8. [147, p. 92] *Let X be a linear space. Then, each seminorm on X is also a paranorm but not conversely.*

1.14 *FK* Spaces

One of the main features of FK-space theory is that it provides easy and short proofs of many classical results of summability theory. It was initiated by Zeller [238] which is the most powerful and widely used tool in the characterization of matrix mappings between sequence spaces, and the most important result was that matrix mappings between FK-spaces are continuous [228, Theorem 4.2.8].

A locally convex space is then defined to be a vector space X along with a family of seminorms $\{p_\alpha\}_{\alpha \in A}$ on X. Note that a seminormed space is locally convex.

Definition 1.14.1. *We give the following:*

(i) *A sequence space X with linear topology is called a K-space if each of the maps $P_n : X \to \mathbb{C}$ defined by $P_n(x) = x_n$ is continuous for all $x = (x_n) \in X$ and every $n \in \mathbb{N}_0$.*

(ii) *A Fréchet space is a complete linear metric space. In other words, a locally convex space is called a Fréchet space if it is metrizable and the underlying metric space is complete.*

(iii) *K-space X is called an FK-space if X is a complete linear metric space, that is, X is an FK-space if X is Fréchet space with continuous coordinates, i.e., if $x^{(n)} \to x$, as $n \to \infty$, in the metric of X then $x_k^{(n)} \to x_k$, as $n \to \infty$, for each fixed $k \in \mathbb{N}_0$.*

(iv) *A normed FK-space is called a BK-space, that is, a BK-space is a Banach sequence space with continuous coordinates.*

Note that some authors include local convexity in the definition of FK-space. But much of the theory can be developed without local convexity.

Remark 1.14.2. *A Fréchet sequence space (X, d_X) is an FK-space if its metric d_X is stronger than the metric $d|_X$ of ω on X. Hence, an FK-space X is continuously embedded in ω, that is, the inclusion map $\iota : (X, d_X) \to (\omega, d)$ defined by $\iota(x) = x$ $(x \in X)$ is continuous.*

Theorem 1.14.3. *[154, Theorem 1.14] Let (X, d_X) be a Fréchet space, (Y, d_Y) an FK-space and $f : X \to Y$ a linear map. Then, $f : (X, d_X) \to (Y, d|_Y)$ is continuous if and only if $f : (X, d_X) \to (Y, d_Y)$ is continuous.*

Proof. First, we assume that $f : (X, d_X) \to (Y, d_Y)$ is continuous. Since Y is an FK-space, its metric d_Y is stronger than the metric $d|_Y$ of ω on Y. So, $f : (X, d_X) \to (Y, d|_Y)$ is continuous.

Conversely, we assume that $f : (X, d_X) \to (Y, d|_Y)$ is continuous. Since (Y, d_Y) is a Hausdorff space and f is continuous, the graph $G(f)$ of f, $G(f) =$

$\{(x, f(x)) : x \in X\}$, is a closed set in $(X, d_X) \times (Y, d|_Y)$ by closed graph theorem, hence a closed set in $(X, d_X) \times (Y, d_Y)$, since the FK metric d_Y is stronger than $d|_Y$. By closed graph theorem, the map $f : (X, d_X) \to (Y, d_Y)$ is continuous.

This completes the proof. □

Corollary 1.14.4. *Let X be a Fréchet space, Y an FK-space, $f : X \to Y$ a linear map and P_n the nth coordinate, that is, $P_n(y) = y_n$ for all $n \in \mathbb{N}_0$ with $y = (y_n) \in Y$. If each map $P_n \circ f : X \to \mathbb{C}$ is continuous, so is $f : X \to Y$.*

Proof. Since $P_n \circ f : X \to \mathbb{C}$ is continuous for each $n \in \mathbb{N}_0$, the map $f : X \to \omega$ is continuous by the equivalence of coordinate-wise convergence and convergence in ω. By Theorem 1.14.3, $f : X \to Y$ is continuous.

This completes the proof. □

We shall frequently make use of the following result.

Theorem 1.14.5. [154, Theorem 1.16] *Let $X \supset \phi$ be an FK-space and $a = (a_k) \in \omega$. If the series $\sum_k a_k x_k$ converges for each $x = (x_k) \in X$, then the linear functional $f_a : X \to \mathbb{C}$ defined by $f_a(x) = \sum_k a_k x_k$ for all $x = (x_k) \in X$ is continuous.*

Proof. For each $n \in \mathbb{N}_0$, we define the linear functional $f_{a,n} : X \to \mathbb{C}$ by $f_{a,n}(x) = \sum_{k=0}^{n} a_k x_k$ for all $x = (x_k) \in X$. Since X is an FK-space, the coordinates $P_k : X \to \mathbb{C}$ are continuous on X for all $k \in \mathbb{N}_0$ and so are the functionals $f_{a,n}(x) = \sum_{k=0}^{n} a_k P_k(x)$ for all $n \in \mathbb{N}_0$. For each $x \in X$, $f_{a,n}(x) \to f_a(x)$, as $n \to \infty$, exists, and so by Banach-Steinhaus theorem, $f_a : X \to \mathbb{C}$ is continuous.

This completes the proof. □

Definition 1.14.6. *An FK-space $X \supset \phi$ is said to have AK if every sequence $x = (x_k) \in X$ has a unique representation $x = \sum_k x_k e^{(k)}$, that is $\sum_{k=0}^{n} x_k e^{(k)} \to x$, as $n \to \infty$. This means that $\{e^{(k)}\}_k$ is a Schauder basis for any FK-space with AK such that every sequence, in an FK-space with AK, coincides with its sequence of coefficients with respect to this basis. X has AD if ϕ is dense in X. If an FK-space has AK or AD we also say that it is an AK- or AD-space. Note that every AK-space has AD but converse is not true, in general.*

An FK-space $X \supset \phi$ is said to have AB if every sequence $(x^{[n]})$ is a bounded set in X for every $x \in X$.

Example 1.14.7. *Let $1 \leq p < \infty$. Then, the following statements hold:*

(i) *The famous example of an FK-space which is not a BK-space is the space (ω, d_ω).*

(ii) *The spaces ℓ_∞, c, c_0, ℓ_p, bs, cs and bv are BK-spaces with their natural norms.*

(iii) The spaces ω, c_0 and ℓ_p have AK, but the spaces c and ℓ_∞ do not have AK.

Definition 1.14.8. *Let H be a Hausdorff space (H is a Hausdorff space if any two distinct points of H can be separated by neighborhoods) and X be a linear space. An FH-space is a Fréchet space X such that*

(i) X is a linear subspace of H.

(ii) The topology of X is stronger than that of H.

Note that an FK-space is a special kind of FH-space in which $H = \omega$ with its topology given by the metric d_ω. A BH-space is an FH-space which is a Banach space.

Note that the letters F, H and B stand for Fréchet, Hausdorff and Banach.

Remark 1.14.9. *The following statements hold:*

(a) By Remark 1.14.2, if X is an FH-space, then the inclusion map $\iota :$ $X \to H$ with $\iota(x) = x$ for all $x \in X$ is continuous. Therefore, X is continuously embedded in H.

(b) Since convergence in (ω, d_ω) and coordinate-wise convergence are equivalent [228, Theorem 4.1.1, p. 54], convergence in an FK-space implies coordinate-wise convergence.

Example 1.14.10. *The sequence spaces c, c_0, ℓ_∞ and ℓ_p with $p > 0$, are FH-spaces.*

Chapter 2

Geometric Properties of Some Sequence Spaces

Keywords. *Geometric properties, Orlicz sequence spaces, Cesàro sequence spaces, sequence spaces related to ℓ_p spaces.*

2.1 Introduction

Recently there has been a lot of interest in investigating geometric properties of sequence spaces besides topological and some other usual properties. In the literature, there are many papers concerning geometric properties of various Banach sequence spaces. For example, Cui and Hudzik [65, 66, 67], Cui and Meng [68], Sanhan and Suantai [192], Ng and Lee [177] investigated the geometric properties of the Cesàro sequence space $ces(p)$ equipped with Luxemburg norm. Furthermore, Mursaleen et al. [169] studied some geometric properties of normed Euler sequence spaces. Quite recently, Karakaya [117] has defined a new sequence space involving lacunary sequence spaces equipped with the Luxemburg norm and studied Kadec-Klee (H) and rotundity (R) properties of these spaces. In this chapter, we give some geometric properties of Orlicz sequence spaces, Cesàro sequence spaces and other sequence spaces related to ℓ_p spaces.

2.2 Geometric Properties

We begin with by giving the definition of some geometric concepts for normed spaces.

A Banach space X is called *uniformly convex (UC)* if for each $\varepsilon > 0$, there is $\delta > 0$ such that for $x, y \in S_X$, the inequality $\|x - y\| > \varepsilon$ implies that $\|(x + y)/2\| < 1 - \delta$.

Definition 2.2.1. (see [65] and [71]) *A Banach space X is said to be a Köthe sequence space if X is a subspace of ω such that*

(i) *if $x \in \omega$, $y \in X$ and $|x_i| \leq |y_i|$ for all $i \in \mathbb{N}_0$, then $x \in X$ and $\|x\| \leq \|y\|$;*

(ii) *there exists an element $x = (x_i) \in X$ such that $x_i > 0$ for all $i \in \mathbb{N}_0$.*

An element x of Köthe sequence space X is said to be order continuous if for any sequence (x_n) and any x in X_+ (the positive cone in X) such that $x_n(i) \leq |x_i|$ for all $n \in \mathbb{N}_0$ and $x_n(i) \to 0$, as $n \to \infty$ for each $i \in \mathbb{N}_0$, we have $\|x_n\| \to 0$, as $n \to \infty$.

A Köthe sequence space X is said to be order continuous if all sequences in X are order continuous, i.e., $\|(0, 0, \ldots, x_{n+1}, x_{n+2}, \ldots)\| \to 0$, as $n \to \infty$, for any $x \in X$. In this case, we say that the norm is *absolutely continuous* at $x = (x_i)_{i=1}^{\infty}$.

A Köthe sequence space X is said to have the *Fatou property* if for any real sequence $x \in \omega$ and any (x_n) in X such that $x_n \uparrow x$ coordinatewise (i.e., $|x_n(i)| \uparrow |x_i|$ for all $i \in \mathbb{N}_0$) and $\sup_{n \in \mathbb{N}_0} \|x_n\| < \infty$, we have that $x \in X$ and $\|x_n\| \to \|x\|$, as $n \to \infty$.

A Banach space X is said to have the *Banach-Saks property* if every bounded sequence (x_n) in X admits a subsequence (z_n) such that the sequence $\{t_k(z)\}$ is convergent in X with respect to the norm, where

$$t_k(z) = \frac{1}{k+1}(z_0 + z_1 + z_2 + \cdots + z_k) \quad \text{for all } k \in \mathbb{N}_0.$$

A Banach space X is said to have the *weak Banach-Saks property* whenever given any weakly null sequence (x_n) in X there exists its subsequence (z_n) such that the sequence $\{t_k(z)\}$ strongly converges to zero.

Given any $p \in (1, \infty)$, we say that a Banach space $(X, \|\cdot\|)$ has the *Banach-Saks property of type p* if there exists a constant $c > 0$ such that every weakly null sequence (x_k) has a subsequence (x_{k_l}) such that (see [128])

$$\left\| \sum_{l=0}^{n} x_{k_l} \right\| \leq c(n+1)^{1/p} \quad \text{for all } n \in \mathbb{N}_0.$$

The Banach-Saks property of type $p \in (1, \infty)$ and weak Banach-Saks property for Cesàro sequence spaces have been considered in [65].

We say that a Banach space X has the *(weak) fixed point property* if every nonexpansive self-mapping defined on a non-empty (weakly) compact convex subset of X has a fixed point.

In [88], Garcia-Falset introduced the following coefficient for a Banach space $(X, \|\cdot\|)$:

$$R(X) = \sup \left\{ \liminf_{n \to \infty} \|x_n - x\| : x \in B(X), (x_n) \subset B(X) \text{ and } x_n \overset{weakly}{\longrightarrow} 0 \right\}$$

and proved (see [88, 89]) that a Banach space X with $R(X) < 2$ has the weak fixed point property.

The *Clarkson modulus of convexity* of a normed space $(X, \|\cdot\|)$ is defined (see Clarkson [59]) by the formula

$$\delta_X(\varepsilon) = \inf\left\{1 - \frac{\|x+y\|}{2}; \ x, y \in S(X), \ \|x-y\| = \varepsilon\right\}$$

for any $\varepsilon \in [0, 2]$. The inequality $\delta_X(\varepsilon) > 0$ characterizes the uniform convexity of X for all $\varepsilon \in (0, 2]$ and the equality $\delta_X(2) = 1$ characterizes strict convexity (=rotundity) of X.

The *Gurariĭ modulus of convexity* of a normed space X is defined (see [95]) by

$$\beta_X(\varepsilon) = \inf\left\{1 - \inf_{\alpha \in [0,1]} \|\alpha x + (1-\alpha)y\|; \ x, y \in \delta(X), \ \|x-y\| = \varepsilon\right\}$$

for any $\varepsilon \in [0, 2]$. It is obvious that $\delta_X(\varepsilon) \leq \beta_X(\varepsilon)$ for any Banach space X and any $\varepsilon \in [0, 2]$. It is also known that $\beta_X(\varepsilon) \leq 2\delta_X(\varepsilon)$ for any $\varepsilon \in [0, 2]$ and that X is rotund if and only if $\beta_X(\varepsilon) = 2$ and X is uniformly convex if and only if $\beta_X(\varepsilon) > 0$ for any $\varepsilon \in [0, 2]$. Gurariĭ [94] proved that if $X = c_0$ is renormed by the norm

$$\|x\| = \|x\|_\infty + \left[\sum_n \left(\frac{x_n}{2^n}\right)^2\right]^{1/2} \quad \text{for all} \ \ x = (x_n) \in c_0,$$

then $\beta_X(\varepsilon) = 0$ for any $\varepsilon \in [0, 2)$ and $\beta_X(2) = 1$.

Gurariĭ and Sozonov [96] proved that a normed linear space $(X, \|\cdot\|)$ is an inner product space if and only if for every $x, y \in \delta(X)$

$$\inf_{\alpha \in [0,1]} \|\alpha x + (1-\alpha)y\| = \frac{\|x+y\|}{2}.$$

Zanco and Zucchi [237] gave an example of a normed space X with $\delta_X(2) \neq \beta_X(2)$.

A Banach space X is said to have the *Kadec-Klee property (or H-property)* if every weakly convergent sequence on the unit sphere is convergent in norm.

A sequence $(x_n) \subset X$ is said to be an *ε-separated sequence* for some $\varepsilon > 0$ if $sep(x_n) = \inf\{\|x_n - x_m\| : n \neq m\} > \varepsilon$.

A Banach space X is said to have the *uniform Kadec-Klee property (UKK for short)* if for every $\varepsilon > 0$ there exists $\delta > 0$ such that for every sequence (x_n) in S_X with $sep(x_n) > \varepsilon$ and $x_n \overset{w}{\to} x$, we have $\|x\| < 1 - \delta$. Note that every (UKK) Banach space has H-property.

A Banach space X is said to be the *nearly uniformly convex (NUC)* if for every $\varepsilon > 0$ there exists $\delta > 0$ such that for every sequence $(x_n) \subset B(X)$ with $sep[(x_n)] > \varepsilon$, we have

$$conv[(x_n)] \cap (1-\delta)B(X) \neq \emptyset.$$

Let $k \in \mathbb{N}_2$. A Banach space X is said to be *k-nearly uniformly convex (k-NUC)* if for any $\varepsilon > 0$ there exists $\delta > 0$ such that for every sequence $(x_n) \subset B(X)$ with $sep[(x_n)] > \varepsilon$, there are $n_0, n_1, \ldots, n_k \in \mathbb{N}_0$ such that

$$\left\| \frac{x_{n_0} + x_{n_1} + \cdots + x_{n_k}}{k+1} \right\| < 1 - \delta.$$

Of course, a Banach space X is (NUC) whenever it is $(k - NUC)$ for some integer $k \geq 2$. Clearly, $(k-NUC)$ Banach spaces are (NUC), but the opposite implication does not hold, in general (see [135]).

A point $x \in S(X)$ is called an *extreme point* if for any $y, z \in B(X)$ the equality $2x = y + z$ implies $y = z$. A Banach space X is said to be *rotund* (abbreviated as (R)) if every point of $S(X)$ is an extreme point. A Banach space X is said to be *fully k–rotund* (write kR) if for every sequence $(x_n) \subset B(X)$,

$$\lim_{n_0, n_1, \ldots, n_k \to \infty} \|x_{n_0} + x_{n_1} + \cdots + x_{n_k}\| = k$$

implies that (x_n) is convergent.

It is well known that (UC) implies (kR) and (kR) implies $((k+1)R)$, and (kR) spaces are reflexive and rotund, and it is easy to see that $(k - NUC)$ implies (kR).

The Opial property [181] is important because Banach spaces with this property have the weak fixed point property. Opial has proved that the sequence spaces ℓ_p have this condition but $L_p[0, 2\pi]$ do not with $p \neq 2$, where $1 < p < \infty$.

A Banach space X is said to have the *non-strict Opial property* if every weakly null sequence $(x_n) \subset S_X$ with $\lim_{n\to\infty} \|x_n\| = 1$ satisfies

$$1 \leq \liminf_{n\to\infty} \|x_n + x\|$$

for every $x \in X$.

A Banach space X is said to have the *weak orthogonality property* if every weakly null sequence $(x_n) \subset S_X$ satisfies

$$\lim_{n\to\infty} | \|x_n + x\| - \|x_n - x\| | = 0$$

for every $x \in X$.

A Banach space X is said to have the *Opial property* if every sequence (x_n) that is weakly convergent to x_0 satisfies

$$\liminf_{n\to\infty} \|x_n - x_0\| < \liminf_{n\to\infty} \|x_n - x\|.$$

A Banach space X is said to have the *uniform Opial property* if each $\varepsilon > 0$ there exists $\tau > 0$ such that for any weakly null sequence (x_n) in S_X and every $x \in X$ with $\|x\| \geq \varepsilon$, we have (see [187]) $1 + \tau < \liminf_{n\to\infty} \|x_n + x\|$.

For a real vector space X, a function $\rho : X \to [0, \infty]$ is called a *modular* if it satisfies the following conditions:

(i) $\rho(x) = 0$ implies $x = \theta$.

(ii) $\rho(\alpha x) = \rho(x)$ for all $\alpha \in F$ with $\|\alpha\| = 1$.

(iii) $\rho(\alpha x + \beta y) \leq \rho(x) + \rho(y)$ for all $x, y \in X$ and all $\alpha, \beta \geq 0$ with $\alpha + \beta = 1$. Further the modular ρ is called convex if

(iv) $\rho(\alpha x + \beta y) \leq \alpha \rho(x) + \beta \rho(y)$ holds for all $x, y \in X$ and all $\alpha, \beta \geq 0$ with $\alpha + \beta = 1$.

For any modular ρ on X, the space

$$X_\rho = \left\{ x \in X : \lim_{\lambda \to 0^+} \rho(\lambda x) = 0 \right\}$$

is called the *modular space*. A sequence (x_n) of elements of X_ρ is called *modular convergent* to $x \in X_\rho$ if there exists a $\lambda > 0$ such that $\rho[\lambda(x_n - x)] \to 0$, as $n \to \infty$. If ρ is a convex modular, then we have the following formula:

$$\|x\|_L = \inf \left\{ \lambda > 0 : \rho\left(\frac{x}{\lambda}\right) \leq 1 \right\}$$

define a norm on X_ρ which is called the *Luxemburg norm*.

A modular ρ is said to satisfy the $\Delta_2-condition$ ($\rho \in \Delta_2$) if for any $\varepsilon > 0$ there exists constants $K \geq 2$ and $a > 0$ such that $\rho(2x) \leq K\rho(x) + \varepsilon$ for all $x \in X_\rho$ with $\rho(x) \leq a$.

If ρ satisfies the Δ_2-condition for all $a > 0$ with $K \geq 2$ dependent on a, we say that ρ satisfies the strong Δ_2-condition ($\rho \in \Delta_2^s$).

2.3 Orlicz Sequence Spaces

A map $\Phi : \mathbb{R} \to \mathbb{R}^+$ is said to be an *Orlicz function* if Φ vanishes only at 0, Φ is even, convex, continuous on the whole line $\mathbb{R}^+$ and $\Phi(u)/u \to \infty$, as $u \to \infty$.

The *Orlicz sequence space* ℓ_Φ is defined by

$$\ell_\Phi := \left\{ x \in \omega : I_\Phi(\lambda x) = \sum_i \Phi[\lambda x_i] < \infty \text{ for some } \lambda > 0 \right\}$$

equipped with the *Luxemburg norm*

$$\|x\|_\Phi = \inf \left\{ k > 0 : I_\Phi\left(\frac{x}{k}\right) \leq 1 \right\}$$

or the *Orlicz norm*

$$\|x\|_\Phi^0 = \inf \left\{ \frac{1}{k}[i + I_\Phi(kx)] : k > 0 \right\}.$$

To simplify, we denote $\ell_\Phi = (\ell_\Phi, \|\cdot\|_\Phi)$ or $\ell_\Phi^0 = (\ell_\Phi, \|\cdot\|_\Phi^0)$. Put

$$h_\Phi := \left\{ x \in l_0 : I_\Phi(\lambda x) = \sum_i \Phi[\lambda x_i] < \infty \text{ for any } \lambda > 0 \right\}.$$

We write shortly $h_\Phi = (h_\Phi, \|\cdot\|_\Phi)$ and $h_\Phi^0 = (h_\Phi, \|\cdot\|_\Phi^0)$. It is well known that $x \in h_\Phi$ if and only if the norm is also continuous at x.

An Orlicz function Φ is said to satisfy the δ_2–condition ($\Phi \in \delta_2$) if there exist $u_0 > 0$, $K \geq 2$ such that $\Phi(2u) \leq K\Phi(u)$ whenever $|u| \leq u_0$.

Theorem 2.3.1. *Let X be a Köthe sequence space with the Fatou property. Then, the following statements are equivalent:*

(i) X has a weak orthogonality property.

(ii) X has the non-strict Opial property.

(iii) X has absolute continuous norm.

Proof. (i) $\Rightarrow$ (ii). See Proposition 2 in [203].

(ii) $\Rightarrow$ (iii). Suppose that X does not have an absolutely continuous norm. Then, there exist $\epsilon_0 > 0$ and $x_0 \in S(X)$ such that

$$\left\| \sum_{i=n+1}^\infty x_0(i)e^{(i)} \right\| \geq \epsilon_0$$

for any $n \in \mathbb{N}_0$, where $e^{(i)} = (0,0,\dots,1^{i^{th}} \text{ term},0,\dots)$ for all $i \in \mathbb{N}_0$.

Take $n = 0$. Since X has Fatou property, there is $n_1 \in \mathbb{N}_0$ such that

$$\left\| \sum_{i=0}^{n_1} x_0(i)e^{(i)} \right\| \geq \frac{3\epsilon_0}{4}.$$

Notice that

$$\lim_{m \to \infty} \left\| \sum_{i=n_1+1}^m x_0(i)e^{(i)} \right\| \geq \epsilon_0,$$

so there exists $n_2 > n_1$ such that

$$\left\| \sum_{i=n_1+1}^{n_2} x_0(i)e^{(i)} \right\| \geq \frac{3\epsilon_0}{4}.$$

In this way, we get a sequence (n_i) of natural numbers such that

$$\left\| \sum_{j=n_i+1}^{n_{i+1}} x_0(j)e^{(j)} \right\| \geq \frac{3\epsilon_0}{4} \text{ for all } i \in \mathbb{N}_1.$$

Put $x_i = \sum_{j=n_i+1}^{n_{i+1}} x_0(j)e^{(j)}$. Then,

(a) $\|x_i\| \geq (3\epsilon_0)/4$ for all $i \in \mathbb{N}_0$;

(b) $x_i \overset{w}{\to} 0$, as $i \to \infty$. It is well known that for any Köthe space X we have $X^* = X^\beta \oplus S$, where S is the space of all singular functionals over X, i.e., functionals which vanish on the subspace

$$X_\alpha = \{x \in X : x \text{ has absolutely continuous norm}\}.$$

This means that every $f \in X^*$ is uniquely represented in the form $f = T_y + \varphi$, where $\varphi \in S$ and for $y \in X$ the function T_y is defined by $T_y(x) = \sum_i x_i y_i$ for all $x \in X$.

Suppose $\sum_i x_i y_i$ converges. We have

$$\lim_{n \to \infty} \sum_j x_n(j)y(j) = \lim_{i \to \infty} \sum_{j=n_i+1}^{n_{i+1}} x_i(j)y(j) = 0.$$

Take $i_0 \in \mathbb{N}_0$ large enough so that $\left\| \sum_{i=i_0+1}^{\infty} x_0(i)e^{(i)} \right\| \leq (4\epsilon_0)/3$. Put $z_0 = \sum_{i=i_0+1}^{\infty} x_0(i)e^{(i)}$ and $z_i = -2x_i$ for all $i \in \mathbb{N}_0$. Then,

(c) $\|z_i + z_0\| = \|z_0\| \leq (4\epsilon_0)/3$ for any i large enough and $\|z_i\| = 2\|x_i\| \geq (3\epsilon_0)/2$. This contradicts X having the non-strict Opial property.

(iii) $\Rightarrow$ (i). For any $\epsilon > 0$, $x \in S(X)$ and every weakly null sequence (x_n) in X, there are $n_0, n_1 \in \mathbb{N}_0$ such that

$$\left\| \sum_{i=n_0+1}^{\infty} x_i e^{(i)} \right\| < \frac{\epsilon}{4} \quad \text{and} \quad \left\| \sum_{i=0}^{n_0} x_n(i)e^{(i)} \right\| < \frac{\epsilon}{4} \quad \text{when } n \geq n_0.$$

Put

$$X_n = \sum_{i=0}^{n_0} x_i e^{(i)} + \sum_{i=n_0+1}^{\infty} x_n(i)e^{(i)}$$

$$Y_n = \sum_{i=0}^{n_0} x_i e^{(i)} - \sum_{i=n_0+1}^{\infty} x_n(i)e^{(i)}.$$

Then, $\|X_n\| = \|Y_n\|$ for all $n \in \mathbb{N}_0$ and

$$\|(x + x_n) - X_n\| = \left\| \sum_{i=0}^{n_0} x_n(i)e^{(i)} + \sum_{i=n_0+1}^{\infty} x_i e^{(i)} \right\|$$

$$\leq \left\| \sum_{i=0}^{n_0} x_n(i)e^{(i)} \right\| + \left\| \sum_{i=n_0+1}^{\infty} x_i e^{(i)} \right\|$$

$$\leq \frac{\epsilon}{4} + \frac{\epsilon}{4} = \frac{\epsilon}{2} \quad \text{when } n \geq n_1$$

and

$$\|(x + x_n) - Y_n\| = \left\| \sum_{i=n_0+1}^{\infty} x_i e^{(i)} - \sum_{i=0}^{n_0} x_n(i) e^{(i)} \right\|$$

$$\leq \left\| \sum_{i=0}^{n_0} x_n(i) e^{(i)} \right\| + \left\| \sum_{i=n_0+1}^{\infty} x_i e^{(i)} \right\|$$

$$\leq \frac{\epsilon}{4} + \frac{\epsilon}{4} = \frac{\epsilon}{2} \text{ when } n \geq n_1.$$

Hence, we have

$$\left| \|x + x_n\| - \|x - x_n\| \right| = \left| \|x + x_n\| - \|X_n\| + \|x - x_n\| - \|Y_n\| \right|$$

$$\leq \left| \|x + x_n\| - \|X_n\| \right| + \left| \|x - x_n\| - \|Y_n\| \right|$$

$$\leq \frac{\epsilon}{2} + \frac{\epsilon}{2} = \epsilon \text{ when } n \geq n_1.$$

This means that $\left| \|x + x_n\| - \|x - x_n\| \right| \to 0$, as $n \to \infty$. $\qquad\square$

Theorem 2.3.2. *The Orlicz sequence space h_Φ^0 has the Opial property.*

Proof. Let $x \in h_\Phi^0$ be given. Take $\epsilon = \min\{(1/5)\, I_\Phi(x), 1\}$. Then, there exists an $i_0 \in \mathbb{N}_0$ such that

$$\left\| \sum_{i=i_0+1}^{\infty} x_i \right\|^0 < \epsilon$$

furthermore we have

$$\sum_{i=i_0+1}^{\infty} \Phi(x_i) \leq \left\| \sum_{i=i_0+1}^{\infty} x_i \right\|^0 < \epsilon.$$

For every weakly null sequence $(x_n) \subset S(h_\Phi^0)$, by using $x_n \overset{w}{\to} 0$, as $n \to \infty$, we have $x_n(i) \to 0$ for all $i \in \mathbb{N}_0$. Hence, there exists $n_0 \in \mathbb{N}_0$ such that

$$\left\| \sum_{i=0}^{i_0} x_n(i) e^{(i)} \right\|^0 < \epsilon \text{ when } n > n_0.$$

So, there exists $k_n > 0$ such that

$$\left\| \sum_{i=0}^{i_0} x_i e^{(i)} + \sum_{i=i_0+1}^{\infty} x_n(i) e^{(i)} \right\|^0 = \frac{1}{k_n} \left\{ 1 + \sum_{i=0}^{i_0} \Phi[k_n x_i] + \sum_{i=i_0+1}^{\infty} \Phi[k_n x_n(i)] \right\},$$

by Theorem 1.31 of [56]. Without loss of generality, we may assume that $k_n \geq 1$. Therefore,

$$
\left\| \sum_{i=0}^{i_0} x_i e^{(i)} + \sum_{i=i_0+1}^{\infty} x_n(i) e^{(i)} \right\|^0 = \frac{1}{k_n} \left\{ 1 + \sum_{i=0}^{i_0} \Phi[k_n x_i] + \sum_{i=i_0+1}^{\infty} \Phi[k_n x_n(i)] \right\}
$$

$$
= \frac{1}{k_n} \left\{ 1 + \sum_{i=i_0+1}^{\infty} \Phi[k_n x_n(i)] \right\} + \frac{1}{k_n} \sum_{i=0}^{i_0} \Phi[k_n x_i]
$$

$$
\geq \left\| \sum_{i=i_0+1}^{\infty} x_n(i) e^{(i)} \right\|^0 + \sum_{i=0}^{i_0} \Phi[x_i]
$$

$$
\geq 1 - \epsilon + 4\epsilon = 1 + 3\epsilon \quad \text{when } n > n_0.
$$

Hence,

$$
\|x + x_n\|^0 = \left\| \sum_{i} x_i e^{(i)} + \sum_{i} x_n(i) e^{(i)} \right\|^0
$$

$$
\geq \left\| \sum_{i=0}^{i_0} x_i e^{(i)} + \sum_{i=i_0+1}^{\infty} x_n(i) e^{(i)} \right\|^0 - 2\epsilon
$$

$$
\geq 1 + 3\epsilon - 2\epsilon = 1 + \epsilon,
$$

i.e., h_Φ^0 has the Opial property. $\qquad\square$

Corollary 2.3.3. *The Orlicz sequence space h_Φ^0 has the fixed point property.*

Corollary 2.3.4. *An Orlicz sequence space ℓ_Φ^0 has the Opial property if and only if $\Phi \in \delta_2$.*

Theorem 2.3.5. *The following statements are equivalent:*

(a) ℓ_Φ has the uniform Opial property.

(b) ℓ_Φ has the Opial property.

(c) h_Φ has the Opial property.

(d) $\Phi \in \delta_2$.

Proof. In view of the previous result it is only necessary to show that (c) $\Rightarrow$ (d) and (d) $\Rightarrow$ (a).

(c) $\Rightarrow$ (d). Suppose that $\Phi \notin \delta_2$. Then, there exists a $x \in S(\ell_\Phi)$ with $x(1) \neq 0$ such that $I_\Phi(\lambda x) = \infty$ for any $\lambda > 1$. This means that $\| \sum_{i=n}^{\infty} x_i e^{(i)} \| = 1$ for all $n \in \mathbb{N}_0$. By the same argument as in the proof of Theorem 2.3.1, we get a sequence (n_i) of natural numbers such that

$$
1 \geq \left\| \sum_{j=n_i+1}^{n_{i+1}} x_j e^{(j)} \right\| \geq 1 - \frac{1}{i+1} \quad \text{for all } i \in \mathbb{N}_0.
$$

Put $x_i = \sum_{i=n_i+1}^{n_{i+1}} x_i e^{(i)}$. Then, $x_i \overset{w}{\to} 0$, as $i \to \infty$. Put $x_0 = \{x_1, 0, 0, \ldots\}$. Then,

$$1 = \lim_{i \to \infty} \|x_i\| = \lim_{i \to \infty} \|x_i + x_0\|,$$

a contradiction.

(d) $\Rightarrow$ (a). Let $\epsilon > 0$ be given. Using $\Phi \in \delta_2$, there exists a $\delta > 0$ such that $\|x\| \geq \epsilon/2$ implies $I_\Phi(x) \geq \delta$. By $\Phi \in \delta_2$, there exists a $\delta_1 \in (0, \epsilon)$ such that $I_\Phi(u) \geq 1 + (\delta/2)$ implies $\|u\| \geq 1 + 3\delta_1$. For any $x \in X$ with $\|x\| \geq \epsilon$, there is an $i_0 \in \mathbb{N}_0$ such that

$$\left\| \sum_{i=i_0+1}^{\infty} x_i e^{(i)} \right\| < \delta_1.$$

For every weakly null sequence $(x_n) \subset S(\ell_\Phi)$, there exists an $n_0 \in \mathbb{N}_0$ such that

$$\left\| \sum_{i=0}^{i_0} x_n(i) e^{(i)} \right\| < \frac{\epsilon}{2}$$

when $n > n_0$. Furthermore, we have

$$I_\Phi \left[\sum_{i=0}^{i_0} x_n(i) e^{(i)} \right] \leq \left\| \sum_{i=0}^{i_0} x_n(i) e^{(i)} \right\| < \frac{\epsilon}{2}.$$

Hence,

$$\|x_n + x\| \geq \left\| \sum_{i=i_0+1}^{\infty} x_n(i) e^{(i)} + \sum_{i=0}^{i_0} x_i e^{(i)} \right\| - 2\delta_1.$$

Next, we estimate the norm of $\sum_{i=i_0+1}^{\infty} x_n(i) e^{(i)} + \sum_{i=0}^{i_0} x_i e^{(i)}$:

$$I_\Phi \left[\sum_{i=i_0+1}^{\infty} x_n(i) e^{(i)} + \sum_{i=0}^{i_0} x_i e^{(i)} \right] = \sum_{i=0}^{i_0} \Phi[x_i] + \sum_{i=i_0+1}^{\infty} \Phi[x_n(i)]$$

$$\geq \delta + 1 - \frac{\delta}{2} = 1 + \frac{\delta}{2}.$$

Hence,

$$\left\| \sum_{i=i_0+1}^{\infty} x_n(i) e^{(i)} + \sum_{i=0}^{i_0} x_i e^{(i)} \right\| \geq 1 + 3\delta_1$$

when $n > n_0$. So,

$$\|x_n + x\| \geq \left\| \sum_{i=i_0+1}^{\infty} x_n(i) e^{(i)} + \sum_{i=0}^{i_0} x_i e^{(i)} \right\| - 2\delta_1 \geq 1 + \delta_1$$

when $n > n_0$. This means that ℓ_Φ has the uniform Opial property. $\square$

Remark 2.3.6. *By Theorem 2.3.1 and Theorem 2.3.5, we get that the non-strict Opial property is weaker than the Opial property.*

Theorem 2.3.7. *The following statements are equivalent:*

(i) ℓ_Φ^0 has the uniform Opial property.

(ii) h_Φ^0 has the Opial property.

(iii) $\Phi \in \delta_2$.

Proof. (i) $\Rightarrow$ (ii) is clear.

(ii) $\Rightarrow$ (iii). Suppose that $\Phi \notin \delta_2$. Then, there exists a $x \in S(\ell_\Phi)$ such that $I_\Phi(\lambda x) = \infty$ for any $\lambda > 1$. This means that $\|\sum_{i=n}^{\infty} x_i e^{(i)}\| = 1$ for all $n \in \mathbb{N}_0$. By Theorem 1.43 of [56], we have

$$\lim_{n \to \infty} \left\| \sum_{i=n}^{\infty} x_i e^{(i)} \right\|^0 = \lim_{n \to \infty} \left\| \sum_{i=n}^{\infty} x_i e^{(i)} \right\| = 1,$$

and we get a sequence (n_i) of natural numbers such that

$$1 + \frac{1}{i} \geq \left\| \sum_{j=n_i+1}^{n_{i+1}} x(j) e^{(j)} \right\|^0 \geq 1 - \frac{1}{i} \quad \text{for all } i \in \mathbb{N}_1.$$

Put $x_i = \sum_{i=n_i+1}^{n_{i+1}} x_i e^{(i)}$. Then $x_i \overset{w}{\to} 0$, as $i \to \infty$. For any $\epsilon > 0$, there exists an $i_1 \in \mathbb{N}_0$ such that

$$1 = \left\| \sum_{i=i_1}^{\infty} x_i e^{(i)} \right\| = \left\| \sum_{i=i_1}^{\infty} x_i e^{(i)} \right\|^0 < 1 + \epsilon.$$

Hence, there exists $i_2 > i_1$ such that

$$\frac{1}{2} < \left\| \sum_{i=i_1}^{i_2} x_i e^{(i)} \right\|^0 < 1 + \epsilon.$$

Put $x_\epsilon = \{0, \ldots, 0, x(i_1), x(i_1 + 1), \ldots, x(i_2), 0, \ldots\}$. Then,

$$\lim_{i \to \infty} \|x_i + x_\epsilon\| \leq \left\| \sum_{i=i_1}^{\infty} x_i e^{(i)} \right\|^0 < 1 + \epsilon,$$

a contradiction.

(iii) $\Rightarrow$ (i). Let $\epsilon > 0$ be given. By $\Phi \in \delta_2$, there exists a $\delta > 0$ such that $\|x\| \geq \epsilon/2$ implies $I_\Phi(x) \geq 2\delta$. For any $x \in \ell_\Phi^0$ with $\|x\| \geq \epsilon$, there exists an $i_0 \in \mathbb{N}_0$ such that $\|\sum_{i=i_0+1}^{\infty} x_i e^{(i)}\|^0 < \epsilon/2$. Using $x_n \overset{w}{\to} 0$, we have $x_n(i) \to 0$ for all $i \in \mathbb{N}_0$. Hence, there exists $n_0 \in \mathbb{N}_0$ such that $\|\sum_{i=0}^{i_0} x_n(i) e^{(i)}\|^0 < \delta$

when $n > n_0$. So, there exists $k_n > 0$ such that, by Theorem 1.31 of [56], we have

$$\left\| \sum_{i=0}^{i_0} x_i e^{(i)} + \sum_{i=i_0+1}^{\infty} x_n(i) e^{(i)} \right\|^0 = \frac{1}{k_n} \left\{ 1 + \sum_{i=0}^{i_0} \Phi[k_n x_i] + \sum_{i=i_0+1}^{\infty} \Phi[k_n x_n(i)] \right\}$$

$$= \frac{1}{k_n} \left\{ 1 + \sum_{i=i_0+1}^{\infty} \Phi[k_n x_i] \right\} + \frac{1}{k_n} \sum_{i=0}^{i_0} \Phi[k_n x_i]$$

$$\geq \left\| \sum_{i=i_0+1}^{\infty} x_n(i) e^{(i)} \right\|^0 + \sum_{i=0}^{i_0} \Phi(x_i)$$

$$\geq 1 - \delta + 2\delta$$

$$= 1 + \delta$$

when $n > n_0$. Hence,

$$\|x + x_n\|^0 = \left\| \sum_i x_i e^{(i)} + \sum_i x_n(i) e^{(i)} \right\|^0$$

$$\geq \left\| \sum_{i=0}^{i_0} x_i e^{(i)} + \sum_{i=i_0+1}^{\infty} x_n(i) e^{(i)} \right\|^0 - 2\epsilon$$

$$\geq 1 + 3\epsilon - 2\epsilon$$

$$= 1 + \epsilon,$$

i.e., h_Φ^0 has uniform Opial property. $\qquad\square$

2.4   Cesàro Sequence Spaces

The *Cesàro sequence space* Ces_p was introduced by Shiue [202] which is defined as

$$Ces_p := \left\{ x = (x_k) \in \omega : \|x\|_{Ces_p} = \left[\sum_n \left(\frac{1}{n+1} \sum_{k=0}^{n} |x_k| \right)^p \right]^{1/p} < \infty \right\}, \ (1 < p < \infty).$$

In this section, we will discuss the results of Chui and Hudzik [65]. First, we describe some geometric constants which are very useful in studying various geometric properties of Banach sequence spaces.

Let X be a Banach space. Then,

$$p(X) = \sup\{r > 0 : \exists\, (x_n) \subset B_X \text{ such that } \|x_n\| \leq 1 - r,\ \|x_n - x_m\| \geq 2r,\ n \neq m\}$$

is known as the *packing constant of a Banach space* of X.

$$sep[(x_n)] = \inf\{\|x_n - x_m\| : n \neq m\},$$
$$D(X) = \sup\{sep[(x_n)] : (x_n) \subset S_X\}.$$

For a sequence (x_n) in X, we define

$$
\begin{aligned}
A(x_n) &= \liminf_{n \to \infty}\{\|x_i + x_j\| : i, j \geq n, \ i \neq j\}, \\
A_1(x_n) &= \liminf_{n \to \infty}\{\|x_i - x_j\| : i, j \geq n, \ i \neq j\}, \\
C(X) &= \sup\{A(x_n) : (x_n) \text{ is weakly null sequence in } S_X\}.
\end{aligned}
$$

We start with the following result:

Theorem 2.4.1. *Any Banach space X with $C(X) < 2$ has the weak Banach–Saks property.*

Proof. Take a positive number ε such that $\theta = C(X) + \varepsilon < 2$. For any weakly null sequence (x_n) in S_X, there exists a subsequence (x_{n_k}) of (x_n) such that $\|x_{n_i} + x_{n_j}\| < 0$ for $i \neq j$. By the result due to Kakutani [109], we get that the Banach space has the weak Banach–Saks property. $\square$

Theorem 2.4.2. *For any Köthe sequence space X, the inequality $C(X) \leq D(X) \leq 2$ holds. If $D(X) < 2$, then X has the Banach–Saks property.*

Proof. Since $D(X) < 2$ implies that the reflexivity of Köthe sequence space X (see [65]), this means that X has an absolutely continuous norm. So, we get that $\{e^{(i)}\}_{i \in \mathbb{N}_0}$ is a basis of X. Now, in virtue of Theorem 2.4.1, we only need to prove that $C(X) \leq D(X)$. Put

$$K = \sup\left\{ A(\{u_n\}) : u_n = \sum_{i=i_{n-1}+1}^{i_n} u_n(i)e^{(i)} \in S_X, \ 0 = i_0 < i_1 < \cdots, \ u_n \xrightarrow{w} 0 \right\}.$$

We will first prove that $C(X) = K$. Clearly $C(X) \geq K$. For any $\varepsilon > 0$, there exists a sequence (x_n) in S_X with $x_n \xrightarrow{w} 0$, as $n \to \infty$, such that $A(x_n) + \varepsilon > C(X)$. By the definition of $A((x_n))$, there exists a subsequence (y_n) of (x_n) such that $\|y_n + y_m\| + 2\varepsilon \geq C(X)$ for $n \neq m$. Define $v_1 = y_1$. There exists $i_1 \in \mathbb{N}_0$ such that

$$\left\| \sum_{i=i_1+1}^{\infty} v_1(i)e^{(i)} \right\| < \varepsilon.$$

Put $z_1 = \sum_{i=0}^{i_1} v_1(i)e^{(i)}$. Then, $\|z_1 + y_m\| + 2\varepsilon \geq C(X)$ for any $m \in \mathbb{N}_2$. Since $y_n \to 0$, as $n \to \infty$, coordinatewise, there exists $n_2 \in \mathbb{N}_0$ such that

$$\left\| \sum_{i=0}^{i_1} y_n(i)e^{(i)} \right\| < \varepsilon$$

for all $n \geq n_2$. Define $v_2 = y_{n_2}$. Then, there is $i_2 > i_1$ such that

$$\left\| \sum_{i=i_2+1}^{\infty} v_2(i)e^{(i)} \right\| < \varepsilon.$$

Put $z_2 = \sum_{i=i_1+1}^{i_2} v_2(i)e^{(i)}$. Then, $\|z_1 + z_2\| + 4\varepsilon \geq C(X)$. In virtue of the inequality $\|z_i + y_n\| + 2\varepsilon \geq C(X)$ for $i = 1, 2$ and $y_n \to 0$, as $n \to \infty$, coordinatewise, there exists $n_3 \in \mathbb{N}_0$ such that

$$\left\| \sum_{i=0}^{i_2} y_n(i)e^{(i)} \right\| < \varepsilon$$

for all $n \geq n_3$. Define $v_3 = y_{n_3}$. Then, there is $i_3 > i_2$ such that

$$\left\| \sum_{i=i_3+1}^{\infty} v_3(i)e^{(i)} \right\| < \varepsilon.$$

Put $z_3 = \sum_{i=i_2+1}^{i_3} v_3(i)e^{(i)}$. Then, $\|z_i + z_3\| + 4\varepsilon \geq C(X)$ for $i = 1, 2$. In the same way, we can find a sequence (z_n) such that

(1) $z_n = \sum_{i=i_{n-1}+1}^{i_n} v_n(i)e^{(i)}$, where $0 = i_0 < i_1 < \cdots$;

(2) $\|z_n + z_m\| + 4\varepsilon \geq C(X)$ for $m, n \in \mathbb{N}_0$;

(3) $\|z_n\| \leq 1$ for $n \in \mathbb{N}_0$;

(4) $z_n \overset{w}{\to} 0$, as $n \to \infty$.

Properties (3) and (4) follow by the fact $(x_n) \subset S_X$, $x_n \overset{w}{\to} 0$, as $n \to \infty$, $|z_n| \leq |x_n|$ for all $n \in \mathbb{N}_0$ and X is a Köthe sequence space. Put $u_n = z_n/\|z_n\|$. Then, $u_n \in S_X$ and note that z_n and z_m are orthogonal for $n \neq m$, whence by $\|z_n\| \leq 1$ and $\|z_m\| \leq 1$, we get

$$\|u_n + u_m\| = \left\| \frac{z_n}{\|z_n\|} + \frac{z_m}{\|z_m\|} \right\| \geq \|z_n + z_m\| \geq C(X) - 5\varepsilon$$

for any $n, m \in \mathbb{N}_0$. Since $\varepsilon > 0$ was arbitrary, we get $C(X) = K$.

Obviously, $D(X) = \sup\{A_1((x_n)) : (x_n) \subset S_X\}$. Since $K = \sup A((u_n))$, where the supremum is taken over the sequences (u_n) as above, we have $C(X) \leq D(X) < 2$.

This completes the proof of the theorem. $\qquad\square$

Corollary 2.4.3. *Let $1 < p < \infty$. Then, $C(\ell_p) = 2^{1/p}$ for the sequence spaces ℓ_p.*

Proof. For any $u = (u_1, u_2, \ldots, u_m, 0, 0, \ldots)$ with $\|u\|_p = 1$, define

$$x_n = (\overbrace{0, 0 \ldots, 0}^{m \times n \ \text{terms}}, u_1, u_2 \ldots, u_m, 0, 0 \ldots)$$

Then, $x_n \overset{w}{\to} 0$ and $\|x_k + x_l\|_p^p = 2\sum_{i=0}^m |u_i|^p = 2$, i.e., $\|x_k + x_l\|_p = 2^{1/p}$ for $k \neq l$. This means that $A(x_n) = 2^{1/p}$ and consequently $C(\ell_p) \geq 2^{1/p}$. Since ℓ_p is a Köthe sequence space and by Theorem 2.4.2, $C(X) \leq D(X)$ for any Köthe sequence space, we have $C(\ell_p) \leq D(\ell_p) = 2^{1/p}$. Hence, $C(\ell_p) = 2^{1/p}$.

This completes the proof of the theorem. $\qquad\square$

Theorem 2.4.4. *The Cesàro sequence space Ces_p has the Banach–Saks property of type p, where $p > 1$.*

Proof. Let (ε_n) be a sequence of positive numbers for which $\sum_n \varepsilon_n \leq 1/2$. Let (x_n) be a weakly null sequence in $B(Ces_p)$. Set $r_0 = 0$ and $t_1 = x_1$. Then, there exists $r_1 \in \mathbb{N}_0$ such that

$$\left\| \sum_{i=r_1+1}^{\infty} t_1(i)e^{(i)} \right\|_{Ces_p} < \varepsilon_1.$$

Since the fact that (x_n) is a weakly null sequence implies that $x_n \to 0$, as $n \to \infty$, coordinatewise, there is an $n_2 \in \mathbb{N}_0$ such that

$$\left\| \sum_{i=0}^{r_1} x_n(i)e^{(i)} \right\|_{Ces_p} < \varepsilon_1$$

for all $n \geq n_2$. Set $t_2 = x_{n_2}$. Then, there exists an $r_2 > r_1$ such that

$$\left\| \sum_{i=r_2+1}^{\infty} t_2(i)e^{(i)} \right\|_{Ces_p} < \varepsilon_2.$$

By using again the fact that $x_n \to 0$, as $n \to \infty$, coordinatewise, there exists an $n_3 > n_2$ such that

$$\left\| \sum_{i=0}^{r_2} x_n(i)e^{(i)} \right\|_{Ces_p} < \varepsilon_2$$

for all $n \geq n_3$. Continuing this process, we can find by induction two increasing subsequences (r_i) and (n_i) of natural numbers such that

$$\left\| \sum_{i=0}^{r_j} x_n(i)e^{(i)} \right\|_{Ces_p} < \varepsilon_j$$

for all $n \geq n_{j+1}$ and

$$\left\| \sum_{i=r_j+1}^{\infty} t_j(i)e^{(i)} \right\|_{Ces_p} < \varepsilon_j,$$

where $t_j = x_{n_j}$. This yields, by the inequality $\varepsilon_{j-1} + \varepsilon_j < 1$, that

$$\sum_s \left[\frac{1}{s} \sum_{i=0}^{s} |t_j(i)| \right]^p \leq (\varepsilon_{j-1} + \varepsilon_j)^p < 1$$

for all $j \in \mathbb{N}_0$. Hence,

$$\left\| \sum_{j=0}^{n} t_j \right\|_{Ces_p} = \left\| \sum_{j=0}^{n} \left[\sum_{i=0}^{r_{j-1}} t_j(i)e^{(i)} + \sum_{i=r_{j-1}+1}^{r_j} t_j(i)e^{(i)} + \sum_{i=r_j+1}^{\infty} t_j(i)e^{(i)} \right] \right\|_{Ces_p}$$

$$\leq \left\| \sum_{j=0}^{n} \left[\sum_{i=r_{j-1}+1}^{r_j} t_j(i)e^{(i)} \right] \right\|_{Ces_p} + \left\| \sum_{j=0}^{n} \left[\sum_{i=0}^{r_{j-1}} t_j(i)e^{(i)} \right] \right\|_{Ces_p}$$

$$+ \left\| \sum_{j=0}^{n} \left[\sum_{i=r_j+1}^{\infty} t_j(i)e^{(i)} \right] \right\|_{Ces_p}$$

$$\leq \left\| \sum_{j=0}^{n} \left[\sum_{i=r_{j-1}+1}^{r_j} t_j(i)e^{(i)} \right] \right\|_{Ces_p} + 2 \sum_{j=0}^{n} \varepsilon_j,$$

and

$$\left\| \sum_{j=0}^{n} \left[\sum_{i=r_{j-1}+1}^{r_j} t_j(i)e^{(i)} \right] \right\|_{Ces_p}^p = \sum_{j=0}^{n} \sum_{s=r_{j-1}+1}^{\infty} \left[\frac{1}{s} \sum_{i=0}^{s} \|t_j(i)\| \right]^p$$

$$\leq \sum_{j=0}^{n} \sum_{s=1}^{\infty} \left[\frac{1}{s} \sum_{i=0}^{s} |t_j(i)| \right]^p$$

$$\leq n.$$

Therefore,

$$\left\| \sum_{j=0}^{n} t_j \right\|_{Ces_p} \leq n^{1/p} + 1 \leq 2n^{1/p}.$$

Hence, the space Ces_p has the Banach–Saks type p.
     This completes the proof of the theorem.   □

2.5   Sequence Spaces Related to ℓ_p Spaces

Sargent [193] defined the following sequence spaces.

Let $\mathcal{C}$ denote the space whose elements are finite sets of distinct positive integers. Given any set $\sigma \in \mathcal{C}$, we define the sequence $c(\sigma) = \{c_n(\sigma)\}$ by

$$c_n(\sigma) := \begin{cases} 1 & , \quad n \in \sigma, \\ 0 & , \quad n \notin \sigma. \end{cases}$$

Further,

$$\mathcal{C}_s := \left\{ \sigma \in \mathcal{C} : \sum_{n=1}^{\infty} c_n(\sigma) \le s \right\},$$

that is, $\mathcal{C}_s$ is the set of those σ whose support has cardinality at most s. The set Φ consists of all real sequences (ϕ_k) such that

$$\phi_1 > 0, \ \Delta \phi_k \ge 0 \ \text{ and } \ \Delta \left(\frac{\phi_k}{k} \right) \le 0 \ \text{ for all } \ k \in \mathbb{N}_1,$$

that is,

$$\Phi := \{ \phi = (\phi_k) \in \omega : 0 < \phi_1 \le \phi_n \le \phi_{n+1} \text{ and } (n+1)\phi_n \ge n\phi_{n+1} \}.$$

For $\phi \in \Phi$, the sequence spaces $m(\phi)$ and $n(\phi)$ were defined by

$$m(\phi) \ := \ \left\{ x = (x_k) \in \omega : \sup_{s \ge 1} \sup_{\sigma \in \mathcal{C}_s} \left(\frac{1}{\phi_s} \sum_{k \in \sigma} |x_k| \right) < \infty \right\},$$

$$n(\phi) \ := \ \left\{ x = (x_k) \in \omega : \sup_{u \in S(x)} \left(\sum_{k=1}^{\infty} |u_k| \Delta \phi_k \right) < \infty \right\}$$

and studied by Sargent [193], and further studied in [153] which are BK-spaces with their natural norms defined by

$$\|x\|_{m(\phi)} = \sup_{s \in \mathbb{N}_0} \sup_{\sigma \in \mathcal{C}_s} \left(\frac{1}{\phi_s} \sum_{k \in \sigma} |x_k| \right) \ \text{ and } \ \|x\|_{n(\phi)} = \sup_{u \in S(x)} \left(\sum_k |u_k| \Delta \phi_k \right),$$

where $\Delta \phi_k = \phi_k - \phi_{k-1}$, $\phi_{-1} = 0$ and $S(x)$ denotes the set of all sequences that are rearrangements of x.

Recently in [166], some of the geometric properties of $m(\phi)$ have been investigated. In [222], Tripathy and Sen have extended the space $m(\phi)$ to $m(\phi, p)$, as follows:

$$m(\phi, p) := \left\{ x = (x_n) \in \omega : \sup_{\tau \in \mathbb{N}_0} \sup_{\sigma \in \mathcal{C}_\tau} \left(\frac{1}{\phi_\tau} \sum_{n \in \sigma} |x_n|^p \right) < \infty \right\}$$

for $\phi \in \Phi$ and $p > 0$. It has been proved in [222] that $m(\phi, p)$ is a Banach space if it is endowed with the norm

$$\|x\|_{m(\phi, p)} = \sup_{\tau \in \mathbb{N}_0} \sup_{\sigma \in \mathcal{C}_\tau} \frac{1}{\phi_\tau} \left(\sum_{n \in \sigma} |x_n|^p \right)^{1/p},$$

where $1 \leq p < \infty$. It is easy to see that $m(\phi, p)$ is a Köthe sequence space, indeed a BK-space with respect to its natural norm (see [193, Lemma 11]). Note that throughout the present section, we study the space $m(\phi, p)$ except the case $\phi_n = n$ for which it is reduced to the space ℓ_∞.

Remark 2.5.1. *The following statements hold:*

(i) *If $\phi_n = 1$ for all $n \in \mathbb{N}_0$, then $m(\phi) = \ell_1$, $n(\phi) = \ell_\infty$; and if $\phi_n = n$ for all $n \in \mathbb{N}_0$, then $m(\phi) = \ell_\infty$, $n(\phi) = \ell_1$.*

(ii) *If $x \in m(\phi)$ $[x \in n(\phi)]$ and $u \in S(x)$, then $u \in m(\phi)$ $[u \in n(\phi)]$ and $\|u\| = \|x\|$.*

(iii) *If $x \in m(\phi)$ $[x \in n(\phi)]$ and $|u_n| \leq |x_n|$ for all $n \in \mathbb{N}_0$, then $u \in m(\phi)$ $[u \in n(\phi)]$ and $\|u\| \leq \|x\|$.*

(iv) *$[m(\phi)]^\beta$ and $n(\phi)$, $[(n(\phi)]^\beta$ and $m(\phi))$ are norm isomorphic, i.e., $[m(\phi)]^\beta \cong n(\phi)$ and $[n(\phi)]^\beta \cong m(\phi)$.*

(v) *If $\phi_n = 1$ for all $n \in \mathbb{N}_0$, then $m(\phi, p) = \ell_p$. Moreover, $\ell_p \subseteq m(\phi, p) \subseteq \ell_\infty$.*

(vi) *If $p = 1$, then $m(\phi, p) = m(\phi)$. Also, for any $p \geq 1$, $m(\phi) \subseteq m(\phi, p)$.*

Now, we investigate some geometric properties of the spaces $m(\phi)$ and $m(\phi, p)$.

Let us start with the following lemma:

Lemma 2.5.2. *If an FK-space X containing ϕ has the property AK, then it is order continuous, i.e., $\|(0, 0 \ldots, x_n, x_{n+1}, \ldots)\| \to 0$, as $n \to \infty$, for any $x \in X$.*

Proof. From the definition of property AK, we have that every $x = (x_i) \in X$ has a unique representation $x = \sum_i x_i e^{(i)}$, i.e., $x^{[n]} = \sum_{i=0}^n x_i e^{(i)} \to x$, as $n \to \infty$. Hence, $\|x - x^{[n]}\|_X \to 0$, as $n \to \infty$, i.e., $\|(0, 0 \ldots, x_n, x_{n+1}, \ldots)\| \to 0$, as $n \to \infty$, which means that X is order continuous. $\qquad \square$

Corollary 2.5.3. *The space $m(\phi, p)$ is order continuous.*

Proof. It is easy to see that $m(\phi, p)$ contains ϕ and that every $x = \{x_i\} \in m(\phi, p)$ has a unique representation $x = \sum_i x_i e^{(i)}$, i.e., $x^{[n]} = \sum_{i=0}^n x_i e^{(i)} \to x$, as $n \to \infty$, which means that $m(\phi, p)$ has AK. Hence, $m(\phi, p)$ is order continuous by Lemma 2.5.2. $\qquad \square$

Theorem 2.5.4. *The space $m(\phi, p)$ has the Fatou property.*

Proof. Let x be any real sequence from ω_+ and (x_n) be any non-decreasing sequence of non-negative elements from $m(\phi, p)$ be such that $x_n(i) \to x_i$, as $n \to \infty$, coordinatewise and $\sup_{n \in \mathbb{N}_0} \|x_n\|_{m(\phi, p)} < \infty$.

Let us denote $s = \sup_{n \in \mathbb{N}_0} \|x_n\|_{m(\phi,p)}$. Then, since the supremum is homogeneous, we have

$$\frac{1}{s} \sup_{\tau \in \mathbb{N}_0} \left\{ \sup_{\sigma \in \zeta_\tau} \left[\frac{1}{\phi_\tau} \sum_{i \in \sigma} \|x_n(i)\|^p \right]^{1/p} \right\} = \sup_{\tau \in \mathbb{N}_0} \left\{ \sup_{\sigma \in \zeta_\tau} \left[\frac{1}{\phi_\tau} \sum_{i \in \sigma} \left\| \frac{x_n(i)}{s} \right\|^p \right]^{1/p} \right\}$$

$$\leq \sup_{\tau \in \mathbb{N}_0} \left\{ \sup_{\sigma \in \zeta_\tau} \left[\frac{1}{\phi_\tau} \sum_{i \in \sigma} \left\| \frac{x_n(i)}{\|x_n\|_{m(\phi,p)}} \right\|^p \right]^{1/p} \right\}$$

$$= \frac{1}{\|x_n\|_{m(\phi,p)}} \|x_n\|_{m(\phi,p)} = 1.$$

Moreover, (x_n) is non-decreasing and converges to x coordinatewisely, and by the Beppo-Levi theorem we have

$$\frac{1}{s} \lim_{n \to \infty} \left[\sup_{\tau \in \mathbb{N}_0} \left(\sup_{\sigma \in \zeta_\tau} \left(\frac{1}{\phi_\tau} \sum_{i \in \sigma} \|x_n(i)\|^p \right)^{1/p} \right) \right] = \sup_{\tau \in \mathbb{N}_0} \left(\sup_{\sigma \in \zeta_\tau} \left(\frac{1}{\phi_\tau} \sum_{i \in \sigma} \left\| \frac{x_i}{s} \right\|^p \right)^{1/p} \right)$$

$$= \left\| \frac{x}{s} \right\|_{m(\phi,p)} \leq 1,$$

whence

$$\|x\|_{m(\phi,p)} \leq s = \sup_{n \in \mathbb{N}_0} \|x_n\|_{m(\phi,p)} = \lim_{n \to \infty} \|x_n\|_{m(\phi,p)} < \infty.$$

Therefore, $x \in m(\phi,p)$. On the other hand, since $0 \leq x_n \leq x$ for any natural number n and the sequence (x_n) is non-decreasing, we have that the sequence $\left\{ \|x_n\|_{m(\phi,p)} \right\}$ is non-decreasing from above by $\|x\|_{m(\phi,p)}$. In consequence $\lim_{n \to \infty} \|x_n\|_{m(\phi,p)} \leq \|x\|_{m(\phi,p)}$ which together with the opposite inequality proved already yields that $\|x_n\|_{m(\phi,p)} \to \|x\|_{m(\phi,p)}$, as $n \to \infty$.

This completes the proof of the theorem. $\square$

Theorem 2.5.5. *The space $m(\phi,p)$ has the weak fixed point property, if $K > 2^{1-p}$, where $K = \sup_{\tau \in \mathbb{N}_0} \phi_\tau < \infty$ and $1 < p < \infty$.*

Proof. If $\psi_\tau = 1$ for all $\tau \in \mathbb{N}_0$, then it follows that $m(\phi,p) \subseteq \ell_p$ if and only if $\sup_{\tau \in \mathbb{N}_0} (\phi_\tau)^{1/p} < \infty$. Again, we have

$$\|x\|_{m(\phi,p)} = \sup_{\tau \in \mathbb{N}_0} \left(\frac{1}{\phi_\tau} \right)^{1/p} \|x\|_p.$$

Hence, since $R(\ell_p) = 2^{1/p}$, we have

$$R[m(\phi,p)] = \sup_{\tau \in \mathbb{N}_0} \left(\frac{1}{\phi_\tau} \right)^{1/p} R(\ell_p) = \left(\frac{2}{K} \right)^{1/p} < 2,$$

where $R(X)$ stands for the Garcia-Falset coefficient of X. Therefore, $m(\phi,p)$ has the weak fixed point property. $\square$

Theorem 2.5.6. *The space $m(\phi)$ has the weak Banach–Saks property.*

Proof. Let (ϵ_n) be a sequence of natural numbers such that $\sum_n \epsilon_n \leq 1/2$. Let (x_n) be a weakly null sequence in $B[m(\phi)]$. Set $x_0 = 0$ and $z_1 = x_{n_1} = x_1$. Then, there exists $s_1 \in \mathbb{N}_0$ such that

$$\left\| \sum_{i \in \tau_1} z_1(i) e^{(i)} \right\|_{m(\phi)} < \epsilon_1,$$

where τ_1 consists of the elements of σ which exceed s_1. Since $x_n \overset{\omega}{\to} 0$, as $n \to \infty$, implies that $x_n \to 0$, as $n \to \infty$, coordinatewise, there is $n_2 \in \mathbb{N}_0$ such that

$$\left\| \sum_{i=0}^{s_1} x_n(i) e^{(i)} \right\|_{m(\phi)} < \epsilon_1$$

for all $n \geq n_2$. Set $z_2 = x_{n_2}$. Then, there exists $s_2 > s_1$ such that

$$\left\| \sum_{i \in \tau_2} z_2(i) e^{(i)} \right\|_{m(\phi)} < \epsilon_2,$$

where τ_2 consists of all elements of σ which exceed s_2. Using again the fact that $x_n \to 0$, as $n \to \infty$, coordinatewise there exists $n_3 > n_2$ such that

$$\left\| \sum_{i=0}^{s_2} x_n(i) e^{(i)} \right\|_{m(\phi)} < \epsilon_2$$

for all $n \geq n_3$. Continuing this process, we can find two increasing sequences (s_i) and (n_i) such that

$$\left\| \sum_{i=0}^{s_j} x_n(i) e^{(i)} \right\|_{m(\phi)} < \epsilon_j \quad \text{for each } n \geq n_{j+1} \text{ and } \left\| \sum_{i \in \tau_j} z_j(i) e^{(i)} \right\|_{m(\phi)} < \epsilon_j,$$

where $z_j = x_{n_j}$ and τ_j consist of the elements of σ which exceed s_j. Since $\epsilon_{j-1} + \epsilon_j < 1$, we have $\dfrac{1}{\phi_s} \sum_{n \in \sigma} |z_j(n)| \leq \epsilon_{j-1} + \epsilon_j < 1$ for all $j, s \in \mathbb{N}_0$. Hence,

$$\left\| \sum_{j=0}^{n} z_j \right\|_{m(\phi)} = \left\| \sum_{j=0}^{n} \left[\sum_{i=0}^{s_{j-1}} z_j(i) e^{(i)} + \sum_{i=s_{j-1}+1}^{s_j} z_j(i) e^{(i)} + \sum_{i \in \tau_j} z_j(i) e^{(i)} \right] \right\|_{m(\phi)}$$

$$\leq \left\| \sum_{j=0}^{n} \left[\sum_{i=0}^{s_{j-1}} z_j(i) e^{(i)} \right] \right\|_{m(\phi)} + \left\| \sum_{j=0}^{n} \left[\sum_{i=s_{j-1}+1}^{s_j} z_j(i) e^{(i)} \right] \right\|_{m(\phi)}$$

$$\leq \sum_{j=0}^{n} \left\| \left[\sum_{i=s_{j-1}+1}^{s_j} z_j(i) e^{(i)} \right] \right\|_{m(\phi)} + 2 \sum_{j=0}^{n} \epsilon_j,$$

$$\left\| \sum_{j=0}^{n} \left[\sum_{i=s_{j-1}+1}^{s_j} z_j(i)e^{(i)} \right] \right\|_{m(\phi)} \le \sum_{j=0}^{n} \left\| \sum_{i \in \tau_j} z_j(i)e^{(i)} \right\|_{m(\phi)} < \sum_{j=0}^{n} \varepsilon_j.$$

Therefore,

$$\left\| \sum_{j=0}^{n} z_j \right\|_{m(\phi)} \le 3 \sum_{j=0}^{n} \varepsilon_j \quad \text{and} \quad \lim_{n \to \infty} \left\| \frac{1}{n+1} \sum_{j=0}^{n} z_j \right\|_{m(\phi)} \le \lim_{n \to \infty} \frac{3}{n+1} \sum_{j=0}^{n} \varepsilon_j = 0.$$

This completes the proof of the theorem. □

Theorem 2.5.7. *The space $m(\phi, p)$ has the Banach–Saks property of type p.*

Proof. This is obtained on the similar lines as the proof of Theorem 2.4.4. □

2.6 Sequence Spaces $\ell_p(u, v)$ and $\ell_\Delta(u, v, p)$

Let us define the weighted mean matrix $G = G(u, v) = (g_{nk})$ by

$$g_{nk} := \begin{cases} u_n v_k & , \quad 0 \le k \le n, \\ 0 & , \quad k > n \end{cases} \tag{2.6.1}$$

for all $k, n \in \mathbb{N}_0$, where u_n depends only on n and v_k only on k. The following sequence spaces have been introduced and studied in [98]. Let $u = (u_n)$ and $v = (v_n)$ be arbitrary sequences of non-zero reals and let for any $p \in [1, \infty)$,

$$\ell_p(u, v) = \left\{ x = (x_k) \in \omega : \sum_n \left| \sum_{k=0}^{n} u_n v_k x_k \right|^p < \infty \right\}.$$

It is obvious that $\ell_p(u, v)$ is a linear space. It is easy to see that $\ell_p(u, v)$ is a Banach space with the norm

$$\|x\|_{\ell_p(u,v)} = \|G(u, v)x\|_p = \left(\sum_n \left| \sum_{k=0}^{n} u_n v_k x_k \right|^p \right)^{1/p}.$$

Note that

(i) If $v = (1, 1, 1 \ldots)$ and $u = \{1/(n+1)\}_{n \in \mathbb{N}_0}$, then $\ell_p(u, v)$ is reduced to the Cesàro sequence space X_p of non-absolute type (see [177]), and

$$\|x\|_{X_p} = \|C_1 x\|_p = \left(\sum_n \left| \frac{1}{n+1} \sum_{k=0}^{n} x_k \right|^p \right)^{1/p}, \quad (1 \le p < \infty).$$

(ii) If $v = (q_n)$ is a sequence of non-zero reals and $u = (1/Q_n)_{n \in \mathbb{N}_0}$, where $Q_n = q_0 + q_1 + \cdots + q_n$ for all $n \in \mathbb{N}_0$, then $\ell_p(u, v)$ is reduced to the Riesz sequence space r_p^q of non-absolute type (see [10]), and

$$\|x\|_{r_p^q} = \|R^q x\|_p = \left(\sum_n \left| \frac{1}{Q_n} \sum_{k=0}^n q_k x_k \right|^p \right)^{1/p}, \quad (1 \le p < \infty).$$

(iii) Let u and Q_n be as in Part (ii), and $v = (q_{n-k})_{k \in \mathbb{N}_0}$ for each $n \in \mathbb{N}_0$. Then, $\ell_p(u, v)$ is reduced to the Nörlund sequence space $X_{a(p)}$ of non-absolute type (see [223]), and

$$\|x\|_{N_q} = \|N_q x\|_p = \left(\sum_k \left| \frac{1}{Q_n} \sum_{k=0}^n q_{n-k} x_k \right|^p \right)^{1/p} ; \quad (1 \le p < \infty).$$

Theorem 2.6.1. *The Gurariĭ modulus of convexity for the normed space $\ell_p(u, v)$ satisfies the inequality*

$$\beta_{\ell_p(u,v)}(\varepsilon) \le 1 - \left[1 - \left(\frac{\varepsilon}{2} \right)^p \right]^{1/p} \quad \text{for any } 0 \le \varepsilon \le 2. \tag{2.6.2}$$

Proof. Let $x \in \ell_p(u, v)$. By using (2.6.1), we have

$$\|x\|_{\ell_p(u,v)} = \|G(u, v)x\|_p = \left(\sum_n |\{G(u, v)x\}_n|^p \right)^{1/p}.$$

Let $0 \le \varepsilon \le 2$. Then, by using (2.6.2) let us consider the following sequences:

$$x = (x_n) = \left\{ H \left[\left(1 - \left(\frac{\varepsilon}{2} \right)^p \right)^{1/p} \right], H \left(\frac{\varepsilon}{2} \right), 0, 0 \ldots \right\},$$

$$t = (t_n) = \left\{ H \left[\left(1 - \left(\frac{\varepsilon}{2} \right)^p \right)^{1/p} \right], H \left(-\frac{\varepsilon}{2} \right), 0, 0 \ldots \right\}.$$

Since $y_n = \{G(u, v)x\}_n$ and $z_n = \{G(u, v)t\}_n$, we have

$$y = (y_n) = \left\{ \left[1 - \left(\frac{\varepsilon}{2} \right)^p \right]^{1/p}, \left(\frac{\varepsilon}{2} \right), 0, 0 \ldots \right\},$$

$$z = (z_n) = \left\{ \left[1 - \left(\frac{\varepsilon}{2} \right)^p \right]^{1/p}, \left(-\frac{\varepsilon}{2} \right), 0, 0 \ldots \right\}.$$

By using the sequences x and t, given previously, we obtain the following equalities:

$$\|x\|_{\ell_p(u,v)}^p = \|G(u,v)x\|_p^p = \left|\left[1-\left(\frac{\varepsilon}{2}\right)^p\right]^{1/p}\right|^p + \left|\frac{\varepsilon}{2}\right|^p$$

$$= 1-\left(\frac{\varepsilon}{2}\right)^p + \left(\frac{\varepsilon}{2}\right)^p = 1$$

$$\|t\|_{\ell_p(u,v)}^p = \|G(u,v)t\|_p^p = \left|\left[1-\left(\frac{\varepsilon}{2}\right)^p\right]^{1/p}\right|^p + \left|-\frac{\varepsilon}{2}\right|^p$$

$$= 1-\left(\frac{\varepsilon}{2}\right)^p + \left(\frac{\varepsilon}{2}\right)^p = 1$$

$$\|x-t\|_{\ell_p(u,v)} = \|G(u,v)x - G(u,v)t\|_p$$

$$= \left\{\left|\left[1-\left(\frac{\varepsilon}{2}\right)^p\right]^{1/p} - \left[1-\left(\frac{\varepsilon}{2}\right)^p\right]^{1/p}\right|^p + \left|\frac{\varepsilon}{2}-\left(-\frac{\varepsilon}{2}\right)\right|^p\right\}^{1/p}$$

$$= \varepsilon.$$

To complete the upper estimate of the Gurariĭ modulus of convexity, it remains to calculate the infimum of $\|\alpha x + (1-\alpha)t\|_{\ell_p(u,v)}$ for $0 \leq \alpha \leq 1$. We have

$$\inf_{0\leq\alpha\leq1}\|\alpha x + (1-\alpha)t\|_{\ell_p(u,v)} = \inf_{0\leq\alpha\leq1}\|\alpha G(u,v)x + (1-\alpha)G(u,v)t\|_p$$

$$= \inf_{0\leq\alpha\leq1}\left\{\left|\alpha\left[1-\left(\frac{\varepsilon}{2}\right)^p\right]^{1/p} + (1-\alpha)\left[1-\left(\frac{\varepsilon}{2}\right)^p\right]^{1/p}\right|^p\right.$$

$$\left. + \left|\alpha\left(\frac{\varepsilon}{2}\right) + (1-\alpha)\left(-\frac{\varepsilon}{2}\right)\right|^p\right\}^{1/p}$$

$$= \inf_{0\leq\alpha\leq1}\left[1-\left(\frac{\varepsilon}{2}\right)^p + |2\alpha-1|^p\left(\frac{\varepsilon}{2}\right)^p\right]^{1/p} = \left[1-\left(\frac{\varepsilon}{2}\right)^p\right]^{1/p}.$$

Consequently, we get the inequality for $p \geq 1$ that

$$\beta_{\ell_p(u,v)}(\varepsilon) \leq 1 - \left[1-\left(\frac{\varepsilon}{2}\right)^p\right]^{1/p},$$

which is the desired result. $\qquad\qquad\qquad\qquad\qquad\qquad\qquad\qquad\qquad\square$

Let us define the matrix $V = (v_{nk})$ by

$$v_{nk} := \begin{cases} u_n(v_k - v_{k+1}) & , \quad k < n, \\ u_n v_n & , \quad k = n, \\ 0 & , \quad k > n \end{cases}$$

for all $k, n \in \mathbb{N}_0$. Here, $u_n \neq 0$ and $v_n \neq 0$ for all $n \in \mathbb{N}_0$, and (u_n) depend on n, (v_k) depend on k. The following paranormed sequence space was introduced and studied in [119]:

$$\ell_\Delta(u,v,p) := \left\{(x_n) \in \omega : (y_n) = \left(\sum_{k=0}^{n-1} u_n(v_k - v_{k+1})x_k + u_n v_n x_n\right)_{n\in\mathbb{N}_0} \in \ell(p)\right\}.$$

The forward difference $\triangle x$ of a sequence $x = (x_k)$ is defined by $(\triangle x)_k = x_k - x_{k+1}$ for all $k \in \mathbb{N}_0$. Throughout this study, we use the convention that any term with negative subscript is equal to naught and we assume here and after that $p = (p_k)$ is a bounded sequence of positive real numbers, $H = \sup_{k \in \mathbb{N}_0} p_k$, $M = \max\{1, H\}$. The sequence space $\ell_\triangle(u, v, p)$ is a complete metric space of non-absolute type with respect to the paranorm defined by

$$g(x) = \left[\sum_n \left| \sum_{k=0}^{n-1} u_n (v_k - v_{k+1}) x_k + u_n v_n x_n \right|^{p_k} \right]^{1/M}.$$

In this section, we will give some basic properties of the modular ρ on the space $\ell_{\rho\triangle}(u, v, p)$ and study some geometric properties of this space.

We introduce the modular sequence space $\ell_{\rho\triangle}(u, v, p)$ by

$$\ell_{\rho\triangle}(u, v, p) := \left\{ x = (x_n) \in w : \sum_{k=1}^{\infty} \left(\sum_{i=1}^{k-1} u_k \triangle v_i |x_i| + u_k v_k |x_k| \right)^{p_k} < \infty \right\}.$$

The Luxemburg norm on the sequence space $\ell_{\rho\triangle}(u, v, p)$ is defined as follows:

$$\|x\|_L = \inf \left\{ \lambda > 0 : \rho \left(\frac{x}{\lambda} \right) \leq 1 \right\} \quad \text{for every } x \in \ell_{\rho\triangle}(u, v, p).$$

Here, the modular ρ is defined by

$$\rho(x) = \sum_{k=1}^{\infty} \left(\sum_{i=1}^{k-1} u_k \triangle v_i |x_i| + u_k v_k |x_k| \right)^{p_k},$$

which is a convex modular on $\ell_{\rho\triangle}(u, v, p)$.

Lemma 2.6.2. *The functional ρ is a convex modular on $\ell_{\rho\triangle}(u, v, p)$.*

The following basic properties are easy to verify [212]:

Lemma 2.6.3. *For $x \in \ell_{\rho\triangle}(u, v, p)$, the modular ρ on $\ell_{\rho\triangle}(u, v, p)$ satisfies the following properties:*

(i) *If $0 < a < 1$, then $a^M \rho(x/a) \leq \rho(x)$ and $\rho(ax) \leq a\rho(x)$.*

(ii) *If $a \geq 1$, then $\rho(x) \leq a^M \rho(x/a)$.*

(iii) *If $a \geq 1$, then $\rho(x) \leq a\rho(x) \leq \rho(ax)$.*

Lemma 2.6.4. *For any $x \in \ell_{\rho\triangle}(u, v, p)$,*

(i) *If $\|x\| < 1$, then $\rho(x) \leq \|x\|$.*

(ii) *If $\|x\| > 1$, then $\rho(x) \geq \|x\|$.*

(iii) *$\|x\| = 1$ if and only if $\rho(x) = 1$.*

(iv) If $\|x\| < 1$, then $\rho(x) < 1$.

(v) If $\|x\| > 1$, then $\rho(x) > 1$.

Lemma 2.6.5. *Let (x_n) be a sequence in $\ell_{\rho\Delta}(u, v, p)$. Then, the following statements hold:*

(i) If $\|x_n\| \to 1$, as $n \to \infty$, then $\rho(x_n) \to 1$, as $n \to \infty$.

(ii) If $\rho(x_n) \to 0$, as $n \to \infty$, then $\|x_n\| \to 0$, as $n \to \infty$.

Lemma 2.6.6. *For any $L > 0$ and $\varepsilon > 0$, there exists $\delta > 0$ such that $\|\rho(u + v) - \rho(u)\| < \varepsilon$ whenever $u, v \in \ell_{\rho\Delta}(u, v, p)$ with $\rho(u) \leq L$ and $\rho(v) \leq \delta$.*

Lemma 2.6.7. *For any sequence $(x_n) \in \ell_{\rho\Delta}(u, v, p)$, $\|x_n\| \to 0$, as $n \to \infty$, if and only if $\rho(x_n) \to 0$, as $n \to \infty$.*

Lemma 2.6.8. *For any $x \in \ell_{\rho\Delta}(u, v, p)$ and $\varepsilon \in (0, 1)$, there exists $\delta \in (0, 1)$ such that $\rho(x) \leq 1 - \varepsilon$ implies $\|x\| \leq 1 - \delta$.*

Now, we will show that $\ell_{\rho\Delta}(u, v, p)$ is a Banach space with respect to the Luxemburg norm.

Theorem 2.6.9. *$\ell_{\rho\Delta}(u, v, p)$ is a Banach space with respect to the Luxemburg norm defined by $\|x\| = \inf\left\{\lambda > 0 : \rho\left(\frac{x}{\lambda}\right) \leq 1\right\}$.*

Proof. We will show that every Cauchy sequence in $\ell_{\rho\Delta}(u, v, p)$ is convergent with respect to the Luxemburg norm. Let (x_k^n) be a Cauchy sequence in $\ell_{\rho\Delta}(u, v, p)$ and $\varepsilon \in (0, 1)$. Thus, there exists $n_0(\varepsilon)$ such that $\|x^n - x^m\| < \varepsilon$ for all $m, n \geq n_0$. By Part (i) of Lemma 2.6.5, we obtain that $\rho(x^n - x^m) < \|x^n - x^m\| < \varepsilon$ for all $m, n \geq n_0(\varepsilon)$. That is,

$$\sum_k \left[\sum_{i=0}^{k-1} u_k \Delta v_i \|x_n(i) - x_m(i)\| + u_k v_k \|x_n(k) - x_m(k)\|\right]^{p_k} < \varepsilon.$$

For any fixed $k \in \mathbb{N}_0$, we get that $\|x_n(i) - x_m(i)\| < \varepsilon$. Hence, we obtain that $\{x_n(i)\}$ is a Cauchy sequence in $\mathbb{R}$. Since the real axis $\mathbb{R}$ is complete, $x_m(i) \to x_i$, as $m \to \infty$. Therefore, we have for any fixed $k \in \mathbb{N}_0$ that

$$\left[\sum_{i=0}^{k-1} u_k \Delta v_i \|x_n(i) - x_i\| + u_k v_k \|x_n(k) - x(k)\|\right]^{p_k} < \varepsilon \text{ for all } n \geq n_0(\varepsilon),$$

as $m \to \infty$. So, we obtain for all $n \geq n_0(\varepsilon)$, as $m \to \infty$, that $\rho(x_n - x_m) \to \rho(x_n - x)$. So, for all $n \geq n_0(\varepsilon)$ from Part (i) of Lemma 2.6.5, $\rho(x_n - x) < \|x_n - x\| < \varepsilon$. It can be seen for all $n \geq n_0$ that $x_n \to x$, as $n \to \infty$, and $(x_n - x) \in \ell_{\rho\Delta}(u, v, p)$.

From the linearity of the sequence space $\ell_{\rho\Delta}(u, v, p)$, we can write that $x = (x_n) - (x_n - x) \in \ell_{\rho\Delta}(u, v, p)$. Hence, $\ell_{\rho\Delta}(u, v, p)$ is a Banach space with respect to the Luxemburg norm.

This completes the proof of theorem. $\square$

Lemma 2.6.10. *Let $x \in \ell_{\rho\Delta}(u,v,p)$ and $(x_n) \subseteq \ell_{\rho\Delta}(u,v,p)$. If $\rho(x_n) \to \rho(x)$, as $n \to \infty$, and $x_n(i) \to x_i$, as $n \to \infty$ for all $i \in \mathbb{N}_0$, then $x_n \to x$, as $n \to \infty$.*

Now, we give the theorems of this section involving the geometric properties of the space $\ell_{\rho\Delta}(u,v,p)$.

Theorem 2.6.11. *The space $\ell_{\rho\Delta}(u,v,p)$ has the Kadec-Klee property.*

Proof. Let $x \in S[\ell_{\rho\Delta}(u,v,p)]$ and $\{x_n(i)\} \subseteq \ell_{\rho\Delta}(u,v,p)$ such that $\|x_n(i)\| \to 1$ and $x_n(i) \xrightarrow{w} x_i$, as $n \to \infty$. From Part (iii) of Lemma 2.6.4, we get $\rho(x) = 1$. So, from Part (i) of Lemma 2.6.5, it follows that $\rho(x_n) \to \rho(x)$, as $n \to \infty$. Since the mapping $\pi_i : \ell_{\rho\Delta}(u,v,p) \to \mathbb{R}$ defined by $\pi_i(y) = y_i$ is a continuous linear functional on $\ell_{\rho\Delta}(u,v,p)$. It follows that $x_n(i) \to x_i$, as $n \to \infty$, for all $i \in \mathbb{N}_0$. So, from Lemma 2.6.10, $x_n \to x$, as $n \to \infty$. □

Theorem 2.6.12. *The space $\ell_{\rho\Delta}(u,v,p)$ is $k - NUC$ for any integer $k \geq 2$, where $1 < p < \infty$.*

Proof. Let $\varepsilon > 0$ and $(x_n) \subseteq B[\ell_{\rho\Delta}(u,v,p)]$ with $sep(x_n) \geq \varepsilon$. For each $m \in \mathbb{N}_0$, let $x_n^m = \left\{ \underbrace{0,0,\ldots,0}_{m-1 \text{ terms}}, x_n(m), x_n(m+1), \ldots \right\}$. Since $\{x_n(i)\}_{n \in \mathbb{N}_0}$ is bounded for each $i \in \mathbb{N}_0$, by using the diagonal method, we can find a subsequence (x_{n_j}) for each $m \in \mathbb{N}_0$ of (x_n) such that (x_{n_j}) converges for each $j \in \mathbb{N}_0$ with $0 \leq j \leq m$. Therefore, there exists an increasing sequence of positive integers (t_m) such that $sep\{x_{n_j}^m\}_{j>t_m} \geq \varepsilon$. Hence, there is an increasing sequence of positive integers (r_m) such that $\|x_{r_m}^m\| \geq \varepsilon/2$ for all $m \in \mathbb{N}_0$. Then, by Lemma 2.6.8, we may assume that there exists $\mu > 0$ such that

$$\rho\left(x_{r_m}^m\right) \geq \mu \quad \text{for all} \quad m \in \mathbb{N}_0. \tag{2.6.3}$$

Let $\alpha > 0$ be such that $1 < \alpha < \liminf\limits_{n \to \infty} p_n$. For fixed integer $k \geq 2$, let $\varepsilon_1 = \mu(k^{\alpha-1} - 1)/[2k^\alpha(k-1)]$. Then, by Lemma 2.6.7, there is a $\delta > 0$ such that

$$|\rho(u+v) - \rho(u)| < \varepsilon_1 \tag{2.6.4}$$

whenever $\rho(u) \leq 1$ and $\rho(v) \leq \delta$.

Since by Part (i) of Lemma 2.6.5, $\rho(x_n) \leq 1$ for all $n \in \mathbb{N}_0$, there exist positive integers m_i with $m_1 < m_2 < \cdots < m_{k-1}$ such that $\rho(x_i^{m_i}) \leq \delta$ and $\alpha \leq p_j$ for all $j \geq m_{k-1}$. Define $m_k = m_{k-1} + 1$. From (2.6.3), we have $\rho\left(x_{r_{m_k}}^{m_k}\right) \geq \mu$. Let $s_i = i$ for $1 \leq i \leq k-1$ and $s_k = r_{m_k}$. Then, in virtue of

(2.6.3), (2.6.4) and convexity of function $f_i(u) = |u|^{p_i}$ for each $i \in \mathbb{N}_1$, we have

$$\rho\left(\frac{x_{s_1} + x_{s_2} + \cdots + x_{s_k}}{k}\right)$$

$$= \sum_{n=1}^{\infty} \left[\sum_{i=1}^{n-1} u_n \triangle v_i \left|\frac{x_{s_1}(i) + x_{s_2}(i) + \cdots + x_{s_k}(i)}{k}\right| + u_n v_n \left|\frac{x_{s_1}(n) + x_{s_2}(n) + \cdots + x_{s_k}(n)}{k}\right|\right]^{p_n}$$

$$= \sum_{n=1}^{m_1} \left[\sum_{i=1}^{n-1} u_n \triangle v_i \left|\frac{x_{s_1}(i) + x_{s_2}(i) + \cdots + x_{s_k}(i)}{k}\right| + u_n v_n \left|\frac{x_{s_1}(n) + x_{s_2}(n) + \cdots + x_{s_k}(n)}{k}\right|\right]^{p_n}$$

$$+ \sum_{n=m_1+1}^{\infty} \left[\sum_{i=1}^{n-1} u_n \triangle v_i \left|\frac{x_{s_1}(i) + x_{s_2}(i) + \cdots + x_{s_k}(i)}{k}\right| + u_n v_n \left|\frac{x_{s_1}(n) + x_{s_2}(n) + \cdots + x_{s_k}(n)}{k}\right|\right]^{p_n}$$

$$\leq \sum_{n=1}^{m_1} \frac{1}{k} \sum_{j=0}^{k} \left[\sum_{i=0}^{n-1} u_n \triangle v_i \left|x_{s_j}(i)\right| + u_n v_n \left|x_{s_j}(n)\right|\right]^{p_n}$$

$$+ \sum_{n=m_1+1}^{\infty} \left[\sum_{i=1}^{n-1} u_n \triangle v_i \left|\frac{x_{s_2}(i) + x_{s_3}(i) + \cdots + x_{s_k}(i)}{k}\right| + u_n v_n \left|\frac{x_{s_2}(n) + x_{s_3}(n) + \cdots + x_{s_k}(n)}{k}\right|\right]^{p_n} + \varepsilon_1$$

$$= \sum_{n=1}^{m_1} \frac{1}{k} \sum_{j=1}^{k} \left[\sum_{i=1}^{n-1} u_n \triangle v_i \left|x_{s_j}(i)\right| + u_n v_n \left|x_{s_j}(n)\right|\right]^{p_n}$$

$$+ \sum_{n=m_1+1}^{m_2} \frac{1}{k} \sum_{j=2}^{k} \left[\sum_{i=1}^{n-1} u_n \triangle v_i \left|x_{s_j}(i)\right| + u_n v_n \left|x_{s_j}(n)\right|\right]^{p_n}$$

$$+ \sum_{n=m_2+1}^{\infty} \left[\sum_{i=1}^{n-1} u_n \triangle v_i \left|\frac{x_{s_3}(i) + x_{s_4}(i) + \cdots + x_{s_k}(i)}{k}\right| + u_n v_n \left|\frac{x_{s_3}(n) + x_{s_4}(n) + \cdots + x_{s_k}(n)}{k}\right|\right]^{p_n} + 2\varepsilon_1$$

$$\leq \sum_{n=1}^{m_1} \frac{1}{k} \sum_{j=1}^{k} \left[\sum_{i=1}^{n-1} u_n \triangle v_i \left|x_{s_j}(i)\right| + u_n v_n \left|x_{s_j}(n)\right|\right]^{p_n}$$

$$+ \sum_{n=m_1+1}^{m_2} \frac{1}{k} \sum_{j=2}^{k} \left[\sum_{i=1}^{n-1} u_n \triangle v_i \left|x_{s_j}(i)\right| + u_n v_n \left|x_{s_j}(n)\right|\right]^{p_n}$$

$$+ \sum_{n=m_2+1}^{m_3} \frac{1}{k} \sum_{j=3}^{k} \left[\sum_{i=1}^{n-1} u_n \triangle v_i \left|x_{s_j}(i)\right| + u_n v_n \left|x_{s_j}(n)\right|\right]^{p_n}$$

$$+$$

$$\vdots$$

$$+ \sum_{n=m_{k-2}+1}^{m_{k-1}} \frac{1}{k} \sum_{j=k-1}^{k} \left[\sum_{i=1}^{n-1} u_n \triangle v_i \left|x_{s_j}(i)\right| + u_n v_n \left|x_{s_j}(n)\right|\right]^{p_n}$$

$$+ \sum_{n=m_{k-1}+1}^{\infty} \left[\sum_{i=1}^{n-1} u_n \triangle v_i \left| \frac{x_{s_k}(i)}{k} \right| + u_n v_n \left| \frac{x_{s_k}(n)}{n} \right| \right]^{p_n} + (k-1)\varepsilon_1$$

$$\leq \frac{\rho(x_{s_1}) + \rho(x_{s_2}) + \cdots + \rho(x_{s_k})}{k} + \frac{1}{k} \sum_{n=1}^{m_k} \left[\sum_{i=1}^{n-1} u_n \triangle v_i \left| x_{s_j}(i) \right| + u_n v_n \left| x_{s_j}(n) \right| \right]^{p_n}$$

$$+ \sum_{n=m_{k-1}+1}^{\infty} \left[\sum_{i=1}^{n-1} u_n \triangle v_i \left| \frac{x_{s_k}(i)}{k} \right| + u_n v_n \left| \frac{x_{s_k}(n)}{n} \right| \right]^{p_n} + (k-1)\varepsilon_1$$

$$\leq \frac{k-1}{k} + \frac{1}{k} \sum_{n=1}^{m_k} \left[\sum_{i=1}^{n-1} u_n \triangle v_i \left| x_{s_j}(i) \right| + u_n v_n \left| x_{s_j}(n) \right| \right]^{p_n}$$

$$+ \frac{1}{k^\alpha} \sum_{n=m_k+1}^{\infty} \left[\sum_{i=1}^{n-1} u_n \triangle v_i \left| x_{s_j}(i) \right| + u_n v_n \left| x_{s_j}(n) \right| \right]^{p_n} + (k-1)\varepsilon_1$$

$$\leq 1 - \frac{1}{k} + \frac{1}{k} \left\{ 1 - \sum_{n=m_k+1}^{\infty} \left[\sum_{i=1}^{n-1} u_n \triangle v_i \left| x_{s_j}(i) \right| + u_n v_n \left| x_{s_j}(n) \right| \right]^{p_n} \right\}$$

$$+ \frac{1}{k^\alpha} \sum_{n=m_k+1}^{\infty} \left[\sum_{i=1}^{n-1} u_n \triangle v_i \left| x_{s_j}(i) \right| + u_n v_n \left| x_{s_j}(n) \right| \right]^{p_n} + (k-1)\varepsilon_1$$

$$= 1 - \frac{k^\alpha - 1}{k^\alpha} \sum_{n=m_k+1}^{\infty} \left[\sum_{i=1}^{n-1} u_n \triangle v_i \left| x_{s_j}(i) \right| + u_n v_n \left| x_{s_j}(n) \right| \right]^{p_n} + (k-1)\varepsilon_1$$

$$\leq 1 + (k-1)\varepsilon_1 - \frac{k^{\alpha-1} - 1}{k^\alpha} \mu = 1 - \frac{(k^{\alpha-1} - 1)\mu}{2k^\alpha}$$

By Lemma 2.6.10, there exists $\gamma > 0$ such that

$$\left\| \frac{x_{s_1} + x_{s_2} + \cdots + x_{s_k}}{k} \right\| < 1 - \gamma.$$

Therefore, $\ell_{\rho\triangle}(u, v, p)$ is $k - NUC$.

This completes the proof of the theorem.   □

Chapter 3

Infinite Matrices

Keywords. *Matrix transformations, regular matrices, conservative matrices, Schur matrices.*

3.1 Introduction

Let $A = (a_{nk})_{n,k \in \mathbb{N}_0}$ be an infinite matrix with real or complex elements. We write $A_n = (a_{nk})_{k \in \mathbb{N}_0}$ to denote the sequence in the nth row of A for every $n \in \mathbb{N}_0$. For $x = (x_k) \in \omega$, the *A-transform* of x is defined as the sequence $Ax = \{(Ax)_n\}_{n \in \mathbb{N}_0}$, where

$$(Ax)_n = \sum_k a_{nk} x_k$$

provided the series on the right side converges for each $n \in \mathbb{N}_0$. Further, the sequence x is said to be *A-summable* to the number l if $(Ax)_n \to l$, as $n \to \infty$. In this case, we write $x \to l(A)$; where l is called the A-limit of x.

Let X and Y be subsets of ω, and A be an infinite matrix. Then, we say that A defines a *matrix transformation* from X into Y if Ax exists and is in Y for every $x \in X$. By $(X : Y)$, we denote the class of all infinite matrices that map X into Y. Thus, $A = (a_{nk}) \in (X : Y)$ if and only if $A_n \in X^\beta$ for all $n \in \mathbb{N}_0$ and $Ax \in Y$ for all $x \in X$.

The theory of matrix transformations deals with establishing necessary and sufficient conditions on the elements of a matrix to map a sequence space X into a sequence space Y. This is a natural generalization of the problem to characterize all summability methods given by infinite matrices that preserve convergence. In this chapter, we characterize some classes of matrix transformations.

3.2   Matrix Transformations Between Some FK-Spaces

Now, we give here some results on matrix transformations by using the theory of FK- and BK-spaces.

Let (X, d) be a metric space, $\delta > 0$ and $x_0 \in X$. Then, we write $S[x_0, \delta] = \{x \in X : d(x, x_0) \leq \delta\}$ for the closed ball of radius δ with its center in x_0. If $X \subset \omega$ is a linear metric space and $a = (a_k) \in \omega$, then we write

$$\|a\|_\delta^* = \|a\|_{\delta,X}^* = \sup_{x \in S[x_0,\delta]} \left| \sum_k a_k x_k \right| \tag{3.2.1}$$

provided the expression on the right-hand side exists and is finite which is the case whenever $a \in X^\beta$; if X is a normed space, we write

$$\|a\|_X^* = \sup_{x \in S_X} \left| \sum_k a_k x_k \right|, \tag{3.2.2}$$

where S_X is the unit sphere in X.

Let A be an infinite matrix, D a positive real and X an FK-space. Then, we put $M_{A,D}^*(X : \ell_\infty) = \sup_{n \in \mathbb{N}_0} \|A_n\|_D^*$ and, if X is a BK-space, then we write $M_A^*(X : \ell_\infty) = \sup_{n \in \mathbb{N}_0} \|A_n\|^*$.

For ready reference, we recall the following:

Remark 3.2.1. *Let X denotes any of the spaces ℓ_∞, c, c_0, ℓ_1 or ℓ_p. Then, we have $\|a\|_X^* = \|a\|_{X^\beta}$ for all $a \in X^\beta$, where $\| \cdot \|_{X^\beta}$ is the natural norm on the dual space X^β.*

The following result is one of the most important in matrix transformations:

Theorem 3.2.2. [228, Theorem 4.2.8] *Any matrix map between FK-spaces is continuous.*

Theorem 3.2.3. [154, Theorem 1.23 (b)] *Let X be an FK-space. Then, we have $A = (a_{nk}) \in (X : \ell_\infty)$ if and only if*

$$\|A\|_\delta^* = \sup_{n \in \mathbb{N}_0} \|A_n\|_\delta^* < \infty \quad \text{for some } \delta > 0. \tag{3.2.3}$$

Proof. First, we assume that (3.2.3) is satisfied. Then, the series $\sum_k a_{nk} x_k$ converge for all $x \in B_\delta[0]$ and for all $n \in \mathbb{N}_0$, and $Ax \in \ell_\infty$ for all $x \in B_\delta[0]$. Since $B_\delta[0]$ is absorbing, we conclude that the series $\sum_k a_{nk} x_k$ converge for all $n \in \mathbb{N}_0$ and all $x \in X$, and $Ax \in \ell_\infty$ for all $x \in X$, i.e., $A \in (X : \ell_\infty)$.

Conversely, suppose that $A = (a_{nk}) \in (X : \ell_\infty)$. Then, the map $L_A : X \to \ell_\infty$ defined by $L_A(x) = Ax$ for all $x \in X$ is continuous by Theorem 3.2.2. Hence, there exists a neighborhood U of 0 in X and a real $\delta > 0$ such that $B_\delta[0] \subset U$ and $\|L_A(x)\|_\infty < 1$ for all $x \in X$. This implies (3.2.3). This completes the proof. $\qquad \square$

Theorem 3.2.4. [151, Theorem 3.20] *Let X and Y be BK-spaces. Then, the following statements hold:*

(a) $(X : Y) \subset \mathcal{B}(X : Y)$, *that is, every matrix $A = (a_{nk}) \in (X : Y)$ defines an operator $L_A \in \mathcal{B}(X : Y)$ by $L_A(x) = Ax$ for all $x \in X$.*

(b) *If X has AK, then $\mathcal{B}(X : Y) \subset (X : Y)$, that is, for every operator $L \in \mathcal{B}(X : Y)$ there exists a matrix $A = (a_{nk}) \in (X : Y)$ such that $L(x) = Ax$ for all $x \in X$.*

(c) $A = (a_{nk}) \in (X : \ell_\infty)$ *if and only if*

$$\|A\|_{(X:\ell_\infty)} = \sup_{n \in \mathbb{N}_0} \|A_n\|_X^* < \infty. \qquad (3.2.4)$$

If $A = (a_{nk}) \in (X : \ell_\infty)$, then

$$\|A\|_{(X:\ell_\infty)} = \|L_A\|. \qquad (3.2.5)$$

Proof. (a) This is Theorem 3.2.2.

(b) Let $L : X \to Y$ be a continuous linear operator. We write $L_n = P_n \circ L$ for all $n \in \mathbb{N}_0$, and put $a_{nk} = L_n(e^{(k)})$ for all $n \in \mathbb{N}_0$ and $k \in \mathbb{N}_0$. Let $x = (x_k)$ be given. Since X has AK, we have $x = \sum_k x_k e^{(k)}$ and since Y is a BK–space, it follows that L_n is a continuous linear functional on X for all $n \in \mathbb{N}_0$. Hence, we obtain $L_n(x) = \sum_k x_k L_n(e^{(k)}) = \sum_k a_{nk} x_k = (Ax)_n$ for all $n \in \mathbb{N}_0$, and so $L(x) = Ax$.

(c) This follows immediately from Theorem 3.2.3 and the definition of $\|A\|_{(X:\ell_\infty)}$.

That is, if X is a BK-space, then $L_A \in \mathcal{B}(X : Y)$ implies

$$\|Ax\|_\infty = \sup_{n \in \mathbb{N}_0} |(Ax)_n| = \|L_A(x)\|_\infty \le \|L_A\|$$

for all $x \in X$ with $\|x\| = 1$. Thus, $|(Ax)_n| \le \|L_A\|$ for all $n \in \mathbb{N}_0$ and for all $x \in X$ with $\|x\| = 1$, and by the definition of the norm $\|A\|_{(X:\ell_\infty)}$,

$$\|A\|_{(X:\ell_\infty)} = \sup_{n \in \mathbb{N}_0} \|A_n\|_X^* \le \|L_A\|. \qquad (3.2.6)$$

Further, given $\varepsilon > 0$, there is $x \in X$ with $\|x\| = 1$, $\|Ax\|_\infty \ge \|L_A\| - \varepsilon/2$, and there is $n \in \mathbb{N}_0$ with $|(Ax)_n| \ge \|Ax\|_\infty - \varepsilon/2$, consequently $|(Ax)_n| \ge \|L_A\| - \varepsilon$. Therefore, $\|A\|_{(X:\ell_\infty)} = \sup_{n \in \mathbb{N}_0} \|A_n\|_X^* \ge \|L_A\| - \varepsilon$. Since $\varepsilon > 0$ is arbitrary, $\|A\|_{(X:\ell_\infty)} \ge \|L_A\|$ which together with (3.2.6) gives $\|A\|_{(X:\ell_\infty)} = \|L_A\|$.

This completes the proof. $\qquad \square$

Theorem 3.2.5. [228, 8.3.6 and 8.3.7, p. 123] *The following statements hold:*

(a) *Let Y and Y_1 be FK-spaces with a closed subspace Y_1 of Y. If $\{b^{(k)}\}_{k \in \mathbb{N}_0}$ is a Schauder basis for X, then $A = (a_{nk}) \in (X : Y_1)$ if and only if $A = (a_{nk}) \in (X : Y)$ and $Ab^{(k)} \in Y_1$ for all $k \in \mathbb{N}_0$.*

(b) *Let X be an FK-space, $X_1 = X \oplus e = \{x_1 = x + \lambda e : x \in X, \lambda \in \mathbb{C}\}$ and Y be a linear subspace of ω. Then, $A \in (X_1 : Y)$ if and only if $A = (a_{nk}) \in (X : Y)$ and $Ae \in Y$.*

Proof. (a) The necessity of the conditions for $A = (a_{nk}) \in (X : Y_1)$ is trivial.

Conversely, if $A = (a_{nk}) \in (X : Y)$, then $L_A \in \mathcal{B}(X : Y)$. Since Y_1 is a closed subspace of Y, the FK metrics of Y_1 and Y are the same. Consequently, if S is any subset in Y_1, then for its closures $\overline{S(Y_1)}$ and $\overline{S(Y_{|Y_1})}$ with respect to the metrics d_{Y_1} and $d_{Y}\big|_{Y_1}$, we have

$$\overline{S(Y_1)} = \overline{S(Y_{|Y_1})}. \tag{3.2.7}$$

Let $x \in X$ and $E = \{\sum_{k=0}^{m} \lambda_k b^{(k)} : m \in \mathbb{N}_0, \lambda_k \in \mathbb{C} \text{ for all } k \in \mathbb{N}_0\}$ denotes the span of $\{b^{(k)}\}_{k \in \mathbb{N}_0}$. Since $L_A(b^{(k)}) \in Y_1$ for all $k \in \mathbb{N}_0$ and the metrics d_{Y_1} and $d_{Y_{|Y_1}}$ are equivalent, the map $L_{A|_E} : (X, d_X) \to (Y_1, d_{Y_1})$ is continuous. Further, since $\{b^{(k)}\}_{k \in \mathbb{N}_0}$ is a basis of X, we have $\bar{E} = X$. Therefore, by (3.2.7) and the continuity of L_A, we have

$$L_A(X) = L_A(\bar{E}) = clos_{Y_1}(L_{A|_E}(E)) = clos_{Y_{|Y_1}}(L_{A|_E}(E)) \subset clos_{Y_{|Y_1}}(Y_1) = Y_1.$$

Hence, $A = (a_{nk}) \in (X : Y_1)$.

(b) First, we assume $A = (a_{nk}) \in (X : Y_1)$. Then, $X \subset Y_1$ implies $A \in (X : Y)$ and $e \in X_1$ implies $Ae \in Y$.

Conversely, we assume $A = (a_{nk}) \in (X : Y)$ and $Ae \in Y$. Let $x_1 \in X_1$ be given. Then, there are $x \in X$ and $\lambda \in \mathbb{C}$ such that $x_1 = x + \lambda e$, and it follows that $Ax_1 = A(x + \lambda e) = Ax + \lambda Ae \in Y$.

This completes the proof. □

Theorem 3.2.6. *Let $X \supset \phi$ be a BK-space. Then, $A = (a_{nk}) \in (X : \ell_1)$ if and only if $A_n \in X^\beta$ for all $n \in \mathbb{N}_0$ and*

$$\sup_{N \in \mathcal{F}} \left\| \sum_n A_n \right\|_X^* < \infty. \tag{3.2.8}$$

If $A = (a_{nk}) \in (X : \ell_1)$, then

$$\|A\|_{(X:\ell_1)} \leq \|L_A\| \leq 4\|A\|_{(X:\ell_1)}, \tag{3.2.9}$$

where $\|A\|_{(X:\ell_1)} = \sup_{N \in \mathcal{F}} \|\sum_n A_n\|_X^*$, and $\mathcal{F}$ denotes the collection of all non-empty and finite subsets of $\mathbb{N}_0$.

Proof. For (3.2.8), we refer to [150].

To show (3.2.9), let $A = (a_{nk}) \in (X : \ell_1)$ and $m \in \mathbb{N}_0$ be given. Then, for all $N \subset \{1, 2, \ldots, m\}$ and for all $x \in X$ with $\|x\| = 1$,

$$\left| \sum_n (Ax)_n \right| \leq \sum_{n=0}^{m} |(Ax)_n| \leq \|L_A\|$$

and this implies that

$$\|A\|_{(X:\ell_1)} \leq \|L_A\|. \tag{3.2.10}$$

Furthermore, given $\varepsilon > 0$, there is $x \in X$ with $\|x\| = 1$ such that

$$\|Ax\|_1 = \sum_n |(Ax)_n| \geq \|L_A\| - \frac{\varepsilon}{2},$$

and there is an integer $m(x)$ such that

$$\sum_{n=0}^{m(x)} |(Ax)_n| \geq \|Ax\|_1 - \frac{\varepsilon}{2}.$$

Consequently,

$$\sum_{n=0}^{m(x)} |(Ax)_n| \geq \|L_A\| - \varepsilon.$$

By Lemma 4.9 of [154],

$$4 \left[\max_{N \subset \{0,1...,m(x)\}} \left| \sum_n (Ax)_n \right| \right] \geq \sum_{n=0}^{m(x)} |(Ax)_n| \geq \|L_A\| - \varepsilon$$

and so, $4\|A\|_{(X:\ell_1)} \geq \|L_A\| - \varepsilon$. Since $\varepsilon > 0$ was arbitrary, we have $4\|A\|_{(X:\ell_1)} \geq \|L_A\|$ which together with (3.2.10) yields (3.2.9).

This completes the proof. □

Consequently, we have the following:

Corollary 3.2.7. *We have* $(c_0 : \ell_1) = (c : \ell_1) = (\ell_\infty : \ell_1)$. *Further,* $A = (a_{nk}) \in (c_0 : \ell_1)$ *if and only if*

$$\sup_{K \in \mathcal{F}} \left(\sum_n \left| \sum_{k \in K} a_{nk} \right| \right) < \infty. \tag{3.2.11}$$

Remark 3.2.8. *Since the BK-spaces* c_0 *and* c *are closed subspaces of* ℓ_∞, *the matrix classes* $(X : c_0)$ *and* $(X : c)$ *can be characterized by combining Part (c) of Theorem 3.2.4 and Part (a) of Theorem 3.2.5, where* X *is a BK-space with Schauder basis. On the other hand, we may note that if* X, *in Part (c) of Theorem 3.2.4 or Part (a) of Theorem 3.2.5, is any of the classical sequence spaces, then any of the conditions (3.2.4), (3.2.5) or (3.2.8) implies the condition* $A_n \in X^\beta$ *for all* $n \in \mathbb{N}_0$ *by Remark 3.2.1. Thus, this condition is redundant in such cases. Also, if* X *is a BK-space with AK then we obtain the following result which is immediate by Propositions 4.2 and 4.3 of [156].*

Theorem 3.2.9. *Let X be a BK-space with AK. Then, the following statements hold:*

(a) $A = (a_{nk}) \in (X : \ell_\infty)$ *if and only if (3.2.4) holds.*

(b) $A = (a_{nk}) \in (X : c)$ *if and only if (3.2.4) and $\lim_{n\to\infty} a_{nk}$ exists for every $k \in \mathbb{N}_0$.*

(c) $A = (a_{nk}) \in (X : c_0)$ *if and only if (3.2.4) and $\lim_{n\to\infty} a_{nk} = 0$ for all $k \in \mathbb{N}_0$.*

(d) $A = (a_{nk}) \in (X : \ell_1)$ *if and only if (3.2.8) holds.*

3.3   Conservative Matrices

Regular matrices play a very important role in summability theory and in the study of sequence spaces obtained by their domain. These matrices transform the convergent sequences into a convergent sequence leaving the limit invariant.

Definition 3.3.1. *A matrix A is called conservative if $Ax \in c$ for all $x \in c$. If in addition $A - \lim x = \lim x$ for all $x \in c$, then A is called a regular matrix or a T-matrix . The classes of conservative and regular matrices will be denoted by $(c : c)$ and $(c : c; p)$ or $(c : c)_{reg}$, respectively.*

Definition 3.3.2. *A matrix belonging to the class $(cs : c)$ is called a β-matrix. Let us suppose that a matrix $A = (a_{nk})$ satisfies the following conditions:*

$$\sup_{n \in \mathbb{N}_0} \sum_k |a_{nk} - a_{n,k+1}| < \infty \tag{3.3.1}$$

$$\lim_{n\to\infty} a_{nk} = \alpha_k \text{ for all } k \in \mathbb{N}_0. \tag{3.3.2}$$

It is known (cf. [228, Exercise 8.4.5B]) that a matrix $A = (a_{nk})$ is a β-matrix if and only if (3.3.1) and (3.3.2) hold.

Let $u = (u_k) \in cs$ with $\sum_k u_k = \alpha$. Then, a β-matrix A such that $(Au)_n \to \alpha$, as $n \to \infty$, is called a γ-matrix. By $(cs : c; p)$, we denote the class of all γ-matrices.

A matrix $A = (a_{nk})$ is a γ-matrix if and only if (3.3.1) holds and (3.3.2) also holds with $\alpha_k = 1$ for all $k \in \mathbb{N}_0$.

Suppose that $A = (a_{nk})$ and $B = (b_{nk})$ are dual matrices, that is, the elements of matrices connected with the relation

$$a_{nk} = \sum_{j=k}^{\infty} b_{nj} \text{ or equivalently } b_{nk} = a_{nk} - a_{n,k+1} \text{ for all } k, n \in \mathbb{N}_0.$$

Then, one can easily show that the following statements hold:

(i) A is a β-matrix if and only if B is a K-matrix, where a matrix belonging to the class $(c : c)$ is called a K-matrix.

(ii) A is a γ-matrix if and only if B is a T-matrix.

Definition 3.3.3. *A matrix A belonging to the class $(cs : cs)$ such that $\sum_n (Au)_n = \sum_k u_k$ is called an α-matrix. By $(cs : cs; p)$, we denote the class of α-matrices.*

Let us suppose that a matrix $A = (a_{nk})$ satisfies the following conditions:

$$\sup_{n \in \mathbb{N}_0} \sum_k |a_{nk} - a_{n,k+1}| < \infty, \tag{3.3.3}$$

$$\sum_n a_{nk} = \alpha_k \quad \text{for each} \quad k \in \mathbb{N}_0. \tag{3.3.4}$$

Then, $A \in (cs : cs)$ if and only if (3.3.3) and (3.3.4) hold, (see [228, Exercise 8.4.6B]). Additionally, $A = (a_{nk})$ is an α-matrix if and only if (3.3.3) holds and (3.3.4) also holds with $\alpha_k = 1$ for all $k \in \mathbb{N}_0$.

Let us suppose that the elements of the matrices $A = (a_{nk})$ and $B = (b_{nk})$ are connected with the relation

$$b_{nk} = \sum_{j=0}^{n} a_{jk} \quad \text{for all} \quad k, n \in \mathbb{N}_0.$$

Then, it is not hard to establish the following statements:

(i) $A \in (cs : cs)$ if and only if $B \in (cs : c)$.

(ii) A is an α-matrix if and only if B is a γ-matrix.

Definition 3.3.4. *The characteristic $\chi(A)$ of a matrix $A = (a_{nk}) \in (c : c)$ is defined by*

$$\chi(A) = \lim_{n \to \infty} \sum_k a_{nk} - \sum_k \left(\lim_{n \to \infty} a_{nk} \right)$$

which is a multiplicative linear functional. The numbers $\lim_{n \to \infty} a_{nk}$ and $\lim_{n \to \infty} \sum_k a_{nk}$ are called the characteristic numbers of A. A matrix A is called coregular if $\chi(A) \neq 0$ and is called conull if $\chi(A) = 0$.

From Part (c) of Theorem 3.2.9, we have the following:

Corollary 3.3.5. $A = (a_{nk}) \in (c_0 : c_0)$ *if and only if*

$$M = \sup_{n \in \mathbb{N}_0} \sum_k |a_{nk}| < \infty, \tag{3.3.5}$$

$$\lim_{n \to \infty} a_{nk} = 0 \quad \text{for all} \quad k \in \mathbb{N}_0. \tag{3.3.6}$$

Further, since c_0 has AK, using Theorem 3.2.4, we have $A \in (c_0 : c_0)$ if and only if $L_A \in \mathcal{B}(c_0 : c_0)$ with $L_A(x) = Ax$ for all $x \in c_0$ and $\|A\|_{(\ell_\infty : \ell_\infty)} = M$.

The well-known Silverman-Toeplitz conditions for the regularity of A are given by the following theorem, (cf. [32, 60, 147]):

Theorem 3.3.6. (Silverman-Toeplitz) A *is regular, i.e.,* $A = (a_{nk}) \in (c : c)_{reg}$ *if and only if (3.3.5) and (3.3.6) hold, and*

$$\lim_{n \to \infty} \sum_k a_{nk} = 1. \tag{3.3.7}$$

Proof. Sufficiency. Let the conditions (3.3.5)-(3.3.7) hold and $x = (x_k) \in c$ with $x_k \to l$, as $k \to \infty$. Write

$$\sum_k a_{nk} x_k = \sum_k a_{nk}(x_k - l) + l \sum_k a_{nk}. \tag{3.3.8}$$

Using the condition (3.3.7), we get $\sum_k a_{nk} x_k \to l$, as $n \to \infty$. Now, replacing x_k by $x_k - l$ in Corollary 3.3.5, we have $Ay \in c_0$ for all $y = (y_k) \in c_0$, that is, $\sum_k a_{nk}(x_k - l) \to 0$, as $n \to \infty$. Therefore, by (3.3.8), we get $\sum_k a_{nk} x_k \to l = \lim x$, as $n \to \infty$. Hence, A is regular.

Necessity. Let $A = (a_{nk}) \in (c : c)_{reg}$. Since $x = e^{(k)} \in c_0$, we have $Ax = Ae^{(k)} \in c_0$. That is, $\left(Ae^{(k)}\right)_n = a_{nk} \to 0$, as $n \to \infty$, for each $k \in \mathbb{N}_0$. Hence, (3.3.6) holds. Also, since $x = e \in c$, we have $Ax = Ae \in c$ with $Ae = \left(\sum_k a_{nk}\right)$ and $(Ae)_n = \sum_k a_{nk} \to 1$, as $n \to \infty$, since $e \to 1$. Hence, (3.3.7) holds.

The existence of $(Ax)_n = \sum_k a_{nk} x_k$ for each $n \in \mathbb{N}_0$ and for all $x = (x_k) \in c$ implies that $A_n \in c^\beta = \ell_1$ for each $n \in \mathbb{N}_0$. That is, $\|A_n\|_1 = \sum_k |a_{nk}| < \infty$ for each $n \in \mathbb{N}_0$. Hence, $\|A\|_{(\ell_\infty : \ell_\infty)} = \sup_{n \in \mathbb{N}_0} \|A_n\|_1 = \sup_{n \in \mathbb{N}_0} \sum_k |a_{nk}| < \infty$. Condition (3.3.5) holds.

This completes the proof of the theorem. $\square$

We include a classical gliding hump proof of the Silverman-Toeplitz theorem (see Boos [52]). This method of argument is still very useful. For instance, there are situations where the sliding hump method works, but FK techniques cannot be applied for all results. Many summability students are unfamiliar with this method of argument and that this occasionally hampers their research.

Another Proof of Necessity of Silverman-Toeplitz Theorem. Since $e^{(k)}$ and e are convergent sequences, the necessity of the conditions (3.3.6) and (3.3.7) follows easily. We will now prove the necessity of (3.3.5). We have $Ae^{(k)} \in c \subset \ell_\infty$ for each $k \in \mathbb{N}_0$. Thus for each $r \in \mathbb{N}_0$ we may choose $M(r) > 0$ with

$$\sum_{k=0}^{r} |a_{nk}| < M(r) \quad \text{for all} \quad n \in \mathbb{N}_0. \tag{3.3.9}$$

Also by hypothesis, we have

$$\sum_k |a_{nk}| < \infty \quad \text{for all} \quad n \in \mathbb{N}_0. \tag{3.3.10}$$

On the contrary, suppose that (3.3.5) does not hold, i.e.,

$$\|A\| = \sup_{n \in \mathbb{N}} \sum_k |a_{nk}| = \infty. \tag{3.3.11}$$

We show that $Ax \notin c$ for an $x \in c$. Choose the index sequences (n_p) and (k_p). Put $k_{-1} = 0$ and take an $n \in \mathbb{N}_0$ such that $\sum_k |a_{n_0 k}| > M(k_{-1}) + 1$ by (3.3.11) and then a $k_0 \in \mathbb{N}_0$ with $k_0 > k_{-1}$ and $\sum_{k=k_0+1}^{\infty} |a_{n_0 k}| < 1$ by (3.3.10). Once n_{p-1} and k_{p-1} having been chosen, we may choose $n_p \in \mathbb{N}_0$ with $n_p > n_{p-1}$ and

$$\sum_k |a_{n_p k}| > (p+1)M(k_{p-1}) + p^2 + 1 \tag{3.3.12}$$

by (3.3.10). After that we can choose $k_p \in \mathbb{N}_0$ with $k_p > k_{p-1}$ and

$$\sum_{k=k_p+1}^{\infty} |a_{n_p k}| < 1 \tag{3.3.13}$$

by (3.3.10). Applying (3.3.9), (3.3.12) and (3.3.13), for any $p \in \mathbb{N}$ we get

$$
\begin{aligned}
\sum_{k=k_{p-1}+1}^{k_p} |a_{n_p k}| &= \sum_k |a_{n_p k}| - \sum_{k=k_p+1}^{\infty} |a_{n_p k}| \sum_{k=0}^{k_{p-1}} |a_{n_p k}| \tag{3.3.14} \\
&> (p+1)M(k_{p-1}) + p^2 + 1 - 1 - M(k_{p-1}) \\
&= pM(k_{p-1}) + p^2.
\end{aligned}
$$

Now, the chosen index sequences (n_p) and (k_p) enable us to define $x \in c_0$ with $Ax \notin \ell_\infty$. Define $x = (x_k)$ by

$$
x_k := \begin{cases} 1 & , \quad 0 \le k \le k_0, \\ \dfrac{|a_{n_p k}|}{p a_{n_p k}} & , \quad k_{p-1} < k \le k_p \end{cases} ; \quad a_{n_p k} \ne 0 \text{ for each } p \in \mathbb{N}_1.
$$

Note that $x \in c_0$. Since A is regular, $Ax \in c_0$. However, for $p \ge 1$, we have

$$
\begin{aligned}
\left| \sum_k a_{n_p k} x_k \right| &\ge \left| \sum_{k=k_{p-1}+1}^{k_p} a_{n_p k} x_k \right| - \sum_{k=k_p+1}^{\infty} |a_{n_p k} x_k| - \sum_{k=0}^{k_{p-1}} |a_{n_p k} x_k| \\
&\ge \frac{1}{p} \sum_{k=k_{p-1}+1}^{k_p} |a_{n_p k}| - \sum_{k=k_p+1}^{\infty} |a_{n_p k}| - \sum_{k=0}^{k_{p-1}} |a_{n_p k}| \\
&\ge M(k_{p-1}) + p - 1 - M(k_{p-1}) = p - 1
\end{aligned}
$$

by (3.3.9), (3.3.13) and (3.3.14). That is, $Ax \notin c_0$, which contradicts $A \in (c : c)_{reg}$. Consequently, (3.3.5) must hold.

This completes the proof of the necessity part of Silverman-Toeplitz theorem. $\qquad \square$

Theorem 3.3.7. (Kojima-Schur) $A = (a_{nk})$ *is conservative, i.e.,* $A = (a_{nk}) \in (c : c)$ *if and only if (3.3.5) and (3.3.2) hold, and*

$$\lim_{n \to \infty} \sum_k a_{nk} = \alpha. \qquad (3.3.15)$$

Proof. Sufficiency. Let the conditions (3.3.5), (3.3.2) and (3.3.15) hold, and $x = (x_k) \in c$. Then, $(Ax)_n = \sum_k a_{nk} x_k$ exists for each $n \in \mathbb{N}_0$ and we have

$$|(Ax)_n| \leq \sum_k |a_{nk}||x_k| \leq \left(\sum_k |a_{nk}| \right) \left(\sup_{k \in \mathbb{N}_0} |x_k| \right) \leq \|A\|_{(\ell_\infty : \ell_\infty)} \|x\|_\infty.$$

Since A_n is linear, it follows that $A_n \in c'$ for each $n \in \mathbb{N}_0$ and $\|A_n\| \leq \|A\|_{(\ell_\infty : \ell_\infty)}$.

Now, $(Ae)_n = \sum_k a_{nk}$ and $\left\{ Ae^{(k)} \right\}_n = a_{nk}$, and

$$\lim_{n \to \infty} (Ae)_n = \lim_{n \to \infty} \sum_k a_{nk} = \alpha, \text{ and}$$

$$\lim_{n \to \infty} \left\{ Ae^{(k)} \right\}_n = \lim_{n \to \infty} a_{nk} = \alpha_k \text{ for each } k \in \mathbb{N}_0.$$

Since $\{e, e^{(0)}, e^{(1)}, \ldots\}$ is a Schauder basis for c, $\|A_n\|$ is finite for each $n \in \mathbb{N}_0$. Also, it is well-known that every $x \in c$ can be written uniquely as $x = (\lim x)e + \sum_k (x_k - \lim x)e^{(k)}$. Hence, since $A_n \in c'$,

$$(Ax)_n = (\lim x) \left[(Ae)_n - \sum_k \left\{ Ae^{(k)} \right\}_n \right] + \sum_k x_k \left\{ Ae^{(k)} \right\}_n$$

$$\lim_{n \to \infty} (Ax)_n = (\lim x) \left(\alpha - \sum_k \alpha_k \right) + \sum_k x_k \alpha_k = L(x), \text{ say.}$$

It follows that $\lim_{n \to \infty} (Ax)_n$ exists for all $x \in c$. Hence, $A \in (c : c)$.

Necessity. Suppose that A is conservative. The existence of $(Ax)_n = \sum_k a_{nk} x_k$ for each $n \in \mathbb{N}_0$ and for all $x = (x_k) \in c$ implies that $A_n \in c^\beta = \ell_1$ for each $n \in \mathbb{N}_0$. That is, $\|A_n\|_1 = \sum_k |a_{nk}| < \infty$ for each $n \in \mathbb{N}_0$. Hence, $\|A\|_{(\ell_\infty : \ell_\infty)} = \sup_{n \in \mathbb{N}_0} \|A_n\|_1 = \sup_{n \in \mathbb{N}_0} \sum_k |a_{nk}| < \infty$, i.e., the condition (3.3.5) holds. Furthermore, since $e^{(k)}$ and e are convergent sequences for each $k \in \mathbb{N}_0$, $\lim_{n \to \infty} \left\{ Ae^{(k)} \right\}_n$ and $\lim_{n \to \infty} (Ae)_n$ must exist. Hence, the conditions (3.3.2) and (3.3.15) hold, respectively. □

Remark 3.3.8. *If we take $\alpha_k = 0$ for all $k \in \mathbb{N}_0$ and $\alpha = 1$, then Theorem 3.3.6 directly follows from Theorem 3.3.7.*

Theorem 3.3.9. *The following statements hold:*

(a) *We have* $(c_0 : \ell_\infty) = (c : \ell_\infty) = (\ell_\infty : \ell_\infty)$; *furthermore* $A = (a_{nk}) \in (\ell_\infty : \ell_\infty)$ *if and only if (3.3.5) holds.*

(b) We have $A = (a_{nk}) \in (c_0 : c)$ if and only if (3.3.5) and (3.3.2) hold. If $A = (a_{nk}) \in (c_0 : c)$; then

$$\lim_{n \to \infty} (Ax)_n = \sum_k \alpha_k x_k. \tag{3.3.16}$$

(c) $A = (a_{nk}) \in (c_0 : c_0)$ if and only if (3.3.5) and (3.3.2) hold with $\alpha_k = 0$ for all $k \in \mathbb{N}_0$.

Proof. (a) We have $A = (a_{nk}) \in (c_0 : \ell_\infty)$, if and only if (3.3.5) holds by Theorem 3.2.4, and since $c_0^\beta = \ell_1$ and c_0^* and ℓ_1 are norm isomorphic.

Furthermore, $c_0 \subset c \subset \ell_\infty$ implies $(c_0 : \ell_\infty) \subset (c : \ell_\infty) \subset (\ell_\infty : \ell_\infty)$. Also, $(c_0 : \ell_\infty) = (c_0^{\beta\beta} : \ell_\infty) = (\ell_1^\beta : \ell_\infty) = (\ell_\infty : \ell_\infty)$ by the first part of Theorem 2.39 of [29].

(b) Since c is a closed subspace of ℓ_∞, the characterization of the class $(c_0 : c)$ is an immediate consequence of Parts (a) of Theorem 3.2.5 and of the present theorem.

Now, we assume $A = (a_{nk}) \in (c_0 : c)$ and write $\|A\| = \|A\|_{(\ell_\infty : \ell_\infty)}$, for short. Let m be a given non–negative integer. Then, it follows from (3.3.5) and (3.3.2) that $\sum_{k=0}^m |\alpha_k| = \lim_{n \to \infty} \sum_{k=0}^m |a_{nk}| \leq \|A\|$. Since m was arbitrary, we have $(\alpha_k)_{k \in \mathbb{N}_0} \in \ell_1$,

$$\sum_k |\alpha_k| \leq \|A\| \text{ and } \sum_k |\alpha_k x_k| \leq \|A\| \|x\|_\infty \text{ for all } x \in c. \tag{3.3.17}$$

Now, let $x \in c_0$ and $\varepsilon > 0$ be given. Then, we can choose an integer $k(\varepsilon)$ such that $|x_k| \leq \varepsilon/(4\|A\| + 1)$ for all $k > k(\varepsilon)$. By (3.3.2), we can choose an integer $n(\varepsilon)$ such that $\sum_{k=0}^{k(\varepsilon)} |a_{nk} - \alpha_k| |x_k| < \varepsilon/2$. Let $n > n(\varepsilon)$. Then, (3.3.5) and (3.3.17) implies

$$\left| (Ax)_n - \sum_k \alpha_k x_k \right| \leq \sum_{k=0}^{k(\varepsilon)} |a_{nk} - \alpha_k| |x_k| + \sum_{k=k(\varepsilon)+1}^{\infty} (|a_{nk}| - |\alpha_k|) |x_k|$$

$$\leq \frac{\varepsilon}{2} + \frac{\varepsilon}{4\|A\| + 1} \left(\sum_k |a_{nk}| + \sum_k |\alpha_k| \right) \leq \frac{\varepsilon}{2} + \frac{\varepsilon}{2} = \varepsilon.$$

Hence, (3.3.16) holds.

(c) This directly follows from Part (b).

This completes the proof. □

Theorem 3.3.10. [151, Example 5.5] *We have $(\ell_1 : \ell_1) = \mathcal{B}(\ell_1 : \ell_1)$ and $A \in (\ell_1 : \ell_1)$ if and only if*

$$\|A\|_{(\ell_1 : \ell_1)} = \sup_{k \in \mathbb{N}_0} \sum_n |a_{nk}| < \infty. \tag{3.3.18}$$

If A is in any of the classes, above, then

$$\|L_A\| = \|A\|. \tag{3.3.19}$$

Proof. Since ℓ_1 has AK, Part (b) of Theorem 3.2.4 yields the first part.

We apply the second part of Theorem 2.39 of [29] with $X = \ell_1, Z = c_0$, BK-spaces with AK, and $Y = Z^\beta = \ell_1$ to obtain $A = (a_{nk}) \in (\ell_1 : \ell_1)$ if and only if $A^T \in (\ell_1 : \ell_1)$; by Part (a) of Theorem 2.40 of [29], this is the case if and only if (3.3.18) is satisfied.

Furthermore, if $A = (a_{nk}) \in (\ell_1 : \ell_1)$ then

$$\|L_A(x)\|_1 = \sum_n \left| \sum_k a_{nk} x_k \right| \leq \sum_k \sum_n |a_{nk} x_k| \leq \|A\|_{(\ell_1 : \ell_1)} \|x\|_1$$

implies $\|L_A\| \leq \|A\|_{(\ell_1 : \ell_1)}$. Also, $L_A \in \mathcal{B}(\ell_1 : \ell_1)$ implies $\|L_A(x)\|_1 = \|Ax\|_1 \leq \|L_A\| \|x\|_1$, and it follows from $\left\| e^{(k)} \right\|_1 = 1$ for all $k \in \mathbb{N}_0$ that

$$\|A\|_{(\ell_1 : \ell_1)} = \sup_{k \in \mathbb{N}_0} \sum_n |a_{nk}| = \sup_{k \in \mathbb{N}_0} \left\| L(e^{(k)}) \right\|_1 \leq \|L_A\|.$$

Hence, $\|L_A\| = \|A\|_{(\ell_1 : \ell_1)}$.

This completes the proof. $\qquad\qquad\qquad\qquad\qquad\qquad\qquad\qquad\square$

Furthermore, we prove:

Theorem 3.3.11. [151, Theorem 6.11] *We have $L \in \mathcal{B}(c : c)$ if and only if there exists a matrix $A = (a_{nk}) \in (c_0 : c)$ and a sequence $b = (b_n) \in \ell_\infty$ with*

$$\lim_{n \to \infty} \left(b_n + \sum_k a_{nk} \right) = \widetilde{\alpha} \text{ exists} \qquad (3.3.20)$$

such that

$$L(x) = b \lim_{k \to \infty} x_k + Ax \text{ for all } x \in c. \qquad (3.3.21)$$

Furthermore, we have

$$\|L\| = \sup_{n \in \mathbb{N}_0} \left(|b_n| + \sum_k |a_{nk}| \right). \qquad (3.3.22)$$

Proof. First, we assume that $L \in \mathcal{B}(c : c)$. We write $L_n = P_n \circ L$ for all $n \in \mathbb{N}_0$, where P_n is the nth coordinate with $P_n(x) = x_n$ for all $x = (x_n) \in \omega$. Since the space c is a BK-space, we have $L_n \in c^*$ for all $n \in \mathbb{N}_0$,

$$L_n(x) = b_n \lim_{k \to \infty} x_k + \sum_k a_{nk}, \quad x = (x_k) \in c \qquad (3.3.23)$$

with $b_n = L_n(e) - \sum_k L_n(e^{(k)})$ and $a_{nk} = L_n(e^{(k)})$ for all $k \in \mathbb{N}_0$ and

$$\|L_n\| = |b_n| + \sum_k |a_{nk}|. \qquad (3.3.24)$$

Now, (3.3.23) yields (3.3.21). Since $L(x_0) = Ax_0$ for all x_0, we have $A = (a_{nk}) \in (c_0 : c)$ and so $\|A\| = \sup_{n \in \mathbb{N}_0} \sum_k |a_{nk}| < \infty$ by Part (b) of Theorem 3.3.9. Also $L(e) = b + Ae$ implies (3.3.6), and we obtain $\|b\|_\infty \leq \|L(e)\|_\infty + \|A\|$, that is, $b \in \ell_\infty$. Consequently, we have $C = \sup_{n \in \mathbb{N}_0} (|b_n| + \sum_k |a_{nk}|) < \infty$. Now, $\|L(x)\|_\infty = \sup_{n \in \mathbb{N}_0} |b_n \lim_{k \to \infty} x_k + \sum_k a_{nk} x_k| \leq (\sup_{n \in \mathbb{N}_0} (|b_n| + \sum_k |a_{nk}|) < \infty$ implies $\|L\| \leq C$. We also have $|L_n(x)| \leq \|L(x)\|_\infty$ for all $x \in \overline{B}_c$ and all $n \in \mathbb{N}_0$, and so $\sup_{n \in \mathbb{N}_0} \|L_n\| = C \leq \|L\|$. Thus, (3.3.22) is proved.

Conversely, we assume that $A = (a_{nk}) \in (c_0 : c)$ and $b \in \ell_\infty$ satisfy (3.3.20). Since $A = (a_{nk}) \in (c_0 : c)$ and $b \in \ell_\infty$, we obtain $C < \infty$ by (3.3.5), and so $L \in \mathcal{B}(c : \ell_\infty)$. Finally, let $x \in c$ be given and $x_k \to \xi$, as $k \to \infty$. Then, we have $x - \xi e \in c_0, L_n(x) = b_n \xi + \sum_k a_{nk} x_k = (b_n + \sum_k a_{nk})\xi + \{A(x - \xi e)\}_n$ for all $n \in \mathbb{N}_0$, and it follows from (3.3.20) and $A = (a_{nk}) \in (c_0 : c)$ that $\lim_{n \to \infty} L_n(x)$ exists. Since $x \in c$ was arbitrary, we have $L \in \mathcal{B}(c : c)$.

This completes the proof. $\qquad\square$

3.4 Schur Matrices

A matrix A is called a *Schur matrix* or *coercive matrix* if A-transforms of every bounded sequence are convergent, that is, if $A \in (\ell_\infty : c)$ then we say that A is a Schur matrix.

First, we state the following lemma which is needed in proving Schur's theorem. To the best of our knowledge it seems that the functional analytic proof of Schur's theorem does not exist yet.

Lemma 3.4.1. [32, Theorem 3.3.7] *Let $B = (b_{nk})$ be an infinite matrix such that $\sum_k |b_{nk}| < \infty$ for each $n \in \mathbb{N}_0$ and $\sum_k |b_{nk}| \to 0$, as $n \to \infty$. Then, $\sum_k |b_{nk}|$ converges uniformly in n.*

Proof. $\sum_k |b_{nk}| \to 0$, as $n \to \infty$ implies that $\sum_k |b_{nk}| < \infty$ for $n \geq N(\varepsilon)$. Since $\sum_k |b_{nk}| < \infty$ for $0 \leq n \leq N(\varepsilon)$, there exists $m = M(\varepsilon, n)$ such that $\sum_{k \geq M} |b_{nk}| < \infty$ for all $n \in \mathbb{N}_0$ which means that $\sum_k |b_{nk}|$ converges uniformly in n.

This completes the proof. $\qquad\square$

Theorem 3.4.2. (Schur) [147, Theorem 10]
$A = (a_{nk}) \in (\ell_\infty : c)$ *if and only if (3.3.5) holds and*

$$\sum_k |a_{nk}| \quad \text{converges uniformly in } n. \qquad (3.4.1)$$

Proof. Suppose that the conditions (3.3.5) and (3.4.1) hold, and $x = (x_k) \in \ell_\infty$. Then, $\sum_k a_{nk} x_k$ is absolutely and uniformly convergent in $n \in \mathbb{N}_0$. Hence, $\sum_k a_{nk} x_k \to \sum_k \alpha_k x_k$, as $n \to \infty$ which gives that $A = (a_{nk}) \in (\ell_\infty : c)$.

Conversely, suppose that $A = (a_{nk}) \in (\ell_\infty : c)$ and $x \in c$. Then, necessity of (3.3.5) follows easily by taking $x = e^{(k)}$ for each $k \in \mathbb{N}_0$. Define $b_{nk} = a_{nk} - \alpha_k$ for all $k, n \in \mathbb{N}_0$. Since $\sum_k |\alpha_k| < \infty$, $(\sum_k b_{nk} x_k)_n$ converges whenever $x = (x_k) \in \ell_\infty$. Now, if we can show that this implies

$$\lim_{n \to \infty} \sum_k |b_{nk}| = 0, \tag{3.4.2}$$

then by using Lemma 3.4.1, we obtain the desired result. Suppose to the contrary that $\sum_k |b_{nk}| \not\to 0$, as $n \to \infty$. Then, it follows that $\sum_k |b_{nk}| \to l > 0$, as $n \to \infty$, through some subsequence of the positive integers. Also we have $b_{mk} \to 0$, as $m \to \infty$, for each $k \in \mathbb{N}_0$. Hence, we may determine $m(1)$ such that $|\sum_k |b_{m(1),k}| - l| < \dfrac{1}{2}$ and $b_{m(1),1} < \dfrac{1}{2}$. Since $\sum_k |b_{m(1),k}| < \infty$, we can choose $k(2) > 1$ such that

$$\sum_{k=k(2)+1}^{\infty} |b_{m(1),k}| < \frac{1}{2}.$$

It follows that

$$\left| \sum_{k=2}^{k(2)} |b_{m(1),k}| - l \right| < \frac{1}{2}.$$

For our convenience we use the notation $\sum_{k=p}^{q} |b_{mk}| = B(m, p, q)$.

Now, we choose $m(2) > m(1)$ such that $|B(m(2), 1, \infty) - l| < l/10$ and $B(m(2), 1, k(2)) < l/10$. Then, choose $k(3) > k(2)$ such that $|B(m(2), k(3) + 1, \infty) - l| < l/10$. It follows that $|B(m(2), k(2) + 1, k(3)) - l| < 3l/10$. Continuing in this way and find $m(1) < m(2) < \cdots$, $1 = k(1) < k(2) < \cdots$ such that

$$\begin{cases} B(m(r), 1, k(r)) &< \dfrac{1}{10}, \\ B(m(r), k(r+1)+1, \infty) &< \dfrac{1}{10}, \\ B(m(r), k(r)+1, k(r+1)) - l &< \dfrac{3l}{10}. \end{cases} \tag{3.4.3}$$

Let us define $x = (x_k) \in \ell_\infty$ such that $\|x\| = 1$ by

$$x_k := \begin{cases} 0 & , \quad k = 1, \\ (-1)^r sgn(b_{m(r),k}) & , \quad k(r) < k \le k(r+1) \end{cases} \tag{3.4.4}$$

for $r \in \mathbb{N}_1$. Then, write $\sum_k b_{m(r),k} x_k$ as $\sum_1 + \sum_2 + \sum_3$, where $\sum_1$ is over $1 \le k \le k(r)$, $\sum_2$ is over $k(r) \le k \le k(r+1)$ and $\sum_3$ is over $k > k(r+1)$. It follows immediately from (3.4.3) with the sequence x given by (3.4.4) that $|\sum_k b_{m(r),k} - (-1)^r l| < \dfrac{1}{2}$. Consequently, it is clear that the sequence $Bx =$

$(\sum_k b_{nk} x_k)$ is not a Cauchy sequence and so is not convergent. Thus, we have proved that Bx is not convergent for all $x \in \ell_\infty$ which contradicts the fact that $A = (a_{nk}) \in (\ell_\infty : c)$. Hence, (3.4.2) must hold. Now, it follows by Lemma 3.4.1 that $\sum_k |b_{nk}|$ converges uniformly in n. Therefore, $\sum_k |a_{nk}| = \sum_k |b_{nk} + \alpha_k|$ converges uniformly in n.

This completes the proof. $\qquad\square$

We get the following corollary:

Corollary 3.4.3. $A = (a_{nk}) \in (\ell_\infty : c_0)$ *if and only if*

$$\lim_{n\to\infty} \sum_k |a_{nk}| = 0. \tag{3.4.5}$$

Remark 3.4.4. *The Silverman-Toeplitz theorem yields for a regular matrix A that $\chi(A) = 1$ which leads us to the fact that regular matrices form a subset of coregular matrices. One can easily see for a Schur matrix A that $\chi(A) = 0$ which says us that coercive matrices form a subset of conull matrices. Hence, we have the following result which is known as Steinhaus's theorem.*

Theorem 3.4.5. (Steinhaus) [32, Theorem 3.3.14] *For every regular matrix A, there is a bounded sequence which is not summable by A.*

Proof. We assume that a matrix $A = (a_{nk}) \in (c : c; p) \cap (\ell_\infty : c)$. Then, it follows from the condition (3.3.7) of Theorem 3.3.6 and Theorem 3.4.2 that

$$1 = \lim_{n\to\infty} \sum_k a_{nk} = \sum_k \left(\lim_{n\to\infty} a_{nk} \right) = 0,$$

a contradiction. That is to say that the classes $(c : c; p)$ and $(\ell_\infty : c)$ are disjoint.

This completes the proof. $\qquad\square$

We observe the following application of Corollary 3.4.3.

Theorem 3.4.6. [147, Corollary, p. 225] *Weak and strong convergence coincide in ℓ_1.*

Proof. We assume that the sequence $\{x^{(n)}\}_{n\in\mathbb{N}_0}$ is weakly convergent to x in ℓ_1, that is, $|f(x^{(n)}) - f(x)| \to 0$, as $n \to \infty$ for every $f \in \ell_1^*$. Since ℓ_1^* and ℓ_∞ are norm isomorphic, to every $f \in \ell_1^*$ there corresponds a sequence $a = (a_k) \in \ell_\infty$ such that $f(y) = \sum_k a_k y_k$. We define the matrix $B = (b_{nk})$ by $b_{nk} = x_k^{(n)} - x_k$ for all $k, n \in \mathbb{N}_0$. Then, we have $f(x^{(n)}) - f(x) = \sum_k \left(x_k^{(n)} - x_k \right) a_k = \sum_k b_{nk} a_k \to 0$, as $n \to \infty$ for all $a = (a_k) \in \ell_\infty$, that is, $B \in (\ell_\infty : c_0)$, and it follows from Corollary 3.4.3 that

$$\lim_{n\to\infty} \|x^{(n)} - x\|_1 = \lim_{n\to\infty} \sum_k |x_k^{(n)} - x_k| = \lim_{n\to\infty} \sum_k |b_{nk}| = 0.$$

This completes the proof. $\qquad\square$

3.5 Examples of Regular Matrices

We give here some special and most important examples of regular summability methods (cf. [32, pp. 47-50], [52, p. 23], [60, (4.1, II), p. 64], [147, pp. 226-227]).

3.5.1 Cesàro Matrix

The *Cesàro matrix* $C_1 = (c_{nk})$ of order one is defined by

$$c_{nk} := \begin{cases} \dfrac{1}{n+1} & , \quad 0 \leq k \leq n, \\ 0 & , \quad k > n \end{cases}$$

for all $k, n \in \mathbb{N}_0$. The inverse matrix $C_1^{-1} = (d_{nk})$ of the matrix C_1 is given by

$$d_{nk} := \begin{cases} (-1)^{n-k}(k+1) & , \quad n-1 \leq k \leq n, \\ 0 & , \quad 0 \leq n \leq n-2 \text{ or } k > n \end{cases}$$

for all $k, n \in \mathbb{N}_0$.

Let $r > -1$ and define A_n^r by

$$A_n^r := \begin{cases} \dfrac{(r+1)(r+2) \cdots (r+n)}{n!} & , \quad n \in \mathbb{N}_1, \\ 1 & , \quad n = 0. \end{cases} \tag{3.5.1}$$

It is known that $A_n^r \cong n^r / \Gamma(n+1)$. Then, the *Cesàro matrix* $C_r = (c_{nk}^r)$ of order r is defined by

$$c_{nk}^r := \begin{cases} \dfrac{A_{n-k}^{r-1}}{A_n^r} & , \quad 0 \leq k \leq n, \\ 0 & , \quad k > n \end{cases} \tag{3.5.2}$$

for all $k, n \in \mathbb{N}_0$. The Cesàro matrix of order r is regular if $r \geq 1$.

3.5.2 Euler Matrix

Let $\binom{n}{k} = n! / [k!(n-k)!]$ for all $k, n \in \mathbb{N}_0$, as usual. The *Euler matrix* $E_1 = (e_{nk})$ of order 1 is given by

$$e_{nk} := \begin{cases} \binom{n}{k} 2^{-n} & , \quad 0 \leq k \leq n, \\ 0 & , \quad k > n \end{cases}$$

for all $k, n \in \mathbb{N}_0$ whose generalization $E_q = (b_{nk}^q)$ of order $q > 0$ was defined by

$$b_{nk}^q := \begin{cases} \binom{n}{k}(q+1)^{-n} q^{n-k} & , \quad 0 \leq k \leq n, \\ 0 & , \quad k > n \end{cases}$$

for all $k, n \in \mathbb{N}_0$.

Let $0 < r < 1$. Then, the *Euler matrix* $E^r = (e^r_{nk})$ of order r is defined by

$$
e^r_{nk} := \left\{ \begin{array}{ll} \binom{n}{k}(1-r)^{n-k}r^k & , \quad 0 \le k \le n, \\ 0 & , \quad k > n \end{array} \right.
$$

for all $k, n \in \mathbb{N}_0$. It is clear that E^r corresponds to E_q for $r = (q+1)^{-1}$. Much of the works on the Euler means of order r was done by Knopp [129, 130]. So, some authors refer to E^r as the *Euler-Knopp matrix*. The original Euler means $E_1 = E^{1/2}$ was given by L. Euler in 1755. E^r is invertible such that $(E^r)^{-1} = E^{1/r}$ with $r \ne 0$. The Euler matrix E^r of order r is regular.

3.5.3   Riesz Matrix

Let $t = (t_k)$ be a sequence of non-negative real numbers with $t_0 > 0$ and write $T_n = \sum_{k=0}^n t_k$ for all $n \in \mathbb{N}_0$. Then, the *Riesz matrix* $R^t = (r^t_{nk})$ associated with the sequence $t = (t_k)$ is defined by

$$
r^t_{nk} := \left\{ \begin{array}{ll} t_k/T_n & , \quad 0 \le k \le n, \\ 0 & , \quad k > n \end{array} \right.
$$

for all $k, n \in \mathbb{N}_0$. For $t = e$, the Riesz matrix R^t is reduced to the matrix C_1. The inverse matrix $S^t = (s^t_{nk})$ of the matrix R^t is given by

$$
s^t_{nk} := \left\{ \begin{array}{ll} (-1)^{n-k}T_k/t_n & , \quad n - 1 \le k \le n, \\ 0 & , \quad 0 \le k \le n - 2 \text{ or } k > n \end{array} \right.
$$

for all $k, n \in \mathbb{N}_0$. The Riesz matrix R^t is regular if and only if $T_n \to 0$, as $n \to \infty$.

3.5.4   Nörlund Matrix

Following Peyerimhoff [185, pp. 17–19] and Mears [161], we give short survey on the properties of Nörlund means. Let (t_k) be a sequence of non-negative real numbers with $t_0 > 0$ and write $T_n = \sum_{k=0}^n t_k$ for all $n \in \mathbb{N}_0$. Then, the Nörlund mean associated with the sequence $t = (t_k)$ is defined by the matrix $N^t = (a^t_{nk})$ which is given by

$$
a^t_{nk} := \left\{ \begin{array}{ll} t_{n-k}/T_n & , \quad 0 \le k \le n, \\ 0 & , \quad k > n \end{array} \right.
$$

for all $k, n \in \mathbb{N}_0$. It is known that the Nörlund matrix N^t is regular if and only if $t_n/T_n \to 0$, as $n \to \infty$ [228, Corollary 2.5.8, p. 34], and is reduced in the case $t = e$ to the matrix C_1 of arithmetic mean. Additionally, for $t_n = A_n^{r-1}$ for all $n \in \mathbb{N}_0$, the method N^t is reduced to the Cesàro method C_r of order

$r > -1$, where A_n^r is defined by (3.5.1). Let $t_0 = D_0 = 1$ and define D_n for $n \in \mathbb{N}_1$ by

$$D_n = \begin{vmatrix} t_1 & 1 & 0 & 0 & \cdots & 0 \\ t_2 & t_1 & 1 & 0 & \cdots & 0 \\ t_3 & t_2 & t_1 & 1 & \cdots & 0 \\ \vdots & \vdots & \vdots & \vdots & \ddots & \vdots \\ t_{n-1} & t_{n-2} & t_{n-3} & t_{n-4} & \cdots & 1 \\ t_n & t_{n-1} & t_{n-2} & t_{n-3} & \cdots & t_1 \end{vmatrix}.$$

The inverse matrix $U^t = (u_{nk}^t)$ of the matrix N^t is given by Mears in [161], as follows;

$$u_{nk}^t := \begin{cases} (-1)^{n-k} D_{n-k} T_k & , \quad 0 \le k \le n, \\ 0 & , \quad k > n \end{cases}$$

for all $k, n \in \mathbb{N}_0$.

3.5.5   Borel Matrix

The semi-continuous *Borel matrix* $A = \{a_k(t)\}$ is defined by $a_k(t) = \dfrac{t^k}{e^t k!}$ for all $k \in \mathbb{N}_0$ and all $t > 0$. Since $a_k(t) \to 0$, as $t \to \infty$, for each $k \in \mathbb{N}_0$ and $\sum_k |a_k(t)| = \sum_k a_k(t) = 1$ whence $\sup_{t>0} \sum_k |a_k(t)| < \infty$, the Borel matrix is regular. This shows the essential property of the Borel matrix that it maps convergent sequences into a convergent sequence of functions, leaving the limit unchanged.

3.5.6   Abel Matrix

(cf. Peyerimhoff [185, p. 24]) A sequence (s_k) is called limitable by the *Abel method* A to l if $\sum_k x^k s_k$ exists for $|x| < 1$, and if

$$\lim_{x \to 1-0} \frac{\sum_k x^k s_k}{\sum_k x^k} = \lim_{x \to 1-0} (1-x) \sum_k x^k s_k = l.$$

This shows that the Abel method A is regular.

It is natural that one can derive the corresponding γ-matrices of the examples of Toeplitz matrices mentioned in the present chapter above, by using the relation between the terms of a series and its sequence of partial sums.

Chapter 4

Almost Convergence and Classes of Related Matrix Transformations

Keywords. *Almost convergence, almost conservativeness, almost regularity, strong regularity, almost coercivity, absolute almost convergence, invariant mean of sequences.*

4.1 Introduction

In the theory of sequence spaces, an application of the well-known Hahn-Banach extension theorem gives rise to the notion of Banach limit, which further leads to a beautiful concept of almost convergence. That is, the lim functional defined on c can be extended to the whole of ℓ_∞, and this extended functional is known as the Banach limit [28]. In 1948, Lorentz [140] used this notion of weak limit to define a new type of convergence, known as the almost convergence. Since then a huge amount of literature has appeared concerning various generalizations, extensions and applications of this method.

In this chapter, we study the notion of almost convergence, absolute almost convergence, invariant mean, σ-bounded variation and the related matrix transformations with their applications.

4.2 Almost Convergence

First, we define almost convergence, which will be used to define almost conservative and almost regular matrices.

Definition 4.2.1. *A continuous linear functional L on ℓ_∞ is said to be a Banach limit if it has the following properties:*

(i) $L(x) \geq 0$ if $x \geq 0$.

(ii) $L(e) = 1$, where $e = (1, 1, 1, \ldots)$.

(iii) $L(Sx) = L(x)$, *where S is the shift operator defined by $(Sx)_n = x_{n+1}$ for all $n \in \mathbb{N}_0$.*

Remark 4.2.2. *One can immediately see that for every $x \in \ell_\infty$, we have*

$$\liminf x \leq L(x) \leq \limsup x.$$

We define the following functionals on ℓ_∞:

$$q(x) = q(x_n) = \inf_{n_1,n_2,\ldots,n_p} \limsup_{k\to\infty} \frac{1}{p+1} \sum_{j=0}^{p} x_{n_j+k}$$

$$q'(x) = q'(x_n) = -q(-x) = \sup_{n_1,n_2,\ldots,n_p} \liminf_{k\to\infty} \frac{1}{p+1} \sum_{j=0}^{p} x_{n_j+k}.$$

It is easy to see that q is a sublinear functional and $|q(x)| \leq \|x\|$. Moreover,

$$q'(x) \leq L(x) \leq q(x). \tag{4.2.1}$$

Definition 4.2.3. *A bounded sequence $x = (x_k)$ is said to be almost convergent to the value l if all its Banach limits coincide, i.e., $L(x) = l$ for all Banach limits L.*

Hence from (4.2.1), it follows that the sequence $x = (x_k)$ is almost convergent if and only if $q'(x) = q(x)$. We denote the space of all almost convergent sequences by f, i.e.,

$$f := \left\{ x \in \ell_\infty : \exists l \in \mathbb{C} \text{ such that } \lim_{m\to\infty} t_{mn}(x) = l \text{ uniformly in } n \right\},$$

where

$$t_{mn}(x) = \frac{1}{m+1} \sum_{j=0}^{m} x_{n+j} \quad \text{for all} \quad m, n \in \mathbb{N}_0. \tag{4.2.2}$$

The sequences which are almost convergent are said to be summable by the method F, i.e., $x \in f$, we mean x is almost convergent and F-lim $x_k = L(x)$.

The subspace f_0 of f consists of all almost null sequences, that is, $f_0 := \{x \in f : F - \lim x = 0\}$. Note that $f = f_0 + \{e\}$ and f and f_0 are closed subspaces of ℓ_∞.

The existence of Banach limits was proved by Banach [28]. We present here the proof given by Bennett and Kalton [48], which offers an insightful proof of the existence of Banach limits and differs slightly from original proof of Lorentz. First we give the following lemmas of Bennett and Kalton [48] which are used in the main theorem.

Lemma 4.2.4. *If L is a continuous linear functional on ℓ_∞ with*

(i) $\|L\| = 1$,

(ii) $L(e) = 1$, *and*

(iii) $L(bs) = 0$,

then L is a Banach limit.

Proof. Since $\phi \subseteq bs$, it follows from (iii) that $L(\phi) = 0$, and by the continuity of L, we have $L(c_0) = 0$. Hence, L is an extended limit. Further, for $x \in \ell_\infty$, $x - Sx \in bs$ and hence $L(x) = L(Sx)$. $\qquad\qquad\square$

Lemma 4.2.5. *If $x \in \ell_\infty \setminus c_0$, then there exists an extended limit L with $L(x) \neq 0$.*

Proof. If $x \in \ell_\infty \setminus c_0$, then we may choose an increasing sequence $(n_k)_{k=1}^\infty$ of positive integers such that $x_{n_k} \to l \neq 0$, as $k \to \infty$. Define L by $Ly = \lim y_{n_k}$, where this limit exists, and extend L to ℓ_∞ by the Hahn-Banach theorem. $\quad\square$

Theorem 4.2.6. (see Lorentz [140]) *For $x = (x_n) \in \omega$ to be almost convergent to s, it is necessary and sufficient that*

$$\lim_{p \to \infty} \frac{x_n + x_{n+1} + \cdots + x_{n+p-1}}{p} = s \qquad (4.2.3)$$

holds uniformly in n.

Proof. There is no loss of generality if we suppose that $s = 0$. Define the matrix map $A : \ell_\infty \to \ell_\infty$ by

$$(Ax)_p = \frac{x_{n_p} + x_{n_p+1} + \cdots + x_{n_p+p-1}}{p}, \quad x \in \ell_\infty;$$

where $(n_p)_{p=1}^\infty$ is any increasing sequence of positive integers. Then $Ae = e$, $A(bs) \subset c_0$ and $\|A\|_{(\ell_\infty : \ell_\infty)} = 1$. For an extended limit L, by Lemma 4.2.4, LA is a Banach limit. Therefore, we get $L(Ax) = 0$ for $x \in f_0$. Using Lemma 4.2.5, we have $Ax \in f_0$. Therefore,

$$\lim_{p \to \infty} \frac{x_{n_p} + x_{n_p+1} + \cdots + x_{n_p+p-1}}{p} = 0. \qquad (4.2.4)$$

Since (4.2.4) holds for any sequence $(n_p)_{p=1}^\infty$, we have

$$\lim_{p \to \infty} \sup_{n \in \mathbb{N}_0} \left| \frac{x_{n_p} + x_{n_p+1} + \cdots + x_{n_p+p-1}}{p} \right| = 0,$$

that is (4.2.3) holds.

Conversely, suppose that (4.2.3) holds. Then,

$$\lim_{p \to \infty} \left\| \frac{x_{n_p} + x_{n_p+1} + \cdots + x_{n_p+p-1}}{p} \right\|_\infty = 0.$$

Hence, $L(x) = 0$ for any Banach limit L, i.e., x is almost convergent to zero. $\qquad\qquad\square$

The sequences which are almost convergent are said to be summable by the method F, i.e., $x \in f$, we mean x is almost convergent and F-lim $x_k = L(x)$.

Definition 4.2.7. *Let $A = (a_{mk})_{m,k \in \mathbb{N}_0}$ be a regular matrix method. A bounded sequence $x = (x_k)$ is said to be F_A-summable to the value l if $y_{mn} = \sum_k a_{mk} x_{k+n} \to l$, as $m \to \infty$, uniformly in n. A sequence $x = (x_k)$ is said to be almost A-summable to the value l if its A-transform is almost convergent to l.*

Note that if A is replaced by the C_1 matrix, then F_A-summability is reduced to the almost convergence.

Theorem 4.2.8. (see Lorentz [140]) *If the matrix $A = (a_{mk})_{m,k \in \mathbb{N}_0}$ is regular, then an F_A-summable sequence $x = (x_n)$ is almost convergent.*

In the following theorem, Schaefer [198] has replaced regularity by almost regularity.

Theorem 4.2.9. (see Schaefer [198]) *If the matrix $A = (a_{mk})_{m,k \in \mathbb{N}_0}$ is almost regular, then an F_A-summable sequence $x = (x_n)$ is almost convergent.*

Proof. Let $x = (x_n)$ be an F_A-summable sequence to the value l and $x^{(k)} = (x_{n+k})_{n \in \mathbb{N}_0}$ for each $k \in \mathbb{N}_0$. Then, $x^{(k)} \in \ell_\infty$; since $\|x^{(k)}\|_\infty \leq \|x\|_\infty$. Let $y_{mk} = l + \alpha_{mk}$. Then, for every $\epsilon > 0$ an m_0 can be found such that $|\alpha_{mk}| < \epsilon$ when $m \geq m_0$ for $k \in \mathbb{N}_0$. Let us write $y^{(m)} = (y_{mk})_k$. Then $y^{(m)} = \sum_n a_{mn} x^{(n)}$ such that each $y^{(m)} \in \ell_\infty$. Thus, if we put $\alpha^{(m)} = (\alpha_{mk})_k$, then we have $y^{(m)} = le + \alpha^{(m)}$. Since $L(x) = L(x^{(n)})$ for all n and since Banach limits are continuous on ℓ_∞, we have for all $m \geq m_0$, $\sum_n a_{mn} L(x) = l + \alpha_m$, where $|\alpha_m| = |L(\alpha^{(m)})| < \epsilon$. Now let $a = (\sum_n a_{mn})$ and $\alpha = (\alpha_m)$. Then, the last expression implies that $L(x)a = le + \alpha$. Once again take any Banach limit L^* of both sides of this equality. Since the matrix A is almost regular, we have $L^*(a) = 1$. Therefore, we get $L(x)L^*(a) = L(x) \cdot 1 = l + L^*(\alpha) = l$ since $\lim_n \alpha_n = 0$. Hence, $L(x) = l$, i.e., x is almost convergent and F-lim $x = l$. $\square$

Examples 4.2.10. [140] *We have the following:*

(i) *For $z \in \mathbb{C}$ on the circumference of $|z| = 1$, $L(z^n) = 0$ holds everywhere except for $z = 1$. The assertion immediately follows from*

$$\left| \frac{1}{k} (z^n + z^{n+1} + \cdots + z^{n+k-1}) \right| = \left| z^n \frac{1 - z^k}{k(1-z)} \right| \leq \frac{2}{k(1 - |z|)}.$$

It is easy to see that the geometric series $\sum_n z^n$ for all $z \in \mathbb{C}$ such that $|z| = 1$ with $z \neq 1$ is almost convergent to $1/(1-z)$. Hence, it follows that the Taylor series of a function $f(z)$, which for $|z| < 1$ is regular and on $|z| = 1$ has simple poles, is almost convergent at every point of the circumference $|z| = 1$ with the limit $f(z)$.

(ii) *A periodic sequence (x_n) for which numbers N and p (the period) exist such that $x_{n+p} = x_n$ holds for $n \geq N$ is almost convergent to the value $L(x_n) = (x_N + x_{N+1} + \cdots + x_{N+p-1})/p$. For example, the periodic sequence $(1, 0, 0, 1, 0, 0, 1, \ldots)$ is almost convergent to $1/3$.*

(iii) *We say that a sequence (x_n) is almost periodic if for every $\varepsilon > 0$, there are two natural numbers N and r such that in every interval $(k, k + r)$ with $k > 0$, at least one "ε-period" p exists. More precisely, $|x_{n+p} - x_n| < \varepsilon$ for $n \geq N$ must hold for this p. Thus, it is easy to see that every almost periodic sequence is almost convergent. But there are almost convergent sequences which are not almost periodic. For example, the sequence $x = (x_k)$ defined by*

$$x_k := \begin{cases} 1 & , \quad k = n^2, \\ 0 & , \quad k \neq n^2; \end{cases} \quad (n \in \mathbb{N}_0)$$

is almost convergent to 0 but is not almost periodic.

Remark 4.2.11. [52] *The following statements hold:*

(i) *Note that $c \subset f$ and for $x \in c$, $F-\lim x_k = \lim x_k$. That is, every convergent sequence is almost convergent to the same limit but not conversely. For example, the sequence $x = (x_k)$ defined by*

$$x_k := \begin{cases} 1 & , \quad k \text{ is odd,} \\ 0 & , \quad k \text{ is even} \end{cases}$$

is not convergent but is almost convergent to $1/2$.

(ii) *In contrast to the well-known fact that c is a separable subspace of $(\ell_\infty, \|\cdot\|_\infty)$, f is a non-separable closed subspace of $(\ell_\infty, \|\cdot\|_\infty)$.*

(iii) *f is a BK-space with $\|\cdot\|_\infty$.*

(iv) *f is nowhere dense in ℓ_∞, dense in itself and closed, and therefore perfect.*

(v) *The method is not strong in spite of the fact that it contains certain classes of matrix methods for bounded sequences.*

(vi) *Most of the commonly used matrix methods contain the method F, e.g., every almost convergent sequence is also (C, α) and (E, α)-summable to its F-limit, where $\alpha > 0$.*

(vii) *The method F is equivalent to none of the matrix methods, i.e., the method F cannot be expressed in the form of a matrix method.*

(viii) *The method F is related to the Cesàro method C_1. In fact the method C_1 can be replaced in this definition by any other regular matrix method A satisfying certain conditions.*

(ix) Since $c \subset f \subset \ell_\infty$, we have $\ell_1 = \ell_\infty^\dagger \subset f^\dagger \subset c^\dagger = \ell_1$. That is, the $\dagger$-dual of f is ℓ_1, where $\dagger$ stands for α, β and γ. Hence, $\|a\|_f^* = \|a\|_1$ for all $a \in \ell_1$.

4.3   Almost Conservative and Almost Regular Matrices

King [124] used the idea of almost convergence to study the almost conservative and almost regular matrices.

Definition 4.3.1. *An infinite matrix A is said be almost conservative if $Ax \in f$ for all $x \in c$, i.e., $A \in (c : f)$. In addition if F-$\lim(Ax)_n = \lim x_k$ for all $x = (x_k) \in c$, then A is said to be almost regular, and in this case, we write $A \in (c : f)_{reg}$.*

Theorem 4.3.2. *The following statements hold:*

(a) $A = (a_{nk})_{n,k\in\mathbb{N}_0}$ is almost conservative, i.e., $A = (a_{nk}) \in (c : f)$ if and only if

(i) $\|A\| < \infty$.

(ii) $t(n, k, p) \to \alpha_k \in \mathbb{C}$, as $p \to \infty$, for each $k \in \mathbb{N}_0$, uniformly in n.

(iii) $\sum_k t(n, k, p) \to \alpha \in \mathbb{C}$, as $p \to \infty$, uniformly in n; where $t(n, k, p) = \left(\sum_{j=n}^{n+p} a_{jk}\right)/(p+1)$ for all $k, n, p \in \mathbb{N}_0$. In this case, the F-limit of Ax is $(\lim x)(\alpha - \sum_k \alpha_k) + \sum_k x_k \alpha_k$ for every $x = (x_k) \in c$.

(b) A is almost regular if and only if the condition (i) holds and the conditions (ii) and (iii) hold with $\alpha_k = 0$ for each $k \in \mathbb{N}_0$ and $\alpha = 1$, respectively.

Proof. (a) Let the conditions (i)–(iii) hold and $x = (x_k) \in c$. Then, we have

$$|t_{pn}(x)| \le \frac{1}{p+1} \sum_k \sum_{j=n}^{n+p} |a_{jk}||x_k| \le \|A\|\|x\|_\infty.$$

Since t_{pn} is obviously linear on c, it follows that $t_{pn} \in c^*$ and that $\|t_{pn}\| \le \|A\|$. Now,

$$t_{pn}(e) = \frac{1}{p+1} \sum_k \sum_{j=n}^{n+p} a_{jk} = \frac{1}{p+1} \sum_{j=n}^{n+p} \sum_k a_{jk},$$

so $\lim_{p\to\infty} t_{pn}(e)$ exists uniformly in n and equals α. Similarly, $t_{pn}(e^{(k)}) \to \alpha_k$, as $p \to \infty$, for each $k \in \mathbb{N}_0$, uniformly in n. Since $\{e, e^{(1)}, e^{(2)}, \ldots\}$ is a

Schauder basis for c, and $\sup_{p \in \mathbb{N}} |t_{pn}(x)| < \infty$ for each $x \in c$, it follows that $t_{pn}(x) \to t_n(x)$, as $p \to \infty$, exists for all $x \in c$. Furthermore, $\|t_n\| \leq \liminf_{p \to \infty} \|t_{pn}\|$ for each $n \in \mathbb{N}_0$, and $t_n \in c^*$. Thus,

$$t_n(x) = (\lim x) \left[t_n(e) - \sum_k t_n\left(e^{(k)}\right) \right] + \sum_k t_n\left(e^{(k)}\right) x_k$$

$$\lim_{n \to \infty} t_n(x) = (\lim x) \left(\alpha - \sum_k \alpha_k \right) + \sum_k \alpha_k x_k,$$

an expression independent of n. Denote this expression by $L(x)$.

In order to see that $t_{pn}(x) \to L(x)$, as $p \to \infty$, uniformly in n, set $G_{pn}(x) = t_{pn}(x) - L(x)$. Then, $G_{pn} \in c^*$, $\|G_{pn}\| \leq 2\|A\|$ for all $p, n \in \mathbb{N}_0$, $G_{pn}(e) \to 0$, as $p \to \infty$, uniformly in n, and $G_{pn}\left(e^{(k)}\right) \to 0$, as $p \to \infty$, uniformly in n for each $k \in \mathbb{N}_0$. Let K be an arbitrary positive integer. Then,

$$x = (\lim x)e + \sum_{k=0}^{K} \left(x_k - \lim x\right) e^{(k)} + \sum_{k=K+1}^{\infty} \left(x_k - \lim x\right) e^{(k)}$$

and we have

$$G_{pn}(x) = (\lim x)G_{pn}(e) + \sum_{k=0}^{K} \left(x_k - \lim x\right) G_{pn}\left(e^{(k)}\right) + G_{pn}\left[\sum_{k=K+1}^{\infty} \left(x_k - \lim x\right) e^{(k)} \right].$$

Now,

$$\left| G_{pn} \left[\sum_{k=K+1}^{\infty} \left(x_k - \lim x\right) e^{(k)} \right] \right| \leq 2\|A\| \cdot \sup_{k \geq K+1} \left| x_k - \lim_{k \to \infty} x_k \right|$$

for all $p, n \in \mathbb{N}_0$. By first choosing a fixed K large enough, it is easy to see that each of the three displayed terms for $G_{pn}(x)$ can be made to be uniformly small in absolute value for all sufficiently large p, so $G_{pn}(x) \to 0$, as $p \to \infty$, uniformly in n. This shows that $t_{pn}(x) \to L(x)$, as $p \to \infty$, uniformly in n. Hence, $Ax \in f$ for all $x \in c$ and the matrix A is almost conservative.

Conversely, suppose that A is almost conservative. If x is any null sequence, then $Ax \in f \subset \ell_\infty$, i.e., $A \in (c : \ell_\infty)$. We know that $A = (a_{nk}) \in (c : \ell_\infty)$ if and only if $\|A\| < \infty$. Hence, (i) follows. Furthermore, since $e^{(k)}$ and e are convergent sequences the limits $\lim_{p \to \infty} t_{pn}\left(e^{(k)}\right)$ for each $k \in \mathbb{N}_0$ and $\lim_{p \to \infty} t_{pn}(e)$ must exist, uniformly in n. Hence, the conditions (ii) and (iii) are also necessary.

(b) If a matrix A satisfies the three conditions of the theorem, then it is an almost conservative matrix. For $x \in c$, the F-limit of Ax is $L(x)$ which reduces to $\lim x$, since $\alpha = 1$ and $\alpha_k = 0$ for each $k \in \mathbb{N}_0$. Hence, A is an almost regular matrix.

Conversely, if A is almost regular, then $F - \lim(Ae)_n = 1$, $F - \lim\left(Ae^{(k)}\right)_n = 0$, and $\|A\| < \infty$, as in the proof of Part (a).

This completes the proof. $\qquad \square$

Remark 4.3.3. *Since* $c \subset f$, *every regular matrix is almost regular, but an almost regular matrix need not be regular. Indeed, if we define the matrix* $C = (c_{nk})$ *by*

$$c_{nk} := \begin{cases} \dfrac{1 + (-1)^n}{n+1} & , \quad 0 \leq k \leq n, \\ 0 & , \quad n < k \end{cases}$$

for all $k, n \in \mathbb{N}_0$. *It is easy to see that the matrix* C *is almost regular but not regular since the limit of* $(\sum_k c_{nk})_{n \in \mathbb{N}_0}$ *does not exist, as* $n \to \infty$.

Remark 4.3.4. *It is known that* E^r *is regular if and only if* $0 < r \leq 1$. *It is natural to ask whether or not there exist values of* r *for which* E^r *is almost regular but not regular which is impossible for the Euler matrix* E^r. *In fact, we have that* E^r *is almost regular if and only if it is regular.*

By fs, we mean the space of the almost convergent series, i.e., fs consists of all series whose sequence of partial sums are in the space f. That is,

$$fs := \left\{ (x_k) \in \omega : \exists l \in \mathbb{C} \text{ such that } \lim_{m \to \infty} \left| \frac{1}{m+1} \sum_{i=0}^{m} \sum_{j=0}^{n+i} x_j - l \right| = 0 \text{ uniformly in } n \right\}.$$

Now, following Duran [72], we can begin with the following lemma characterizing the class $(f_0 : f_0)$ of matrix transformations:

Lemma 4.3.5. [72, Theorem 3] *An infinite matrix* $A = (a_{nk})$ *transforms the space* f_0 *into itself if and only if*

$$\|A\| = \sup_{n \in \mathbb{N}_0} \sum_k |a_{nk}| < \infty, \tag{4.3.1}$$

$$F - \lim a_{nk} = 0 \quad \text{for each fixed } k \in \mathbb{N}_0, \tag{4.3.2}$$

$$\lim_{q \to \infty} \sum_k \frac{1}{q+1} \left| \sum_{j=0}^{q} \triangle a_{n+j,k} \right| = 0 \text{ uniformly in } n, \tag{4.3.3}$$

where $\triangle a_{n+j,k} = a_{n+j,k} - a_{n+j,k+1}$.

Now, following Orhan and Öztürk [182], we can characterize the class $(fs : f; p)$ of almost strongly regular series-to-sequence matrix transformations.

Theorem 4.3.6. [182, Theorem 3.1] *An infinite matrix* $A = (a_{nk})$ *transforms the space* fs *into the space* f *leaving the* F-*limit/sum invariant if and only if*

$$\sup_{n \in \mathbb{N}_0} \sum_k |\triangle a_{nk}| < \infty, \text{ where } \triangle a_{nk} = a_{nk} - a_{n,k+1}, \tag{4.3.4}$$

$$\lim_{k \to \infty} a_{nk} = 0 \quad \text{for each fixed } n \in \mathbb{N}_0, \tag{4.3.5}$$

$$F - \lim a_{nk} = 1 \quad \text{for each fixed } k \in \mathbb{N}_0, \tag{4.3.6}$$

$$\lim_{q \to \infty} \sum_k \frac{1}{q+1} \left| \sum_{j=0}^{q} \triangle^2 a_{n+j,k} \right| = 0 \text{ uniformly in } n, \tag{4.3.7}$$

where $\triangle^2 a_{n+j,k} = \triangle(a_{n+j,k} - a_{n+j,k+1})$.

Proof. Let $A = (a_{nk})$ be a series-to-sequence almost strongly regular matrix, i.e., $A \in (fs : f; p)$ and $x = (x_k) \in fs$. Let s_n also be the nth partial sum of the series $\sum_k x_k$ for all $n \in \mathbb{N}_0$ and $F - \lim s_n = l$. Then, Ax exists and belongs to f with $F - \lim (Ax)_n = l$. Now, the necessity of the condition (4.3.6) is immediate by taking $x = e^{(k)}$, since $e^{(k)} \in fs$ for each $k \in \mathbb{N}_0$.

To show the necessity of the condition (4.3.5), we assume that (4.3.5) does not hold for some $n \in \mathbb{N}_0$ and obtain a contradiction as in Theorem 2.1 of Öztürk [183]. Indeed, under this assumption we can find some $x \in fs$ such that Ax does not belong to the space f. For example, if we choose $x = \{(-1)^k\} \in fs$ then $(Ax)_n = \sum_k a_{nk}(-1)^k$ does not converge for each $n \in \mathbb{N}_0$. That is to say that the A-transform of the series $\sum_k (-1)^k$ belonging to the space fs, does not even exist. But, this contradicts the almost strong regularity of A. Hence, the condition (4.3.5) is necessary.

Let us consider the equality

$$\sum_{k=0}^{m} a_{nk} x_k = \sum_{k=0}^{m-1} \triangle a_{nk}(s_k - l) + l a_{n1} + (s_m - l) a_{nm} \text{ for each } m, n \in \mathbb{N}_0 \quad (4.3.8)$$

obtained by applying Abel's partial summation to the mth partial sums of the series $\sum_k a_{nk} x_k$. Since $(s_m - l) \in \ell_\infty$ and the third term on the right-hand side of (4.3.8) tends to zero by (4.3.5), letting $m \to \infty$ in (4.3.8), we derive that

$$\sum_k a_{nk} x_k = \sum_k \triangle a_{nk}(s_k - l) + l a_{n1} \text{ for each } n \in \mathbb{N}_0. \quad (4.3.9)$$

Then, we must have $\{\sum_k \triangle a_{nk}(s_k - l)\} \in f_0$, since $F - \lim (Ax)_n = l$ by the hypothesis and $F - \lim a_{n1} = 1$ by (4.3.6). Therefore, we conclude that $B = (b_{nk}) \in (f_0 : f_0)$, where $b_{nk} = \triangle a_{nk}$ for all $k, n \in \mathbb{N}_0$. Hence, the conditions (4.3.1) and (4.3.3) of Lemma 4.3.5 are satisfied by the matrix B which are equivalent to the conditions (4.3.4) and (4.3.7), respectively.

Conversely, suppose that the matrix A satisfies the conditions (4.3.4)–(4.3.7) and $x = (x_k) \in fs$ with $s_n = \sum_{k=0}^{n} x_k$ for all $n \in \mathbb{N}_0$, and $F - \lim s_n = l$. Let us reconsider the matrix $B = (b_{nk})$ defined by $b_{nk} = \triangle a_{nk}$ for all $k, n \in \mathbb{N}_0$. The conditions (4.3.4) and (4.3.5) imply that (4.3.9) holds. Then, by (4.3.4), (4.3.6) and (4.3.7), one can easily see that $B \in (f_0 : f_0)$, since the conditions (4.3.1), (4.3.2) and (4.3.3) of Lemma 4.3.5 are satisfied. Therefore, the equality (4.3.9) implies that $F - \lim (Ax)_n = l$. This means that A is an almost strongly regular series-to-sequence matrix, as desired.

This completes the proof. $\square$

4.4   Almost Coercive Matrices

Eizen and Laush [76] (see also Duran [72]) considered the class of almost coercive matrices.

Definition 4.4.1. *An infinite matrix belonging to any of the classes $(\ell_\infty : f)$, $(\ell_\infty : fs)$, $(bs : f)$ or $(bs : fs)$ is called as almost coercive.*

Theorem 4.4.2. [76, Theorem 2.1], [72, Theorem 1] *$A = (a_{nk})$ is almost coercive, i.e., $A = (a_{nk}) \in (\ell_\infty : f)$ if and only if (4.3.1) holds, and*

$$\exists \alpha_k \in \mathbb{C} \text{ such that } F - \lim a_{nk} = \alpha_k \text{ for each fixed } k \in \mathbb{N}_0, \quad (4.4.1)$$

$$\lim_{q \to \infty} \sum_k \frac{1}{q+1} \left| \sum_{j=0}^{q} a_{n+j,k} - \alpha_k \right| = 0 \text{ uniformly in } n. \quad (4.4.2)$$

In this case, the F-limit of Ax is $\sum_k \alpha_k x_k$ for every $x = (x_k) \in \ell_\infty$.

Proof. Suppose that the matrix $A = (a_{jk})$ satisfies the conditions (4.3.1), (4.4.1) and (4.4.2). Then, we have for any positive integer K

$$\begin{aligned}
\sum_{k=0}^{K} |\alpha_k| &= \sum_{k=0}^{K} \lim_{p \to \infty} \frac{1}{p+1} \left| \sum_{j=n}^{n+p} a_{jk} \right| \\
&= \lim_{p \to \infty} \frac{1}{p+1} \sum_{k=0}^{K} \left| \sum_{j=n}^{n+p} a_{jk} \right| \\
&\leq \limsup_{p \to \infty} \frac{1}{p+1} \sum_{j=n}^{n+p} \sum_k |a_{jk}| \leq \|A\|.
\end{aligned}$$

This implies that $\sum_k |\alpha_k|$ converges, which implies that $\sum_k \alpha_k x_k$ is defined for every bounded sequence $x = (x_k) \in \ell_\infty$.

Let $x = (x_k) \in \ell_\infty$. For $p \in \mathbb{N}_0$, define the operator S^p on ℓ_∞ by the composition of the shift operator S with itself p times as $S^p(x) = x_{n+p}$. Then,

$$\left\| \frac{Ax + S(Ax) + \cdots + S^p(Ax)}{p+1} - \left(\sum_k \alpha_k x_k \right) e \right\|$$

$$= \sup_{n \in \mathbb{N}_0} \left| \sum_k \sum_{j=n}^{n+p} \frac{a_{jk} - \alpha_k}{p+1} x_k \right| \leq \|x\|_\infty \cdot \sup_{n \in \mathbb{N}_0} \sum_k \frac{1}{p+1} \left| \sum_{j=n}^{n+p} a_{jk} - \alpha_k \right|.$$

Let $p \to \infty$. By the uniformity of the limits in the condition (4.4.2), it follows that $[Ax + S(Ax) + \cdots + S^p(Ax)]/(p+1) \to (\sum_k \alpha_k x_k) e$, and that $Ax \in f$ with $F - \lim(Ax)_n = \sum_k \alpha_k x_k$. Hence, A is almost coercive.

Conversely, suppose that $A = (a_{nk}) \in (\ell_\infty : f)$. Then, $A = (a_{nk}) \in (\ell_\infty : c)$ and so the conditions (4.3.1) and (4.3.3) follow immediately by Theorem 3.4.2. To prove (4.4.2), let for some $n \in \mathbb{N}_0$,

$$\limsup_{p \to \infty} \frac{1}{p+1} \sum_k \left| \sum_{j=n}^{n+p} a_{jk} - \alpha_k \right| = N > 0.$$

Since $\|A\|$ is finite, N is also finite. We observe that since $\sum_k |\alpha_k| < \infty$, the matrix $B = (b_{nk})$ defined by $b_{nk} = a_{nk} - \alpha_k$ for all $k, n \in \mathbb{N}_0$, is also an almost coercive matrix. If one sets $G_{kp} = |\sum_{j=n}^{n+p}(a_{jk} - \alpha_k)|/(p+1)$, and $E_{kt} = G_{k,p_t}$, one can follow the construction in the proof of Theorem 2.1 in [76] to obtain a bounded sequence whose B-transform is not in the space f. This contradiction shows that the limit in (4.4.2) is zero for every $n \in \mathbb{N}_0$.

To show that this convergence is uniform in n, we invoke the following lemma, which is proved in [199]. $\square$

Lemma 4.4.3. *Let $\{H(n)\}$ be a countable family of matrices $H(n) = \{h_{pk}(n)\}$ such that $\|H(n)\| \leq M < \infty$ for all $n \in \mathbb{N}_0$ and $h_{pk}(n) \to 0$, as $p \to \infty$, for each $k \in \mathbb{N}_0$, uniformly in n. Then, $\sum_k h_{pk}(n) \to 0$, as $p \to \infty$, uniformly in n, for all $x \in \ell_\infty$ if and only if $\sum_k |h_{pk}(n)| \to 0$, as $p \to \infty$, uniformly in n.*

Proof. Define the matrix $H(n) = \{h_{pk}(n)\}$ by $h_{pk}(n) = \sum_{j=n}^{n+p}(a_{jk} - \alpha_k)/(p+1)$ for all $k, n, p \in \mathbb{N}_0$. It is easy to see that $\|H(n)\| \leq 2\|A\|$ for every $n \in \mathbb{N}_0$ and that $h_{pk}(n) \to 0$, as $p \to \infty$, for each $k \in \mathbb{N}_0$, uniformly in n by the condition (4.4.1). For any $x = (x_k) \in \ell_\infty$, $\lim_{p \to \infty} \sum_k h_{pk}(n)x_k = F - \lim(Ax)_n - \sum_k \alpha_k x_k$, and the limit exists uniformly in n since $Ax \in f$. Moreover, this limit is zero, since

$$\sum_k |h_{pk}(n)x_k| \leq \|x\| \frac{1}{p+1} \sum_k \left| \sum_{j=n}^{n+p} a_{jk} - \alpha_k \right|.$$

Thus, $\sum_k |h_{pk}(n)| \to 0$, as $p \to \infty$, uniformly in n, and the matrix A satisfies the condition (4.4.2).

This completes the proof. $\square$

Corollary 4.4.4. *The classes of almost strongly regular and almost coercive sequence-to-sequence matrix transformations are disjoint.*

Now, following Başar and Solak [40], prior to giving a Steinhaus-type theorem related to series-to-sequence almost strongly regular matrix transformations we characterize the classes $(bs : f)$ and $(bs : fs)$ of almost coercive series-to-sequence and series-to-series matrix transformations.

Theorem 4.4.5. [40, Theorem 2.2] *An infinite matrix $B = (b_{nk})$ transforms the space bs into the space f if and only if the conditions (4.3.4) and (4.3.5) hold with b_{nk} instead of a_{nk}, and*

$$\exists \beta_k \in \mathbb{C} \ \ such \ that \ \ F - \lim_n \ b_{nk} = \beta_k \ for \ each \ fixed \ k \in \mathbb{N}_0, \quad (4.4.3)$$

$$\lim_{q \to \infty} \sum_k \frac{1}{q+1} \left| \sum_{j=0}^{q} \Delta(b_{n+j,k} - \beta_k) \right| = 0 \ uniformly \ in \ n. \quad (4.4.4)$$

Proof. Let $B = (b_{nk}) \in (bs : f)$ and $u = (u_k) \in bs$ with $s_n = \sum_{k=0}^{n} u_k$ for all $n \in \mathbb{N}_0$. In order to B be applicable to the elements of the space bs, the rows B_n of B must belong to the beta-dual $bv_0 = bv \cap c_0$ of the space bs. This shows that (4.3.5) is necessary with b_{nk} instead of a_{nk}.

The necessity of the condition (4.4.3) is easily obtained by taking $u = e^{(k)}$, since $e^{(k)} \in bs$ for each $k \in \mathbb{N}_0$.

Let us consider the equality

$$\sum_{k=0}^{m} b_{nk} u_k = \sum_{k=0}^{m-1} \Delta b_{nk} s_k + b_{nm} s_m \ for \ each \ fixed \ m, n \in \mathbb{N}_0 \quad (4.4.5)$$

obtained by applying Abel's partial summation to the mth partial sums of the series $\sum_k b_{nk} u_k$. Letting $m \to \infty$ in (4.4.5), we have

$$\sum_k b_{nk} u_k = \sum_k \Delta b_{nk} s_k \ for \ each \ \ n \in \mathbb{N}_0, \quad (4.4.6)$$

since the second term on the right-hand side of (4.4.5) tends to zero by (4.3.5) and $(s_k) \in \ell_\infty$. It follows by passing to F-limit in (4.4.6) that $C = (c_{nk}) \in (\ell_\infty : f)$, where $c_{nk} = \Delta b_{nk}$ for all $k, n \in \mathbb{N}_0$. Therefore, the conditions (4.3.1) and (4.4.2) of Theorem 4.4.2 are satisfied by the matrix C, which are equivalent to the conditions (4.3.4) and (4.4.4), respectively.

Conversely, suppose that the matrix B satisfies the conditions (4.3.4), (4.3.5), (4.4.3) and (4.4.4), and $u = (u_k) \in bs$. Let us consider the matrix C defined by $c_{nk} = \Delta b_{nk}$ for all $k, n \in \mathbb{N}_0$. Then, the two-sided implication "the conditions (4.3.1), (4.4.1) and (4.4.2) of Theorem 4.4.2 are satisfied by the matrix C if and only if the conditions (4.3.4), (4.4.3) and (4.4.4) are satisfied by the matrix B, respectively" holds. Hence, $C \in (\ell_\infty : f)$ and this gives by passing to F-limit in (4.4.6) that $Bu \in f$. This means that every element of the space bs is almost B-summable, i.e., $B \in (bs : f)$.

This completes the proof. $\qquad \qquad \square$

As an immediate consequence of Theorem 4.4.5, we have the following corollary:

Corollary 4.4.6. *An infinite matrix B transforms the space bs into the space f_0 of almost null sequences if and only if (4.3.4) and (4.3.5) hold with b_{nk} instead of a_{nk}, and (4.4.3), (4.4.4) also hold with $\beta_k = 0$ for all $k \in \mathbb{N}_0$.*

Now, we can give the Steinhaus-type theorem for the almost strongly regular and almost coercive series-to-sequence matrix transformations.

Theorem 4.4.7. [40, Theorem 2.4] *The classes of almost strongly regular and almost coercive series-to-sequence matrix transformations are disjoint.*

Proof. Let us suppose that $(fs : f; p) \cap (bs : f) \neq \emptyset$. Then, there is at least one $B \in (fs : f; p) \cap (bs : f)$. Since $B \in (fs : f; p)$, the series $\sum_k \triangle b_{nk}$ and also $\sum_k 1/(q+1) \sum_{j=0}^{q} \triangle b_{n+j,k}$ are uniformly convergent in n. Therefore, the condition (4.3.6) of Theorem 4.3.6 leads to

$$\lim_{q \to \infty} \sum_k \frac{1}{q+1} \sum_{j=0}^{q} \triangle b_{n+j,k} = \lim_{q \to \infty} \frac{1}{q+1} \sum_{j=0}^{q} \triangle b_{n+j,0} = F - \lim b_{n0} = 1 \quad (4.4.7)$$

and the same condition with the condition (4.4.4) also leads to

$$\lim_{q \to \infty} \sum_k \frac{1}{q+1} \left| \sum_{j=0}^{q} \triangle b_{n+j,k} \right| = 0 \quad \text{uniformly in } n \quad (4.4.8)$$

which yields that

$$\lim_{q \to \infty} \left| \sum_k \frac{1}{q+1} \sum_{j=0}^{q} \triangle b_{n+j,k} \right| = 0 \quad \text{uniformly in } n.$$

This contradicts (4.4.7).

This step completes the proof. □

Theorem 4.4.8. [40, Theorem 3.1] *An infinite matrix $C = (c_{nk})$ transforms the space bs into the space fs if and only if*

$$\sup_{n \in \mathbb{N}_0} \sum_k \left| \sum_{j=0}^{n} \triangle c_{jk} \right| < \infty, \quad (4.4.9)$$

$$\lim_{k \to \infty} c_{nk} = 0 \quad \text{for each fixed } n \in \mathbb{N}_0, \quad (4.4.10)$$

$$\exists \gamma_k \in \mathbb{C} \quad \text{such that} \quad F - \lim \sum_{j=0}^{n} c_{jk} = \gamma_k \text{ for each fixed } k \in \mathbb{N}_0, \quad (4.4.11)$$

$$\lim_{q \to \infty} \sum_k \frac{1}{q+1} \left| \sum_{i=0}^{q} \sum_{j=0}^{n+i} \triangle(c_{jk} - \gamma_k) \right| = 0 \quad \text{uniformly in } n. \quad (4.4.12)$$

Proof. Let $C = (c_{nk}) \in (bs : fs)$ and $u = (u_k) \in bs$ with $s_n = \sum_{k=0}^{n} u_k$ for all $n \in \mathbb{N}_0$. The necessity of the conditions (4.4.10) and (4.4.11) can be established by the similar way used in the proof of Theorem 4.4.5. Now, consider the equality derived by the similar way of the relation (4.4.5)

$$\sum_{j=0}^{n} \sum_{k=0}^{m} c_{jk} u_k = \sum_{k=0}^{m-1} \left(\sum_{j=0}^{n} \triangle c_{jk} \right) s_k + \sum_{j=0}^{n} c_{jm} s_m \quad \text{for all } m, n \in \mathbb{N}_0. \quad (4.4.13)$$

Letting $m \to \infty$ in (4.4.13), we get

$$\sum_{j=0}^{n} \sum_{k} c_{jk} u_k = \sum_{k} \left(\sum_{j=0}^{n} \triangle c_{jk} \right) s_k \quad \text{for each} \quad n \in \mathbb{N}_0. \qquad (4.4.14)$$

Define the matrix $D = (d_{nk})$ by

$$d_{nk} = \sum_{j=0}^{n} \triangle c_{jk} \quad \text{for all} \quad k, n \in \mathbb{N}_0. \qquad (4.4.15)$$

Thus, it is seen by passing to F-limit in (4.4.14) that $D \in (\ell_\infty : f)$. Therefore, the conditions (4.3.1) and (4.4.2) are satisfied by the matrix D which are equivalent to the conditions (4.4.9) and (4.4.12), respectively.

Conversely, suppose that the matrix C satisfies the conditions (4.4.9)-(4.4.12) and $u = (u_k) \in bs$. Let us consider the matrix $D = (d_{nk})$ defined by (4.4.15). Then, it is immediate that "the conditions (4.3.1), (4.4.1) and (4.4.2) of Theorem 4.4.2 are satisfied by the matrix D if and only if the conditions (4.4.9), (4.4.11) and (4.4.12) are satisfied by the matrix C, respectively" holds. Hence, $D \in (\ell_\infty : f)$ and this gives by passing to F-limit in (4.4.14) that $Cu \in fs$. This means that C-transform of every element of the space bs is in the space fs, i.e., $C \in (bs : fs)$.

This completes the proof. $\qquad\qquad\qquad\qquad\qquad\qquad\qquad\qquad\square$

Corollary 4.4.9. *An infinite matrix C transforms the space bs into the space fs_0 of almost converging series to zero if and only if (4.4.9), (4.4.10) hold and (4.4.11), (4.4.12) also hold with $\gamma_k = 0$ for all $k \in \mathbb{N}_0$.*

Finally, we can give the Steinhaus type theorem without proof for the almost strongly regular and almost coercive series-to-series matrix transformations, since one can prove in the similar way used in the proof of Theorem 4.4.7.

Theorem 4.4.10. *The classes of almost strongly regular and almost coercive series-to-series matrix transformations are disjoint.*

4.5 Strongly Regular Matrices

Definition 4.5.1. (cf. [140]) *An infinite matrix $A = (a_{nk})$ is said to be strongly regular if it sums all almost convergent sequences and $A - \lim x_n = F$-$\lim x_k$ for all $x = (x_k) \in f$. In this case, we write $A = (a_{nk}) \in (f : c)_{reg}$.*

Theorem 4.5.2. *An infinite matrix $A = (a_{nk})$ is strongly regular if and only if the following conditions hold:*

(i) A *is regular.*

(ii) $\sum_k |a_{nk} - a_{n,k+1}| \to 0$, *as* $n \to \infty$.

Proof. Suppose that the conditions (i) and (ii) hold, and $x = (x_n)$ be an almost convergent sequence with the generalized limit l. We have to show that Ax is convergent to l. Fix $\epsilon > 0$. Then, from the definition of almost convergence, there exists a natural number p such that

$$\frac{1}{p+1} \sum_{j=n}^{n+p} x_j = l + \alpha_n, \quad \text{where } |\alpha_n| < \epsilon \text{ for all } n \in \mathbb{N}_0. \tag{4.5.1}$$

Multiplying by a_{mn} and adding, we have

$$\frac{1}{p+1} \sum_{n} a_{mn} \sum_{j=n}^{n+p} x_j = l A_m + \sum_{n} a_{mn} \alpha_n, \tag{4.5.2}$$

where $A_m = \sum_n a_{mn} \to 1$, as $m \to \infty$. Since a_{mn} tends to zero, as $m \to \infty$, we have on the other hand:

$$\frac{1}{p+1} \sum_{n} a_{mn} \sum_{j=n}^{n+p} x_j = o(1) + \sum_{n=p}^{\infty} x_n \frac{1}{p+1} (a_{m,n-p} + \cdots + a_{mn}) \tag{4.5.3}$$

$$= y_m + \sum_{n=p}^{\infty} x_n \left[\frac{1}{p+1} (a_{m,n-p} + \cdots + a_{mn}) - a_{mn} \right] + o(1),$$

where y_m is the A-transform of x, and the last term is infinitely small for $m \to \infty$ and the chosen p. Now, the absolute value of the sum on the right-hand side of (4.5.3) is not larger than

$$\frac{1}{p+1} \sum_{n=p}^{\infty} |(a_{m,n-p} + \cdots + a_{mn}) - (p+1)a_{mn}| \|x\|$$

$$\leq \frac{1}{p+1} \|x\| \sum_{\varrho=0}^{p} \sum_{n=p}^{\infty} |a_{m,n-\varrho} - a_{mn}|$$

$$\leq \frac{1}{p+1} \|x\| \sum_{\varrho=0}^{p} \varrho \sum_{n} |a_{m,n} - a_{m,n+1}|$$

$$\leq \frac{p}{2} \|x\| \sum_{n} |a_{m,n} - a_{m,n+1}|.$$

From (4.5.2) and (4.5.3), we have $y_m = l A_m + \sum_n a_{mn} \alpha_n + o(1)$. Now,

$$l A_m = l + o(1), \quad \left| \sum_{n} a_{mn} \alpha_n \right| \leq M\epsilon \text{ with } M = \sup_{m \in \mathbb{N}_0} \sum_{n} |a_{mn}|.$$

Thus, for sufficiently large m we certainly have $|y_m - l| \leq (M+1)\epsilon$. Therefore, $y_m \to l$, as $m \to \infty$.

Conversely, suppose that A is strongly regular. Then, the condition (i) follows immediately, since $(f:c)_{reg} \subset (c:c)_{reg}$. We now assume that (ii) does not hold. We shall construct a sequence (x_n) for which $F - \lim x_n = 0$ which is not A-summable. According to our hypotheses an $\epsilon > 0$ exists, such that for an infinity of m

$$\sum_n |a_{mn} - a_{m,n+1}| > 8\epsilon.$$

For every such m, we either have

$$\sum_l |a_{m,2l} - a_{m,2l+1}| > 4\epsilon \ \text{ or } \ \sum_l |a_{m,2l+1} - a_{m,2l+2}| > 4\epsilon.$$

By recurrence, we now construct three increasing sequences of natural numbers (m_k), (p_k) and (q_k); where $q_{-1} = 0 < p_1 < q_1 < p_2 \cdots$. We first choose m_1, p_1, q_1 such that

$$|a_{m_1,0}| < \frac{\epsilon}{2},$$

$$\sum_{l=0}^{\frac{q_1-p_1-1}{2}} |a_{m_1,p_1+2l} - a_{m_1,p_1+2l+1}| > 2\epsilon,$$

$$\sum_{n=q_1+1}^{\infty} |a_{m_1,n}| < \frac{\epsilon}{2}.$$

If the numbers m_ν, p_ν, q_ν with $\nu = 1, 2, \ldots, k-1$ are already known, m_k, p_k, q_k (where $q_{k-1} < p_k < q_k$ and one of the numbers p_k, q_k are even, the other are odd) are chosen such that

$$\sum_{n=0}^{q_{k-1}} |a_{m_k,n}| < \frac{\epsilon}{2},$$

$$\sum_{l=0}^{\frac{q_k-p_k-1}{2}} |a_{m_k,p_k+2l} - a_{m_k,p_k+2l+1}| > 2\epsilon,$$

$$\sum_{n=q_k+1}^{\infty} |a_{m_k,n}| < \frac{\epsilon}{2}.$$

We now define the sequence $x = (x_n)$ by

$$x_{p_k+2l} = (-1)^k \operatorname{sign}(a_{m_k,p_k+2l+1} - a_{m_k,p_k+2l+1}),$$
$$x_{p_k+2l+1} = -x_{p_k+2l},$$
$$x_n = 0 \text{ for } q_{k-1} < n < p_k \text{ for } k \in \mathbb{N}_1 \text{ and } l = 0, 1, \ldots, \frac{q_k - p_k - 1}{2}.$$

Under these conditions, we have

$$|y_{m_k}| = \left| \sum_n a_{m_k,n} x_n \right| > \sum_{l=0}^{(q_k - p_k - 1)/2} |a_{m_k,p_k+2l} - a_{m_k,p_k+2l+1}| - \frac{\epsilon}{2} - \frac{\epsilon}{2} > \epsilon$$

and sign $y_{m_k} = (-1)^k$. Hence, it follows that the sequence (y_m) diverges. It is further easy to see that x is almost convergent to 0.

This completes the proof. □

Duran [72] considered the class of almost strongly regular matrices.

Definition 4.5.3. *An infinite matrix A is said to be almost strongly regular if A transforms all almost convergent sequences into an almost convergent sequence leaving the $F - \lim$ invariant.*

Theorem 4.5.4. *An infinite matrix $A = (a_{nk})$ is almost strongly regular if and only if the following conditions hold:*

(i) *A is almost regular.*

(ii) *$\sum_k |t(n, k, p) - t(n, k + 1, p)| \to 0$, as $p \to \infty$, uniformly in n.*

Examples 4.5.5. *The following statements hold:*

(i) *The sequence-to-sequence Cesàro method C_r of the order $r > 0$ defined by (3.5.2) is almost strongly regular. Indeed, by putting $A_{-1}^{r-1} = 0$, one can see that*

$$\lim_{n \to \infty} \sum_k |a_{nk} - a_{n,k+1}| = \lim_{n \to \infty} \frac{1}{A_n^r} \sum_{\nu=0}^{n} |A_{n-\nu}^{r-1} - A_{n-\nu-1}^{r-1}|$$

$$= \lim_{n \to \infty} \frac{1}{A_n^r} \left(|1 - A_n^{r-1}| + 1 \right) = 0,$$

as the numbers A_n^{r-1} are monotonous. Thus, every almost convergent sequence is C_r summable to its F-limit; where $r > 0$.

(ii) *We examine the Euler method E_α with $\alpha > 0$ of summation which is given by the transformation*

$$y_n = \frac{1}{2^{\alpha n}} \sum_{\nu=0}^{n} \binom{n}{\nu} (2^\alpha - 1)^{n-\nu} x_\nu = \sum_{\nu=0}^{n} \binom{n}{\nu} \left(\frac{1}{2^\alpha} \right)^\nu \left(1 - \frac{1}{2^\alpha} \right)^{n-\nu} x_\nu.$$

We put $2^{-\alpha} = t$ and use the notation

$$p_{\nu n}(t) := \begin{cases} \binom{n}{\nu} t^\nu (1-t)^{n-\nu} & , \quad \nu = 0, 1, 2, \dots, n, \\ 0 & , \quad \nu > n \end{cases}$$

for all $\nu, n \in \mathbb{N}_0$. Then,

$$\sum_{\nu=0}^{n} |p_{\nu n}(t) - p_{\nu+1,n}(t)| = \frac{1}{1-t} \sum_{\nu=0}^{n} \frac{n+1}{\nu+1} \left| \frac{\nu+1}{n+1} - t \right| p_{\nu n}(t).$$

We split this sum into two parts, let $\sum_1$ be the sum for those ν for which $\left| \frac{\nu}{n} - t \right| < n^{-1/2}$ and let $\sum_2$ be the remainder. For the evaluation of the sums, we use the following known inequality, in which A signifies an absolute constant:

$$\sum_{\substack{0 \le \nu \le n \\ |\frac{\nu}{n} - t| \ge n^{-\frac{1}{2}}}} p_\nu(t) < \frac{A}{n^2}.$$

With the aid of this inequality, we obtain

$$\left| \sum_2 \right| \le \frac{A}{n^2} \max_{0 \le \nu \le n} \left\{ \frac{n+1}{\nu+1} \left| \frac{\nu+1}{n+1} - t \right| \right\} \le \frac{A(n+1)}{n^2}.$$

For the terms of the sum $\sum_1$ we have

$$\left| \frac{\nu+1}{n+1} - t \right| < \frac{1}{\sqrt{n}} + \frac{1}{n} < \frac{2}{\sqrt{n}}, \quad \frac{n+1}{\nu+1} < \frac{1}{t - \frac{2}{\sqrt{n}}} < \frac{2}{t},$$

and therefore

$$\left| \sum_1 \right| < \frac{2}{t\sqrt{n}} \sum_{\nu=0}^{n} p_\nu(t) = \frac{2}{t\sqrt{n}}.$$

Here, also the condition (ii) of Theorem 4.5.2 is fulfilled, i.e., every almost convergent sequence x is E_α-summable to $F - \lim x_k$ for $\alpha > 0$.

4.6   Applications to Approximation

Several mathematicians have worked on extending or generalizing Korovkin's theorems in many ways and to several settings, including function spaces, abstract Banach lattices, Banach algebras, Banach spaces and so on. This theory is very useful in real analysis, functional analysis, harmonic analysis, measure theory, probability theory, summability theory and partial differential equations. In this section, we prove Korovkin-type approximation theorems by applying the notion of almost convergence and show that these results are stronger than original ones (cf. [126, 162]).

Let $C[0, 2\pi]$ be the space of all continuous functions defined on $[0, 2\pi]$. We know that $C[0, 2\pi]$ is a normed linear space with the norm

$$\|f\|_\infty = \max_{x \in [0, 2\pi]} |f(x)|, \ f \in C[0, 2\pi].$$

We denote the space of all 2π-periodic functions f in $C[0, 2\pi]$ by $\overline{C}[0, 2\pi]$, which are the normed linear spaces with the maximum norm $\| \cdot \|_\infty$.

We write $L_n(f; x)$ for $L_n(f(s); x)$; and we say that L is a positive operator if $L(f; x) \geq 0$ for all $f(x) \geq 0$.

Now, we respectively give the classical Korovkin's first and second theorems, as follows (see [132]) :

Theorem 4.6.1. *Let* (T_n) *be a sequence of positive linear operators on* $C[0, 1]$. *Then,* $\|T_n(f, x) - f(x)\|_\infty \to 0$, *as* $n \to \infty$, *for all* $f \in C[0, 1]$ *if and only if* $\|T_n(f_i, x) - e^i(x)\|_\infty \to 0$, *as* $n \to \infty$, *for* $i = 0, 1, 2$, *where* $e^i(x) = x^i$ *with* $i = 0, 1, 2$.

Theorem 4.6.2. *Let* (T_n) *be a sequence of positive linear operators on* $\overline{C}[0, 2\pi]$. *Then,* $\|T_n(f, x) - f(x)\|_\infty \to 0$, *as* $n \to \infty$, *for all* $f \in C[0, 2\pi]$ *if and only if* $\|T_n(f_i, x) - f_i(x)\|_\infty \to 0$, *as* $n \to \infty$, *for* $i = 0, 1, 2$, *where* $f_0(x) = 1$, $f_1(x) = \cos x$ *and* $f_2(x) = \sin x$.

The following result is due to Mohiuddine [162]:

Theorem 4.6.3. *Let* $C_B[a, b]$ *be the space of all functions* f *continuous and bounded on* $[a, b]$ *and* (T_n) *be a sequence of positive linear operators from* $C_B[a, b]$ *into itself. Then, for any function* $f \in C_B[a, b]$

$$\lim_{p \to \infty} \left\| \frac{1}{p} \sum_{n=m}^{m+p-1} [T_n(f, x) - f(x)] \right\|_\infty = 0, \ uniformly \ in \ m$$

if and only if

$$\lim_{p \to \infty} \left\| \frac{1}{p} \sum_{n=m}^{m+p-1} [T_n(1, x) - 1] \right\|_\infty = 0, \ uniformly \ in \ m \qquad (4.6.1)$$

$$\lim_{p \to \infty} \left\| \frac{1}{p} \sum_{n=m}^{m+p-1} [T_n(t, x) - x] \right\|_\infty = 0, \ uniformly \ in \ m \qquad (4.6.2)$$

$$\lim_{p \to \infty} \left\| \frac{1}{p} \sum_{n=m}^{m+p-1} [T_n(t^2, x) - x^2] \right\|_\infty = 0, \ uniformly \ in \ m. \qquad (4.6.3)$$

Proof. Necessity follows obviously. Let the conditions (4.6.1)–(4.6.3) hold. We have $f \in C_B[a, b]$ and f is bounded on the real line. Hence, $|f(x)| \leq M$ for $-\infty < x < \infty$. Therefore,

$$|f(t) - f(x)| \leq 2M, \quad -\infty < t, x < \infty. \tag{4.6.4}$$

Also since $f \in C_B[a, b]$, we have that f is continuous on $[a, b]$, i.e.,

$$|f(t) - f(x)| < \epsilon, \quad \forall |t - x| < \delta. \tag{4.6.5}$$

Using (4.6.4), (4.6.5) and putting $\psi(t) = (t - x)^2$, we get

$$|f(t) - f(x)| < \epsilon + \frac{2M}{\delta^2} \psi(t), \quad \forall |t - x| < \delta.$$

This means

$$-\epsilon - \frac{2M}{\delta^2} \psi < f(t) - f(x) < \epsilon + \frac{2M}{\delta^2} \psi.$$

Now, since $T_n(f, x)$ is monotone and linear, by operating $T_n(1, x)$ to this inequality we have

$$T_n(1, x) \left(-\epsilon - \frac{2M}{\delta^2} \psi \right) < T_n(1, x)[f(t) - f(x)] < T_n(1, x) \left(\epsilon + \frac{2M}{\delta^2} \psi \right).$$

Note that x is fixed and so $f(x)$ is constant number. Therefore,

$$-\epsilon T_n(1, x) - \frac{2M}{\delta^2} T_n(\psi, x) < T_n(f, x) - f(x) T_n(1, x) < \epsilon T_n(1, x) + \frac{2M}{\delta^2} T_n(\psi, x). \tag{4.6.6}$$

But

$$\begin{aligned}
T_n(f, x) - f(x) &= T_n(f, x) - f(x) T_n(1, x) + f(x) T_n(1, x) - f(x) \tag{4.6.7} \\
&= [T_n(f, x) - f(x) T_n(1, x)] + f(x)[T_n(1, x) - 1].
\end{aligned}$$

Using (4.6.6) and (4.6.7), we have

$$T_n(f, x) - f(x) < \epsilon T_n(1, x) + \frac{2M}{\delta^2} T_n(\psi, x) + f(x)[T_n(1, x) - 1]. \tag{4.6.8}$$

Let us estimate $T_n(\psi, x)$

$$\begin{aligned}
T_n(\psi, x) &= T_n((t - x)^2, x) \\
&= T_n(t^2 - 2tx + x^2, x) \\
&= T_n(t^2, x) + 2x T_n(t, x) + x^2 T_n(1, x) \\
&= [T_n(t^2, x) - x^2] - 2x[T_n(t, x) - x] + x^2[T_n(1, x) - 1].
\end{aligned}$$

Using (4.6.8), we obtain

$$\begin{aligned}
T_n(f, x) - f(x) \ < \ & \epsilon T_n(1, x) + \frac{2M}{\delta^2} \{[T_n(t^2, x) - x^2] + 2x[T_n(t, x) - x] \\
& + x^2[T_n(1, x) - 1]\} + f(x)[T_n(1, x) - 1] \\
= \ & \epsilon[T_n(1, x) - 1] + \epsilon + \frac{2M}{\delta^2} \{[T_n(t^2, x) - x^2] + 2x[T_n(t, x) - x] \\
& + x^2[T_n(1, x) - 1]\} + f(x)[T_n(1, x) - 1].
\end{aligned}$$

Since ϵ is arbitrary, we can write

$$T_n(f,x) - f(x) \leq \epsilon[T_n(1,x) - 1] + \frac{2M}{\delta^2}\{[T_n(t^2,x) - x^2] + 2x[T_n(t,x) - x]$$
$$+ x^2[T_n(1,x) - 1]\} + f(x)[T_n(1,x) - 1].$$

Therefore,

$$\left\|\frac{1}{p}\sum_{n=m}^{m+p-1}[T_n(f,x) - f(x)]\right\|_\infty \leq \left(\epsilon + \frac{2Mb^2}{\delta^2} + M\right)\left\|\frac{1}{p}\sum_{n=m}^{m+p-1}[T_n(1,x) - 1]\right\|_\infty$$

$$+ \frac{4Mb}{\delta^2}\left\|\frac{1}{p}\sum_{n=m}^{m+p-1}[T_n(t,x) - x]\right\|_\infty + \frac{2M}{\delta^2}\left\|\frac{1}{p}\sum_{n=m}^{m+p-1}[T_n(t^2,x) - x^2]\right\|_\infty.$$

Letting $p \to \infty$ and using (4.6.1)–(4.6.3), we get

$$\lim_{p\to\infty}\left\|\frac{1}{p}\sum_{n=m}^{m+p-1}[T_n(f,x) - f(x)]\right\|_\infty = 0, \text{ uniformly in } m.$$

$\square$

In the following, we give an example of a sequence of positive linear operators satisfying the conditions of Theorem 4.6.3 but does not satisfy the conditions of Theorem 4.6.1.

Remark 4.6.4. *Consider the sequence of classical Bernstein polynomials*

$$B_n(f,x) := \sum_{k=0}^n f\left(\frac{k}{n}\right)\binom{n}{k}x^k(1-x)^{n-k}; \ 0 \leq x \leq 1.$$

Let the sequence (P_n) be defined by $P_n : C[0,1] \to C[0,1]$ with $P_n(f,x) = (1+z_n)B_n(f,x)$, where z_n is defined by

$$z_n := \begin{cases} 1 & , \quad n \text{ is odd,} \\ 0 & , \quad n \text{ is even} \end{cases} \tag{4.6.9}$$

for all $n \in \mathbb{N}_0$. Then, $B_n(1,x) = 1$, $B_n(t,x) = x$ and $B_n(t^2,x) = x^2 + (x - x^2)/n$, and the sequence (P_n) satisfies the conditions. Also, the sequence (P_n) is almost convergent to 0. On the other hand, we get $P_n(f,0) = (1+z_n)f(0)$, since $B_n(f,0) = f(0)$, and hence,

$$\|P_n(f,x) - f(x)\|_\infty \geq |P_n(f,0) - f(0)| = z_n|f(0)|.$$

We see that (P_n) does not satisfy the classical Korovkin's theorem, since $\lim_{n\to\infty} z_n$ does not exist.

Our next result is an analogue of Theorem 4.6.2.

Theorem 4.6.5. *Let (T_k) be a sequence of positive linear operators defined on $\overline{C}[0, 2\pi]$. The sequence $\{T_k(f)(x)\}$ is almost convergent to $f(x)$ uniformly on $[0, 2\pi]$ for each $f \in C[0, 2\pi]$ if and only if $\{T_k(e^0)(x)\}$, $\{T_k(\cos x)\}$, $\{T_k(\sin x)\}$ are almost convergent to $e^0(x)$, $\cos x$ and $\sin x$, respectively, uniformly on $[0, 2\pi]$, that is,*

$$F\text{-}\lim_{k\to\infty}\left\|T_k(f; x) - f(x)\right\|_{2\pi} = 0 \tag{4.6.10}$$

if and only if

$$F\text{-}\lim_{k\to\infty} \|T_k(1; x) - 1\|_{2\pi} = 0, \tag{4.6.11}$$

$$F\text{-}\lim_{k\to\infty} \|T_k(\cos t; x) - \cos x\|_{2\pi} = 0, \tag{4.6.12}$$

$$F\text{-}\lim_{k\to\infty} \|T_k(\sin t; x) - \sin x\|_{2\pi} = 0. \tag{4.6.13}$$

Proof. Since each f_1, f_2, f_3 belongs to $\overline{C}[0, 2\pi]$, the conditions (4.6.11)–(4.6.13) follow immediately from (4.6.10). Let the conditions (4.6.11)–(4.6.13) hold and $f \in C[0, 2\pi]$. Fix $x \in [0, 2\pi]$. By the continuity of f at x, it follows that for given $\varepsilon > 0$ there is a number $\delta > 0$ such that $|f(t) - f(x)| < \varepsilon$ for all t whenever $|t - x| < \delta$. Since f is bounded, it follows that $|f(t) - f(x)| \le \|f\|_{2\pi}$ for all $t \in [0, 2\pi]$. For all $t \in (x - \delta, 2\pi + x - \delta]$, it is well-known that

$$|f(t) - f(x)| < \varepsilon + \frac{2\|f\|_{2\pi}}{\sin^2 \dfrac{\delta}{2}}\psi(t), \tag{4.6.14}$$

where $\psi(t) = \sin^2[(t - x)/2]$. Since the function $f \in C[0, 2\pi]$ is 2π-periodic, the inequality (4.6.14) holds for $t \in [0, 2\pi]$.

Now, operating $T_k(1; x)$ to this inequality, we obtain

$$|T_k(f; x) - f(x)|$$
$$\le (\varepsilon + |f(x)|)|T_k(1; x) - 1| + \varepsilon + \frac{\|f\|_{2\pi}}{\sin^2 \dfrac{\delta}{2}}\{|T_k(1; x) - 1|$$
$$+ |\cos x||T_k(\cos t; x) - \cos x| + |\sin x||T_k(\sin t; x) - \sin x|\}$$
$$\le \varepsilon + \left[\varepsilon + |f(x)| + \frac{\|f\|_{2\pi}}{\sin^2 \dfrac{\delta}{2}}\right]$$
$$\times \{|T_k(1; x) - 1| + |T_k(\cos t; x) - \cos x| + |T_k(\sin t; x) - \sin x|\}$$

Therefore, we get by taking supremum over $[0, 2\pi]$ that

$$\|T_k(f; x) - f(x)\|_{2\pi} \le \varepsilon + K\big[\|T_k(1; x) - 1\|_{2\pi} + \|T_k(\cos t; x) - \cos x\|_{2\pi}$$
$$+ \|T_k(\sin t; x) - \sin x\|_{2\pi}\big], \tag{4.6.15}$$

where $K = \sup_{x\in[0,2\pi]}\left\{\varepsilon + \|f\|_{2\pi} + \big[\|f\|_{2\pi}/\sin^2(\delta/2)\big]\right\}$.

Now, replacing $T_k(\cdot, x)$ by $\sum_{k=n}^{n+m} T_k(\cdot, x)/(m+1)$ in (4.6.15) on both sides and then taking the limit as $m \to \infty$ uniformly in n, therefore, we get by using the conditions (4.6.11)–(4.6.13) that

$$\lim_{m \to \infty} \left\| \frac{1}{m+1} \sum_{k=n}^{n+m} T_k(f, x) - f(x) \right\|_{2\pi} = 0 \text{ uniformly in } n,$$

i.e., the condition (4.6.10) is satisfied.

This completes the proof of the theorem. $\qquad\square$

In the following, we construct an example of a sequence of positive linear operators satisfying the conditions of Theorem 4.6.5 not satisfying the conditions of Theorem 4.6.2.

Example 4.6.6. *For any $n \in \mathbb{N}_0$, denote the nth partial sum of the Fourier series of f by $S_n(f)$, i.e.,*

$$S_n(f)(x) = \frac{1}{2}a_0(f) + \sum_{k=1}^{n} a_k(f) \cos kx + b_k(f) \sin kx.$$

For any $n \in \mathbb{N}_0$, write

$$F_n(f) = \frac{1}{n+1} \sum_{k=0}^{n} S_k(f).$$

A straightforward calculation gives for every $t \in [0, 2\pi]$ that

$$
\begin{aligned}
F_n(f; x) &= \frac{1}{2\pi} \int_{-\pi}^{\pi} f(t) \frac{1}{n+1} \sum_{k=0}^{n} \frac{\sin \dfrac{(2k+1)(x-t)}{2}}{\sin \dfrac{x-t}{2}} dt \\
&= \frac{1}{2\pi} \int_{-\pi}^{\pi} f(t) \frac{1}{n+1} \sum_{k=0}^{n} \frac{\sin^2 \dfrac{(n+1)(x-t)}{2}}{\sin^2 \dfrac{x-t}{2}} dt \\
&= \frac{1}{2\pi} \int_{-\pi}^{\pi} f(t) \varphi_n(x-t) dt,
\end{aligned}
$$

where

$$\varphi_n(x) := \begin{cases} \dfrac{\sin^2[(n+1)(x-t)/2]}{\sin^2[(x-t)/2]} & , \quad x \text{ is not a multiple of } 2\pi, \\[2mm] n+1 & , \quad x \text{ is a multiple of } 2\pi. \end{cases}$$

The sequence $(\varphi_n)_{n \in \mathbb{N}_0}$ is a positive kernel called the Fejér kernel, and the corresponding operators F_n, $n \geq 1$ are called the Fejér convolution operators.

Note that Theorem 4.6.2 is satisfied for the sequence (F_n). In fact, we have $F_n(f) \to f$, as $n \to \infty$, for every $f \in \overline{C}[0, 2\pi]$.
Let $L_n : \overline{C}[0, 2\pi] \to \overline{C}[0, 2\pi]$ be defined by

$$L_n(f; x) = (1 + z_n)F_n(f; x), \qquad (4.6.16)$$

where the sequence $z = (z_n)$ is defined as in (4.6.9). Now,

$$L_n(1; x) = 1,$$
$$L_n(\cos t; x) = \frac{n}{n+1} \cos x,$$
$$L_n(\sin t; x) = \frac{n}{n+1} \sin x.$$

So, we have

$$F\text{-} \lim_{n \to \infty} \|L_n(1; x) - 1\|_{2\pi} = 0,$$
$$F\text{-} \lim_{n \to \infty} \|L_n(\cos t; x) - \cos x\|_{2\pi} = 0,$$
$$F\text{-} \lim_{n \to \infty} \|L_n(\sin t; x) - \sin x\|_{2\pi} = 0,$$

that is, the sequence (L_n) satisfies the conditions (4.6.11)–(4.6.13). Therefore, one can see by Theorem 4.6.5 that

$$F\text{-} \lim_{n \to \infty} \|L_n(f) - f\|_{2\pi} = 0,$$

i.e., Theorem 4.6.5 holds. But Theorem 4.6.2 does not hold for the operator L_n defined by (4.6.16), since the sequence (L_n) is not convergent.

Chapter 5

Spectrum of Some Triangle Matrices on Some Sequence Spaces

Keywords. *Spectrum of an operator, fine spectrum of an operator, double sequential band matrix, spectral mapping theorem, Goldberg's classification, difference operator, subdivisions of spectrum, point spectrum, continuous spectrum, residual spectrum, approximate point spectrum, defect spectrum and compression spectrum.*

5.1 Preliminaries, Background and Notations

Let X and Y be Banach spaces, and $T : X \to Y$ also be a bounded linear operator. By $R(T)$, we denote *the range of T*, i.e., $R(T) = \{y \in Y : y = Tx,\ x \in X\}$. The *kernel* or *null space* $Ker(T)$ of T consists of all $x \in X$ such that $Tx = \theta$. By $B(X)$, we also denote the *set of all bounded linear operators* on X into itself. If X is any Banach space and $T \in B(X)$, then the *adjoint* T^* of T is a bounded linear operator on the dual X^* of X defined by $(T^*f)(x) = f(Tx)$ for all $f \in X^*$ and $x \in X$.

Let $X \neq \{\theta\}$ be a non-trivial complex normed space and $T : D(T) \to X$ a linear operator defined on subspace $D(T) \subseteq X$, where $D(T)$ denotes the domain of T. We do not assume that $D(T)$ is dense in X, or that T has closed graph $\{(x, Tx) : x \in D(T)\} \subseteq X \times X$. We mean by the expression "*T is invertible*" that there exists a bounded linear operator $S : R(T) \to X$ for which $ST = I$ on $D(T)$ and $\overline{R(T)} = X$; such that $S = T^{-1}$ is necessarily uniquely determined, and linear; the boundedness of S means that T must be *bounded below*, in the sense that there is $k > 0$ for which $\|Tx\| \geq k\|x\|$ for all $x \in D(T)$. Associated with each complex number α is the perturbed operator $T_\alpha = T - \alpha I$, defined on the same domain $D(T)$ as T. The *spectrum* $\sigma(T, X)$ consists of those $\alpha \in \mathbb{C}$ for which T_α is not invertible, and the *resolvent* is the mapping from the complement $\sigma(T, X)$ of the spectrum into the algebra of bounded linear operators on X defined by $\alpha \mapsto T_\alpha^{-1}$.

5.2   Subdivisions of Spectrum

In this section, we give the definitions of the parts point spectrum, continuous spectrum, residual spectrum, approximate point spectrum, defect spectrum and compression spectrum. There are many different ways to subdivide the spectrum of a bounded linear operator. Some of them are motivated by applications to physics, in particular, quantum mechanics.

5.2.1   The Point Spectrum, Continuous Spectrum and Residual Spectrum

The name resolvent is appropriate, since T_α^{-1} helps to solve the equation $T_\alpha x = y$. Thus, $x = T_\alpha^{-1} y$, provided T_α^{-1} exists. More important, the investigation of properties of T_α^{-1} will be basic for an understanding of the operator T itself. Naturally, many properties of T_α and T_α^{-1} depend on α, and spectral theory is concerned with these properties. For instance, we shall be interested in the set of all α's in the complex plane such that T_α^{-1} exists. Boundedness of T_α^{-1} is another property that will be essential. We shall also ask for what α's the domain of T_α^{-1} is dense in X, to name just a few aspects.

Definition 5.2.1. [134, p. 371] *Let $X \neq \{\theta\}$ be a complex normed space and $T : D(T) \to X$ be a linear operator with $D(T) \subseteq X$. A regular value α of T is a complex number such that*

(R1) T_α^{-1} exists.

(R2) T_α^{-1} is bounded.

(R3) T_α^{-1} is defined on a set which is dense in X.

For our investigation of T, T_α and T_α^{-1}, we need some basic concepts in spectral theory which are given, as follows (see [134, pp. 370-371]):

The *resolvent set* $\rho(T, X)$ *of a linear operator* T *on a space* X is the set of all regular values α of T, i.e., $\rho(T, X) = \mathbb{K} \setminus \sigma(T, X)$, where $\mathbb{K}$ denotes the scalar field of $\mathbb{R}$ or $\mathbb{C}$. A number $\alpha \in \mathbb{K}$ is called an *eigenvalue of a linear operator* T *on a space* X if

$$\sigma_e(T, X) := \{x \in X : Tx = \alpha x\} = Ker(\alpha I - T) \neq \{\theta\}.$$

Here and after, we denote the zero vector in X by θ. $\sigma_e(T, X)$ is called the *eigenspace of* T *corresponding to the eigenvalue* α. The elements of $\sigma_e(T, X) \setminus \{\theta\}$ are called *eigenvectors of* T *corresponding to the eigenvalue* α. Furthermore, the spectrum $\sigma(T, X)$ of a linear operator T on a space X is partitioned into the following three disjoint sets:

The *point (discrete) spectrum* $\sigma_p(T, X)$ is the set such that T_α^{-1} does not exist.

The *continuous spectrum* $\sigma_c(T, X)$ of a linear operator T on a space X is the set such that T_α^{-1} exists and is unbounded, and the domain of T_α^{-1} is dense in X.

The *residual spectrum* $\sigma_r(T, X)$ of a linear operator T on a space X is the set such that T_α^{-1} exists (and may be bounded or not) but the domain of T_α^{-1} is not dense in X.

Therefore, these three parts of spectrum form a disjoint subdivisions such that

$$\sigma(T, X) = \sigma_p(T, X) \cup \sigma_c(T, X) \cup \sigma_r(T, X). \tag{5.2.1}$$

To avoid trivial misunderstandings, let us say that some of the sets defined above, may be empty. This is an existence problem which we shall have to discuss. Indeed, it is well-known that $\sigma_c(T, X) = \sigma_r(T, X) = \emptyset$ and the spectrum $\sigma(T, X)$ consists of only the set $\sigma_p(T, X)$ in the finite dimensional case. However, as was seen from the following example, this is not valid for infinite dimensional Banach spaces.

Example 5.2.2. *Consider the operator T defined on the Hilbert space ℓ_2 of absolutely square-summable complex sequences by*

$$T \; : \qquad \ell_2 \qquad \longrightarrow \qquad \ell_2$$
$$x = (x_k)_{k \in \mathbb{N}_1} \quad \longmapsto \quad Tx = \left(\tfrac{1}{k} x_k\right)_{k \in \mathbb{N}_1}$$

It is immediate that T is injective, but is not surjective. Consequently, 0 is not an eigenvalue of T while it belongs to $\sigma(T, \ell_2)$.

The *approximate spectrum* $\sigma_a(T, X)$ of T is the set of all $\alpha \in \mathbb{K}$ such that $T - \alpha I$ is not bounded below. Thus $\alpha \in \sigma_a(T, X)$ if and only if there is a sequence $x = (x_n)$ in X such that $\|x_n\| = 1$ for each n and $\|Tx_n - \alpha x_n\| \to 0$ as $n \to \infty$. Thus α is called an *approximate eigenvalue* of T. If $\alpha \in \sigma_e(T, X)$ and x is corresponding eigenvector, then letting $x_n = x/\|x\|$ for all n, we conclude that $\alpha \in \sigma_a(T, X)$. Hence, we have the inclusion relations

$$\sigma_e(T, X) \subset \sigma_a(T, X) \subset \sigma(T, X).$$

An operator T on a linear space X is said to be of *finite rank* if the range $R(T)$ of T is finite dimensional.

Prior to giving three more examples we should quote the following lemmas without proof.

Lemma 5.2.3. [139, 12.2 Theorem] *Let X be a normed space and $T \in B(X)$ be of finite rank. Then $\sigma_e(T, X) = \sigma_a(T, X) = \sigma(T, X)$.*

Lemma 5.2.4. [139, 12.3 Lemma] *Let X be a Banach space, $A \in B(X)$ and $\|A^p\| < 1$ for some positive integer p. Then the bounded operator $I - A$ is invertible. Also,*

$$(I - A)^{-1} = \sum_{n=0}^{\infty} A^n \quad and \quad \|(I - A)^{-1}\| \leq \frac{1 + \|A\| < + \cdots + \|A^{p-1}\|}{1 - \|A^p\|}.$$

Lemma 5.2.5. [139, 12.6 Theorem] *Let X be a Banach space over $\mathbb{K}$ and $T \in B(X)$. Then the following statements hold:*

(a) (Neumann expansion) Let $\alpha \in \mathbb{K}$ such that $|\alpha|^p < \|T^p\|$ for some positive integer p. Then $\alpha \notin \sigma(T, X)$ and $(T - \alpha I)^{-1} = -\sum_n \frac{1}{\alpha^n} T^n$. Consequently, for every $\alpha \in \sigma(T, X)$ we have $|\alpha| \le \inf_{n \in \mathbb{N}_1} \|T^n\|^{1/n} \le \|T\|$.

(b) $\sigma(T, X)$ is a compact subset of $\mathbb{K}$.

Example 5.2.6. [33, Problem 2, pp. 350–351]
    Consider the linear operator $T : \mathbb{R}^n \to \mathbb{R}^n$ represented by the $n \times n$ matrix $\mathbf{A} = (a_{ij})$ defined as $a_{ij} := \begin{cases} a &, \quad i = j \\ b &, \quad i \ne j \end{cases}$ for all $i, j \in \{1, 2, \dots, n\}$ with $a, b \in \mathbb{R}$. Find the eigenvalues of the linear operator T and obtain the eigenspaces of T corresponding to each eigenvalue.

Solution. The characteristic polynomial $p(\alpha)$ of the matrix $\mathbf{A}$ is

$$p(\alpha) = |\alpha \mathbf{I_n} - \mathbf{A}| = \begin{vmatrix} \alpha - a & -b & -b & \cdots & -b \\ -b & \alpha - a & -b & \cdots & -b \\ -b & -b & \alpha - a & \cdots & -b \\ \vdots & \vdots & \vdots & \ddots & \vdots \\ -b & -b & -b & \cdots & \alpha - a \end{vmatrix}.$$

Since T is a finite rank operator on the linear space $\mathbb{R}^n$, by Lemma 5.2.3 the problem of finding non-zero eigenvalues of T can be reduced to a matrix eigenvalue problem.

By adding the $2^{nd}, 3^{rd}, \dots, n^{th}$ rows to the first row we have

$$|\alpha \mathbf{I_n} - \mathbf{A}| = \begin{vmatrix} \alpha - a & -b & -b & \cdots & -b \\ -b & \alpha - a & -b & \cdots & -b \\ -b & -b & \alpha - a & \cdots & -b \\ \vdots & \vdots & \vdots & \ddots & \vdots \\ -b & -b & -b & \cdots & \alpha - a \end{vmatrix}$$

$$= \begin{vmatrix} \alpha - a - (n-1)b & \alpha - a - (n-1)b & \cdots & \alpha - a - (n-1)b \\ -b & \alpha - a & \cdots & -b \\ -b & -b & \cdots & -b \\ \vdots & \vdots & \ddots & \vdots \\ -b & -b & \cdots & \alpha - a \end{vmatrix}$$

$$= [\alpha - a - (n-1)b] \begin{vmatrix} 1 & 1 & 1 & \cdots & 1 \\ -b & \alpha - a & -b & \cdots & -b \\ -b & -b & \alpha - a & \cdots & -b \\ \vdots & \vdots & \vdots & \ddots & \vdots \\ -b & -b & -b & \cdots & \alpha - a \end{vmatrix}$$

$$= [\alpha - a - (n-1)b] D_1.$$

By adding -1 times of the first column to the $2^{nd}, 3^{rd}, \ldots, n^{th}$ columns we see that

$$
D_1 = \begin{vmatrix}
1 & 1 & 1 & \cdots & 1 \\
-b & \alpha - a & -b & \cdots & -b \\
-b & -b & \alpha - a & \cdots & -b \\
\vdots & \vdots & \vdots & \ddots & \vdots \\
-b & -b & -b & \cdots & \alpha - a
\end{vmatrix}
$$

$$
= \begin{vmatrix}
1 & 0 & 0 & \cdots & 0 \\
-b & \alpha - a + b & 0 & \cdots & 0 \\
-b & 0 & \alpha - a + b & \cdots & 0 \\
\vdots & \vdots & \vdots & \ddots & \vdots \\
-b & 0 & 0 & \cdots & \alpha - a + b
\end{vmatrix}
$$

$$
= (\alpha - a + b)^{n-1}.
$$

Therefore, we derive the characteristic equation of the matrix $\mathbf{A}$ that

$$
|\alpha \mathbf{I_n} - \mathbf{A}| = [\alpha - a - (n-1)b](\alpha - a + b)^{n-1} = 0,
$$

which gives the characteristic roots of the matrix $\mathbf{A}$ as

$$
\alpha_1 = a + (n-1)b \quad \text{and} \quad \alpha_2 = \alpha_3 = \cdots = \alpha_n = a - b.
$$

By solving the homogen system of the linear equations

$$
(\alpha_1 \mathbf{I_n} - \mathbf{A})\mathbf{x} = \theta \quad \text{and} \quad (\alpha_2 \mathbf{I_n} - \mathbf{A})\mathbf{y} = \theta
$$

we obtain the eigenvectors

$$
\mathbf{x} = [c, c, c, \ldots, c]' \quad \text{and} \quad \mathbf{y} = [-(c + d + \cdots + t), c, d, \ldots, t]',
$$

where $c, d, \ldots, t \in \mathbb{R}$. Therefore, the eigenspaces corresponding to the characteristic roots α_1 and α_2 are respectively obtained as

$$
\{[c, c, \ldots, c]' : c \in \mathbb{R}\} \quad \text{and} \quad \{[-(c + d + \cdots + t), c, d, \ldots, t]' : c, d, \ldots, t \in \mathbb{R}\}.
$$

Thus we have

$$
\begin{aligned}
\sigma(T, \mathbb{R}^n) &= \sigma_p(T, \mathbb{R}^n) = \sigma_e(T, \mathbb{R}^n) = \sigma_a(T, \mathbb{R}^n) \\
&= \{\alpha_1 = a + (n-1)b, \alpha_2 = a - b\}.
\end{aligned}
$$

$\square$

Example 5.2.7. [139, 12.7 Examples (a), p. 203] *Let $X = \mathbb{K}^n$ with a given norm, and $\mathbf{M}$ be an $n \times n$ matrix with scalar entries. Then $\mathbf{M}$ defines a continuous operator on X. For $\alpha \in \mathbb{K}$, the matrix $\mathbf{M} - \alpha \mathbf{I}$ is invertible if and only if $\det(\mathbf{M} - \alpha \mathbf{I}) \neq 0$. Thus $\alpha \in \sigma(\mathbf{M}, X)$ if and only if α is a root of the* *characteristic polynomial $p(\alpha) = det(\mathbf{M} - \alpha\mathbf{I})$. By Lemma 5.2.3, a spectral value of $\mathbf{M}$ is just an eigenvalue of $\mathbf{M}$. Since the characteristic polynomial of $\mathbf{M}$ is of degree n, there are at most n distinct eigenvalues of $\mathbf{M}$. On the other hand, if $\mathbb{K} = \mathbb{R}$, then $\mathbf{M}$ may have no eigenvalues, and so $\sigma(\mathbf{M}, X)$ may be empty. A simple example is provided by the matrix $\mathbf{M} = \begin{bmatrix} 0 & 1 \\ -1 & 0 \end{bmatrix}$, whose characteristic polynomial is $p(\alpha) = \alpha^2 + 1$.*

If $\mathbf{M}$ is a triangle matrix and $k_1, k_2, \ldots, k_n$ are the diagonal entries then we have $det(\mathbf{M} - \alpha\mathbf{I}) = (\alpha - k_1)(\alpha - k_2)\cdots(\alpha - k_n)$, and hence $\sigma(\mathbf{M}, X) = \sigma_e(\mathbf{M}, X) = \{k_1, k_2, \ldots, k_n\}$. If $\mathbf{M}$ is not a triangle matrix, then the problem of finding its eigenvalues poses great difficulties. Alogarithms have been developed to reduce $\mathbf{M}$ to an approximately triangle matrix by similarity transformations. The most notable among these is the QR alogaritm. It enables us to obtain approximations of all eigenvalues of $\mathbf{M}$.

If α is an eigenvalue of $\mathbf{M}$, then $|\alpha| \leq \|\mathbf{M}\|$, where $\|\cdot\|$ is the operator norm on $B(\mathbb{K}^n)$ induced by the given norm on $\mathbb{K}^n$. Various choices of these norms yield upper bounds for the eigenspectrum of $\mathbf{M}$. Let $\mathbf{M} = [k_{ij}]$; $i, j = 1, 2, \ldots, n$, α_1 (resp. α_∞) denote the maximum of the column sums (resp. the row sums) of the matrix $[|k_{ij}|]$ and $\beta_2 = \left(\sum_{i=1}^n \sum_{j=1}^n |k_{ij}|^2\right)^{1/2}$. Then $|\alpha| \leq \min\{\alpha_1, \alpha_\infty, \beta_2\}$ for every $\alpha \in \sigma_e(\mathbf{M}, X)$. The fact "(Gershgorin's theorem) If $\mathbf{M}$ is an $n \times n$ matrix, then $\alpha \in \sigma_e(\mathbf{M})$ is contained in $\cup_{j=1}^n E_j$ as well as in $\cup_{i=1}^n F_i$, where $E_j = \{\alpha \in \mathbb{K} : |\alpha - k_{ij}| \leq \sum_{i \neq j} |k_{ij}|\}$ and $F_i = \{\alpha \in \mathbb{K} : |\alpha - k_{ii}| \leq \sum_{j \neq i} |k_{ij}|\}$." gives a well-known inclusion theorem for eigenvalues of $\mathbf{M}$.

Example 5.2.8. [139, 12.7 Examples (b), pp. 205-207] *Let $X = \ell_p$ with the norm $\|\cdot\|_p$, $1 \leq p \leq \infty$. Define the operators T_1 and T_2 on X by*

$$T_1 : \qquad \ell_p \qquad \longrightarrow \qquad \ell_p$$
$$x = (x_1, x_2, x_3, \ldots) \quad \longmapsto \quad T_1 x = \left(x_1, \tfrac{x_2}{2}, \tfrac{x_3}{3}, \ldots\right),$$

$$T_2 : \qquad \ell_p \qquad \longrightarrow \qquad \ell_p$$
$$x = (x_1, x_2, x_3, \ldots) \quad \longmapsto \quad T_2 x = (0, x_1, x_2, \ldots).$$

Find the spectrum, approximate spectrum and eigenspace of the operators T_1 and T_2 on the space X.

Solution. We consider the operator T_1. It is immediate that $T_1 \in B(X)$ and $\|T_1\|_p = 1$. Since $T_1 e_n = e_n/n$, we see that $1/n$ is an eigenvalue of T_1 with e_n as a corresponding eigenvector, $n \in \mathbb{N}_1$. Since $T_1 x = \theta$ implies $x = \theta$, we note that 0 is not an eigenvalue of T_1. However, since $\|Te_n\|_p \to 0$ as $n \to \infty$ and $\|e_n\|_p = 1$ for each n, we see that T_1 is not bounded below, that is, 0 is an approximate eigenvalue of T_1.

Now consider $\alpha \in \mathbb{K}$ such that $\alpha \notin \{0, 1, 1/2, \ldots\}$. Then there is some $\delta > 0$ such that $|\alpha - \frac{1}{n}| \geq \delta$ for all $n \in \mathbb{N}_1$. For $y = (y_1, y_2, \ldots) \in X$, define

$$B_\alpha y = \left(\frac{y_1}{1 - \alpha}, \frac{y_2}{\frac{1}{2} - \alpha}, \ldots \right).$$

Since $\delta |(B_\alpha y)_n| \leq |\frac{1}{n} - \alpha| |(B_\alpha y)_n| = |y_n|$, for all $n \in \mathbb{N}_1$, $B_\alpha y \in X$ for all $y \in X$. It is easy to see that $B_\alpha \in B(X)$ and $(T_1 - \alpha I)B_\alpha = I = B_\alpha(T_1 - \alpha I)$. Hence $T_1 - \alpha I$ is invertible. Thus

$$\sigma_e(T_1, X) := \left\{ 1, \frac{1}{2}, \ldots \right\},$$

$$\sigma_a(T_1, X) := \left\{ 0, 1, \frac{1}{2}, \ldots \right\} = \sigma(T_1, X).$$

Now, we can consider the right shift operator T_2 on X. Then $\|T_2 x\|_p = \|x\|_p$ for all $x \in X$, so that $T_2 \in B(X)$ and $\|T_2\|_p = 1$. For $\alpha \in \mathbb{K}$, we have

$$(T_2 - \alpha I)x = (-\alpha x_1, x_1 - \alpha x_2, x_2 - \alpha x_3, \ldots).$$

By considering the cases $\alpha = 0$ and $\alpha \neq 0$ separately, we see that if $(T_2 - \alpha I)x = \theta$, then $x = \theta$. Hence no α in $\mathbb{K}$ is an eigenvalue of T_2, that is, $\sigma_e(T_2, X) = \emptyset$.

Now suppose that $(T_2 - \alpha I)x = e_1$ for some $x \in X$. Then

$$-\alpha x_1 = 1 \quad \text{and} \quad x_n - \alpha x_{n+1} = 0 \quad \text{for} \quad n \in \mathbb{N}_1,$$

so that $x_n = -1/\alpha^n$ for $n \in \mathbb{N}_1$. If $|\alpha| < 1$, then $x_n \to \infty$ as $n \to \infty$, showing that x cannot belong to X. Thus there is no $x \in X$ such that $(T_2 - \alpha I)x = e_1$ if $|\alpha| < 1$. Hence for every $\alpha \in \mathbb{K}$ with $|\alpha| < 1$, $T_2 - \alpha I$ is not surjective and $\alpha \in \sigma(T_2, X)$. On the other hand, $|\alpha| \leq \|T_2\|_p = 1$ for every $\alpha \in \sigma(T_2, X)$ by Lemma 5.2.5. Thus, $\{\alpha \in \mathbb{K} : |\alpha| < 1\} \subset \sigma(T_2, X) \subset \{\alpha \in \mathbb{K} : |\alpha| \leq 1\}$. Since $\sigma(T_2, X)$ is closed by Part (b) of Lemma 5.2.5, we conclude that

$$\sigma(T_2, X) = \{\alpha \in \mathbb{K} : |\alpha| \leq 1\}.$$

Let $|\alpha| > 1$. We calculate the inverse of $T_2 - \alpha I$. It can be easily seen that for all $n \in \mathbb{N}_1$,

$$T_2^n x = (0, \ldots, 0, x_1, x_2, \ldots), \quad x = (x_k)_{k \in \mathbb{N}_1} \in X,$$

where the first n entries are zero. By the Neumann expansion given Part (a) of Lemma 5.2.5,

$$(T_2 - \alpha I)^{-1} - = \sum_{n=0}^{\infty} \frac{1}{\alpha^{n+1}} T_2^n,$$

so that for every $y = (y_j) \in X$ and $j \in \mathbb{N}_1$, we have

$$\{(T_2 - \alpha I)^{-1} y\}_j = -\left(\frac{y_j}{\alpha} + \cdots + \frac{y_1}{\alpha^j} \right).$$

Finally, we shall show that $\sigma_a(T_2, X) = \{\alpha \in \mathbb{K} : |\alpha| = 1\}$.

Let $\alpha \in \mathbb{K}$ with $|\alpha| < 1$. For $x \in X$, we have

$$\|T_2 x - \alpha x\|_p \geq \|T_2 x\|_p - |\alpha| \|x\|_p = (1 - |\alpha|) \|x\|_p.$$

Thus $T_2 - \alpha I$ is bounded below, that is, $\alpha \notin \sigma_a(T_2, X)$.

Next, let $\alpha \in \mathbb{K}$ with $|\alpha| = 1$. If $1 \leq p < \infty$, let

$$x_n = n^{-1/p} \left(1, \overline{\alpha}, \ldots, (\overline{\alpha})^{n-1}, 0, 0, \ldots\right)$$

for $n \in \mathbb{N}_1$. Then, $\|x_n\|_p = 1$, but

$$\|T_2 x_n - \alpha x_n\|_p = \left\| n^{-1/p} \left(-\alpha, 0, \ldots, 0, (\overline{\alpha})^{n-1}, 0, 0, \ldots\right) \right\|_p = \left(\frac{2}{n}\right)^{1/p},$$

which tends to zero as $n \to \infty$. If $p = \infty$, let

$$x_n = n^{-1} \left(1, 2\overline{\alpha}, \ldots, n(\overline{\alpha})^{n-1}, (n-1)(\overline{\alpha})^n, \ldots, 2(\overline{\alpha})^{2n-3}, (\overline{\alpha})^{2n-2}, 0, 0, \ldots\right)$$

for $n \in \mathbb{N}_1$. Then again $\|x_n\|_\infty = 1$, but

$$\|T_2 x_n - \alpha x_n\|_\infty = \left\| n^{-1} \left(-\alpha, -1, -\overline{\alpha}, \ldots, -(\overline{\alpha})^{n-2}, (\overline{\alpha})^{n-1}, \ldots, (\overline{\alpha})^{2n-2}, 0, 0, \ldots\right) \right\|_\infty$$

which tends to zero as $n \to \infty$. Hence $\alpha \in \sigma_a(T_2, X)$.

Thus we have shown that

$$\begin{aligned}
\sigma_e(T_2, X) &= \varnothing, \\
\sigma_a(T_2, X) &:= \{\alpha \in \mathbb{K} : |\alpha| = 1\}, \\
\sigma(T_2, X) &:= \{\alpha \in \mathbb{K} : |\alpha| \leq 1\}.
\end{aligned}$$

$\square$

5.2.2   The Approximate Point Spectrum, Defect Spectrum and Compression Spectrum

In this subsection, following Appell et al. [21], we give the definitions of three more subdivisions of the spectrum called as the *approximate point spectrum*, *defect spectrum* and *compression spectrum*.

Given a bounded linear operator T in a Banach space X, we call a sequence (x_k) in X as a *Weyl sequence* for T if $\|x_k\| = 1$ and $\|T x_k\| \to 0$, as $k \to \infty$.

In what follows, we call the set

$$\sigma_{ap}(T, X) := \{\alpha \in \mathbb{C} : \text{there exists a Weyl sequence for } \alpha I - T\} \quad (5.2.2)$$

the *approximate point spectrum of a linear operator T on a space X*. Moreover, the subspectrum

$$\sigma_\delta(T, X) := \{\alpha \in \mathbb{C} : \alpha I - T \text{ is not surjective}\} \quad (5.2.3)$$

is called *defect spectrum of a linear operator T on a space X*.

The two subspectra given by (5.2.2) and (5.2.3) form (not necessarily disjoint) subdivisions of the spectrum $\sigma(T, X)$ such that $\sigma(T, X) = \sigma_{ap}(T, X) \cup \sigma_\delta(T, X)$. There is another subspectrum

$$\sigma_{co}(T, X) := \left\{ \alpha \in \mathbb{C} : \overline{R(\alpha I - T)} \neq X \right\}$$

which is often called *compression spectrum of a linear operator T on a space X*, in the literature. The compression spectrum gives rise to another (not necessarily disjoint) decomposition of the spectrum $\sigma(T, X)$ such that $\sigma(T, X) = \sigma_{ap}(T, X) \cup \sigma_{co}(T, X)$. Clearly, $\sigma_p(T, X) \subseteq \sigma_{ap}(T, X)$ and $\sigma_{co}(T, X) \subseteq \sigma_\delta(T, X)$. Moreover, comparing these subspectrum with those in (5.2.1), we note that

$$\sigma_r(T, X) = \sigma_{co}(T, X) \backslash \sigma_p(T, X),$$
$$\sigma_c(T, X) = \sigma(T, X) \backslash [\sigma_p(T, X) \cup \sigma_{co}(T, X)].$$

Sometimes it is useful to relate the spectrum of a bounded linear operator to that of its adjoint. Building on classical existence and uniqueness results for linear operator equations in Banach spaces and their adjoints are also useful.

Proposition 5.2.9. [21, Proposition 1.3, p. 28] *The following statements concerning the spectrum and subspectrum of an operator $T \in B(X)$ and its adjoint $T^* \in B(X^*)$ hold:*

(a) $\sigma(T^*, X^*) = \sigma(T, X)$.

(b) $\sigma_c(T^*, X^*) \subseteq \sigma_{ap}(T, X)$.

(c) $\sigma_{ap}(T^*, X^*) = \sigma_\delta(T, X)$.

(d) $\sigma_\delta(T^*, X^*) = \sigma_{ap}(T, X)$.

(e) $\sigma_p(T^*, X^*) = \sigma_{co}(T, X)$.

(f) $\sigma_{co}(T^*, X^*) \supseteq \sigma_p(T, X)$.

(g) $\sigma(T, X) = \sigma_{ap}(T, X) \cup \sigma_p(T^*, X^*) = \sigma_p(T, X) \cup \sigma_{ap}(T^*, X^*)$.

The relations (c)–(f) show that the approximate point spectrum is in a certain sense dual to defect spectrum, and the point spectrum dual to the compression spectrum. The equality (g) implies, in particular, that $\sigma(T, X) = \sigma_{ap}(T, X)$ if X is a Hilbert space and T is normal. Roughly speaking, this shows that normal (in particular, self-adjoint) operators on Hilbert spaces are most similar to matrices in finite dimensional spaces (see [21]).

5.2.3   Goldberg's Classification of Spectrum

From Goldberg [91, pp. 58-71], if X is a Banach space and $T \in B(X)$, then there are three possibilities for $R(T)$ and T^{-1}:

(I) $R(T) = X$.

(II) $R(T) \neq \overline{R(T)} = X$.

(III) $\overline{R(T)} \neq X$.

Additionally,

(1) T^{-1} exists and is continuous.

(2) T^{-1} exists but is discontinuous.

(3) T^{-1} does not exist.

Applying Golberg's classification to T_α, we have three possibilities for T_α and T_α^{-1};

(I) T_α is surjective.

(II) $R(T_\alpha) \neq \overline{R(T_\alpha)} = X$.

(III) $\overline{R(T_\alpha)} \neq X$.

Additionally,

(1) T_α is injective and T_α^{-1} is continuous.

(2) T_α is injective and T_α^{-1} is discontinuous.

(3) T_α is not injective.

If these possibilities are combined in all possible ways, nine different states are created (see Table 5.1). These are labelled by: I_1, I_2, I_3, II_1, II_2, II_3, III_1, III_2 and III_3. If α is a complex number such that $T_\alpha \in I_1$ or $T_\alpha \in II_1$ then α is in the resolvent set $\rho(T, X)$ of T. That is, $\sigma(T, X)$ can be divided into the subsets $I_2\sigma(T, X) = \emptyset$, $I_3\sigma(T, X)$, $II_2\sigma(T, X)$, $II_3\sigma(T, X)$, $III_1\sigma(T, X)$, $III_2\sigma(T, X)$, $III_3\sigma(T, X)$. For example, if $T_\alpha = \alpha I - T$ is in a given state, III_2 (say), then we write $\alpha \in III_2\sigma(T, X)$. The further classification gives rise to the *fine spectrum* of T. If an operator is in state II_2 for example, then $R(T) \neq \overline{R(T)} = X$ and T^{-1} exists but is discontinuous and we write $\alpha \in II_2\sigma(T, X)$, (see [91]).

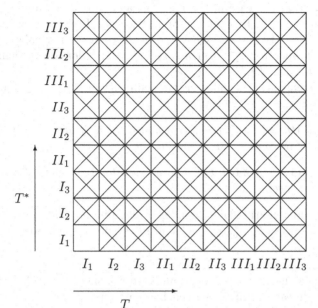

$$\xrightarrow{\hspace{3cm}}$$

T

Table 5.1: State diagram for $B(X)$ and $B(X^*)$ for a non-reflective Banach space X

By the definitions given above, we can illustrate the subdivisions (5.2.1) in the Table 5.2.

		1	2	3
		T_α^{-1} exists and is bounded	T_α^{-1} exists and is unbounded	T_α^{-1} does not exist
I	$R(\alpha I - T) = X$	$\alpha \in \rho(T, X)$	$-$	$\alpha \in \sigma_p(T, X)$ $\alpha \in \sigma_{ap}(T, X)$
II	$\overline{R(\alpha I - T)} = X$	$\alpha \in \rho(T, X)$	$\alpha \in \sigma_c(T, X)$ $\alpha \in \sigma_{ap}(T, X)$ $\alpha \in \sigma_\delta(T, X)$	$\alpha \in \sigma_p(T, X)$ $\alpha \in \sigma_{ap}(T, X)$ $\alpha \in \sigma_\delta(T, X)$
III	$\overline{R(\alpha I - T)} \neq X$	$\alpha \in \sigma_r(T, X)$ $\alpha \in \sigma_\delta(T, X)$ $\alpha \in \sigma_{co}(T, X)$	$\alpha \in \sigma_r(T, X)$ $\alpha \in \sigma_{ap}(T, X)$ $\alpha \in \sigma_\delta(T, X)$ $\alpha \in \sigma_{co}(T, X)$	$\alpha \in \sigma_p(T, X)$ $\alpha \in \sigma_{ap}(T, X)$ $\alpha \in \sigma_\delta(T, X)$ $\alpha \in \sigma_{co}(T, X)$

Table 5.2: Subdivisions of spectrum of a linear operator

Observe that the case in the first row and second column cannot occur in a Banach space X, by Closed Graph Theorem. If we are not in the third column, i.e., if α is not an eigenvalue of T, then we may always consider the resolvent operator T_α^{-1} (on a possibly "thin" domain of definition) as "algebraic" inverse of $\alpha I - T$.

Now, we may give the following two lemmas which are needed in the proof of theorems in the present chapter:

Lemma 5.2.10. [91, p. 59] *T has a dense range if and only if the adjoint operator T^* of the operator T is one to one.*

Lemma 5.2.11. [91, p. 60] *The adjoint operator T^* is onto if and only if T has a bounded inverse.*

We should remark that the index p has different meanings in the notation of the spaces ℓ_p, ℓ_p^*, bv_p, bv_p^* and in the point spectrums $\sigma_p(\Delta^{(1)}, \ell_p)$, $\sigma_p(\Delta^{(1)^*}, \ell_p^*)$, $\sigma_p(\Delta^{(1)}, bv_p)$, $\sigma_p(\Delta^{(1)^*}, bv_p^*)$, $\sigma_p(C_1, bv_p)$, $\sigma_p(C_1^*, bv_p^*)$ which occur in the next sections.

We give a short survey concerning with the spectrum and the fine spectrum of the linear operators defined by some triangle matrices over certain sequence spaces. Wenger [226] examined the fine spectrum of the integer power of the Cesàro operator in c, and Rhoades [190] generalized this result to the weighted mean methods. González [93] studied the fine spectrum of the Cesàro operator on the sequence space ℓ_p, where $1 < p < \infty$. Reade [189], Akhmedov and Başar [1], and Okutoyi [180] investigated the spectrum of the Cesàro operator on the sequence spaces c_0 and bv, respectively. Yıldırım [235] examined the fine spectrum of the Rhally operators on the sequence spaces c_0 and c. Furthermore, Coşkun [64] studied the spectrum and fine spectrum for p-Cesàro operator acting on the space c_0. More recently, de Malafosse [148] and Altay and Başar [8] have, respectively, studied the spectrum and the fine spectrum of the difference operator on the sequence spaces s_r and c_0, c, where s_r denotes the Banach space of all sequences $x = (x_k)$ normed by

$$\|x\|_{s_r} = \sup_{k \in \mathbb{N}_0} \frac{|x_k|}{r^k}, \quad (r > 0).$$

Altay and Karakuş [15] have determined the fine spectrum of the Zweier matrix, which is a band matrix as an operator over the sequence spaces ℓ_p and bv. In 2010, Srivastava and Kumar [204] have determined the spectra and the fine spectra of generalized difference operator Δ_ν on ℓ_1, where Δ_ν is defined by $(\Delta_\nu)_{nn} = \nu_n$ and $(\Delta_\nu)_{n+1,n} = -\nu_n$ for all $n \in \mathbb{N}_0$, under certain conditions on the sequence $\nu = (\nu_n)$. We should note here that the reader can refer to Yeşilkayagil and Başar [234], and references therein for the detail of spectrum of various triangles over some sequence spaces. At this stage, Table 5.3 may be useful:

$\sigma(A,\alpha)$	$\sigma_p(A,\alpha)$	$\sigma_c(A,\alpha)$	$\sigma_r(A,\alpha)$	refer to.
$\sigma(C_1^p,c)$	-	-	-	[226]
$\sigma(W,c)$	-	-	-	[190]
$\sigma(C_1,c_0)$	-	-	-	[189]
$\sigma(C_1,c_0)$	$\sigma_p(C_1,c_0)$	$\sigma_c(C_1,c_0)$	$\sigma_r(C_1,c_0)$	[1]
$\sigma(C_1,bv)$	-	-	-	[180]
$\sigma(R,c_0)$	$\sigma_p(R,c_0)$	$\sigma_c(R,c_0)$	$\sigma_r(R,c_0)$	[235]
$\sigma(R,c)$	$\sigma_p(R,c)$	$,\sigma_c(R,c)$	$\sigma_r(R,c)$	[235]
$\sigma(C_1^p,c_0)$	-	-	-	[64]
$\sigma(\Delta,s_r)$	-	-	-	[148]
$\sigma(\Delta,c_0)$	-	-	-	[148]
$\sigma(\Delta,c)$	-	-	-	[148]
$\sigma(\Delta^{(1)},c)$	$\sigma_p(\Delta^{(1)},c)$	$\sigma_c(\Delta^{(1)},c)$	$\sigma_r(\Delta^{(1)},c)$	[8]
$\sigma(\Delta^{(1)},c_0)$	$\sigma_p(\Delta^{(1)},c_0)$	$\sigma_c(\Delta^{(1)},c_0)$	$\sigma_r(\Delta^{(1)},c_0)$	[8]
$\sigma(B(r,s),\ell_p)$	$\sigma_p(B(r,s),\ell_p)$	$\sigma_c(B(r,s),\ell_p)$	$\sigma_r(B(r,s),\ell_p)$	[50]
$\sigma(B(r,s),bv_p)$	$\sigma_p(B(r,s),bv_p)$	$\sigma_c(B(r,s),bv_p)$	$\sigma_r(B(r,s),bv_p)$	[50]
$\sigma(B(r,s,t),\ell_p)$	$\sigma_p(B(r,s,t),\ell_p)$	$\sigma_c(B(r,s,t),\ell_p)$	$\sigma_r(B(r,s,t),\ell_p)$	[86]
$\sigma(B(r,s,t),bv_p)$	$\sigma_p(B(r,s,t),bv_p)$	$\sigma_c(B(r,s,t),bv_p)$	$\sigma_r(B(r,s,t),bv_p)$	[86]

Table 5.3: Spectrum and fine spectrum of some triangle matrices in certain sequence spaces.

5.3 The Fine Spectrum of the Operator Defined by the Matrix Λ over the Spaces of Null and Convergent Sequences

In this section, following Yeşilkayagil and Başar [232], we determine the fine spectrum with respect to Goldberg's classification of the operator defined by the matrix Λ over the sequence spaces c_0 and c. Throughout this section, the matrix $\Lambda = (\lambda_{nk})$ is defined by

$$\lambda_{nk} := \begin{cases} \dfrac{\lambda_k - \lambda_{k-1}}{\lambda_n} &, \quad 0 \le k \le n, \\ 0 &, \quad k > n \end{cases}$$

for all $k,n \in \mathbb{N}_0$. As a new development, we give the approximate point spectrum, defect spectrum and compression spectrum of the matrix operator Λ on the sequence spaces c_0 and c. Finally, we present a Mercerian theorem.

Lemma 5.3.1. [228, Theorem 1.3.6, p. 6] *The matrix $A = (a_{nk})$ gives rise to a bounded linear operator $T \in B(c)$ from c into itself if and only if*

1. *The rows of A are in ℓ_1 and their ℓ_1 norms are bounded.*

2. *The columns of A are in c.*

3. *The sequence of row sums of A is in c.*

The operator norm of T is the supremum of the ℓ_1 norms of the rows.

Corollary 5.3.2. $\Lambda : c \to c$ *is a bounded linear operator with the norm* $\|\Lambda\|_{(c:c)} = 1$.

Lemma 5.3.3. [228, Example 8.4.5.A, p. 129] *The matrix* $A = (a_{nk})$ *gives rise to a bounded linear operator* $T \in B(c_0)$ *from* c_0 *into itself if and only if*

1. *The rows of* A *are in* ℓ_1 *and their* ℓ_1 *norms are bounded.*

2. *The columns of* A *are in* c_0.

The operator norm of T is the supremum of the ℓ_1 norms of the rows.

Corollary 5.3.4. $\Lambda : c_0 \to c_0$ *is a bounded linear operator with the norm* $\|\Lambda\|_{(c_0:c_0)} = 1$.

5.3.1   The Fine Spectrum of the Operator Λ on the Sequence Space c_0

In this subsection, we examine the spectrum, the point spectrum, the continuous spectrum, the residual spectrum, the fine spectrum, the approximate point spectrum, the defect spectrum and the compression spectrum of the operator Λ on the sequence space c_0. For simplicity in the notation, we write throughout that $c_n = (\alpha_n - \alpha_{n-1})/\alpha_n$ for all $n \in \mathbb{N}_0$, and we use this abbreviation with other letters.

Theorem 5.3.5. [232, Theorem 6] $\sigma(\Lambda, c_0) \subseteq \{\alpha \in \mathbb{C} : |2\alpha - 1| \leq 1\}$.

Proof. Let $|2\alpha - 1| > 1$. Since $\Lambda - \alpha I$ is triangle, $(\Lambda - \alpha I)^{-1}$ exists and solving the matrix equation $(\Lambda - \alpha I)x = y$ for x in terms of y gives the matrix $(\Lambda - \alpha I)^{-1} = B = (b_{nk})$ defined by

$$
b_{nk} := \begin{cases} \dfrac{(-1)^{n-k}(\alpha_k - \alpha_{k-1})}{\alpha_n \alpha^2 \displaystyle\prod_{j=k}^{n} \left(\dfrac{c_j}{\alpha} - 1\right)} & , \quad 0 \leq k \leq n-1, \\[4ex] \dfrac{\alpha_n}{\alpha_n - \alpha_{n-1} - \alpha\alpha_n} & , \quad k = n, \\[2ex] 0 & , \quad k > n \end{cases}
$$

for all $k, n \in \mathbb{N}_0$. Thus, we observe that

$$
\|(\Lambda - \alpha I)^{-1}\|_{(c_0:c_0)} = \sup_{n \in \mathbb{N}} \sum_k |b_{nk}|.
$$

The inequality $|2\alpha - 1| > 1$ is equivalent to $\gamma > -1$, where $-(1/\alpha) = \gamma + i\beta$. For all $\alpha \in \mathbb{C}$,

$$
\left|1 - \frac{c_j}{\alpha}\right| = |1 + (\gamma + i\beta)c_j| \geq 1 + \gamma c_j
$$

holds for all $j \in \mathbb{N}_0$. So, $1/\left|1 - (c_j/\alpha)\right| \leq 1/\left(1 + \gamma c_j\right)$.

Firstly, we take $-1 < \gamma < 0$. Since $0 < c_j \leq 1$, we have $1 + \gamma \leq 1 + \gamma c_j < 1$. Therefore, $1/(1 + \gamma c_j) < 1/(1 + \gamma)$ and $1 < 1/(1 + \gamma) < \infty$ for $0 < 1 + \gamma < 1$.

$$\sum_k |b_{nk}| = \sum_{k=0}^n |b_{nk}| < \frac{\alpha_{n-1}}{\alpha_n |\alpha|^2 (1 + \gamma)^{n+1}} + \frac{1}{|\alpha|(1 + \gamma)} < \infty.$$

Secondly, we get $0 \leq \gamma$. Since $1 < 1 + \gamma c_j \leq 1 + \gamma$, $1/(1 + \gamma c_j) < 1$. So,

$$\sum_k |b_{nk}| = \sum_{k=0}^n |b_{nk}| < \frac{\alpha_{n-1}}{\alpha_n |\alpha|^2} + \frac{1}{|\alpha|} < \infty.$$

Therefore, we have

$$\|(\Lambda - \alpha I)^{-1}\|_{(c_0 : c_0)} = \sup_{n \in \mathbb{N}_0} \sum_{k=0}^n |b_{nk}| < \infty,$$

that is, $(\Lambda - \alpha I)^{-1} \in (c_0 : c_0)$. But, for $|2\alpha - 1| \leq 1$, $\|(\Lambda - \alpha I)^{-1}\|_{(c_0 : c_0)} = \infty$. This means that $(\Lambda - \alpha I)^{-1}$ is not in $B(c_0)$.

This completes the proof. $\qquad\square$

Theorem 5.3.6. [232, Theorem 7] *Define* μ *by* $\mu = \limsup_{j \to \infty} c_j$. *Then,*

$$\left\{ \alpha \in \mathbb{C} : \left| \alpha - \frac{1}{2 - \mu} \right| \leq \frac{1 - \mu}{2 - \mu} \right\} \cup S \subseteq \sigma(\Lambda, c_0); \text{ where } S = \overline{\left\{ \frac{\alpha_j - \alpha_{j-1}}{\alpha_j}, j \geq 0 \right\}}.$$

Proof. Let $|\alpha - 1/(2 - \mu)| < (1 - \mu)/(2 - \mu)$ and $\alpha \neq c_j$ for any $j \in \mathbb{N}_0$. Then,

$$
\begin{aligned}
1 - \frac{c_j}{\alpha} &= \frac{\alpha_j \alpha_{j-1}}{\alpha_j \alpha_{j-1}} - \frac{c_j}{\alpha} \\
&= \frac{\alpha_{j-1}}{\alpha_j} \left[\frac{\alpha_j}{\alpha_{j-1}} - \frac{\alpha_j - \alpha_{j-1}}{\alpha_{j-1}} + \left(1 - \frac{1}{\alpha}\right) \frac{\alpha_j - \alpha_{j-1}}{\alpha_{j-1}} \right] \\
&= \frac{\alpha_{j-1}}{\alpha_j} \left[1 + \left(1 - \frac{1}{\alpha}\right) \frac{\alpha_j - \alpha_{j-1}}{\alpha_{j-1}} \right].
\end{aligned}
$$

So, we have

$$|b_{nk}| = \frac{\alpha_k - \alpha_{k-1}}{\alpha_{k-1} |\alpha|^2 \prod_{j=k}^n \left| 1 + \left(1 - \frac{1}{\alpha}\right) \frac{\alpha_j - \alpha_{j-1}}{\alpha_{j-1}} \right|}. \tag{5.3.1}$$

Note that $\left| 1 + \left(1 - \frac{1}{\alpha}\right)(\alpha_j - \alpha_{j-1})/\alpha_{j-1} \right| \leq 1$ if and only if

$$\left[1 + (1 + \gamma) \frac{\alpha_j - \alpha_{j-1}}{\alpha_{j-1}} \right]^2 + \left(\beta \frac{\alpha_j - \alpha_{j-1}}{\alpha_{j-1}} \right)^2 \leq 1, \tag{5.3.2}$$

where $-(1/\alpha) = \gamma + i\beta$. Therefore, we see that

$$2(1 + \gamma)\frac{\alpha_j - \alpha_{j-1}}{\alpha_{j-1}} + \left[(1 + \gamma)^2 + \beta^2\right]\left(\frac{\alpha_j - \alpha_{j-1}}{\alpha_{j-1}}\right)^2 \leq 0$$

which is equivalent to the inequality

$$2(1 + \gamma) + \left[(1 + \gamma)^2 + \beta^2\right]\left(\frac{\alpha_j - \alpha_{j-1}}{\alpha_{j-1}}\right) \leq 0. \tag{5.3.3}$$

In order for the inequality (5.3.3) to hold for all sufficiently large j, it is sufficient to have

$$\limsup_{j \to \infty}\left\{2(1 + \gamma) + \left[(1 + \gamma)^2 + \beta^2\right]\frac{\alpha_j - \alpha_{j-1}}{\alpha_{j-1}}\right\} < 0. \tag{5.3.4}$$

We can write $(\alpha_j - \alpha_{j-1})/\alpha_{j-1} = \left[(\alpha_j - \alpha_{j-1})\alpha_j\right]/(\alpha_j\alpha_{j-1})$ and $\alpha_j/\alpha_{j-1} = 1/(1 - c_j)$. Therefore,

$$\frac{\alpha_j - \alpha_{j-1}}{\alpha_{j-1}} = \frac{c_j}{1 - c_j},$$

and

$$\limsup_{j \to \infty}\frac{\alpha_j - \alpha_{j-1}}{\alpha_{j-1}} = \frac{\mu}{1 - \mu}, \tag{5.3.5}$$

since the function g defined by $g(x) = x/(1 - x)$ is monotone increasing in x for $0 < x < 1$.

In order to the relation (5.3.4) to be valid for all sufficiently large j, it is sufficient to have μ satisfying

$$2(1 + \gamma) + \left[(1 + \gamma)^2 + \beta^2\right]\frac{\mu}{1 - \mu} < 0$$

which is equivalent to

$$\left|\alpha - \frac{1}{2 - \mu}\right| < \frac{1 - \mu}{2 - \mu}.$$

Therefore, for all $n \geq N$, for some fixed N,

$$\sum_{k=N}^{n-1}|b_{nk}| = \sum_{k=N}^{n-1}\frac{\alpha_k - \alpha_{k-1}}{\alpha_{k-1}|\alpha|^2\prod_{j=k}^{n}\left|1 + \left(1 - \frac{1}{\alpha}\right)\frac{\alpha_j - \alpha_{j-1}}{\alpha_{j-1}}\right|}$$

$$\geq \frac{1}{|\alpha|^2}\sum_{k=N}^{n-1}\frac{\alpha_k - \alpha_{k-1}}{\alpha_{k-1}}$$

which diverges in the light of (5.3.5).

If $\alpha = c_j$ for any $j \in \mathbb{N}_0$, then clearly α lies in the spectrum of Λ. This completes the proof. $\square$

Theorem 5.3.7. [232, Theorem 8] *Define η by $\eta = \liminf_{j \to \infty} c_j$ and μ be defined as in Theorem 5.3.6. Then,*

$$\sigma(\Lambda, c_0) \subseteq \{\alpha \in \mathbb{C} : |\alpha - 1/(2 - \eta)| \leq (1 - \eta)/(2 - \eta)\} \cup S.$$

Proof. Let α be fixed and satisfy the inequality

$$\left| \alpha - \frac{1}{2 - \eta} \right| > \frac{1 - \eta}{2 - \eta}, \tag{5.3.6}$$

and $\alpha \neq c_j$ for any $j \in \mathbb{N}_0$. We shall show that $\alpha \in \rho(\Lambda, c_0)$. From Theorem 5.3.5, we need consider only those values of α satisfying $|2\alpha - 1| > 1$; i.e., $\gamma > -1$. Under the assumption on α, we wish to verify that

$$\left| 1 + \left(1 - \frac{1}{\alpha} \right) \frac{\alpha_j - \alpha_{j-1}}{\alpha_{j-1}} \right| > 1$$

for all sufficiently large j. It will be sufficient to show that

$$\liminf_{j \to \infty} \left\{ 2(1 + \gamma) + \left[(1 + \gamma)^2 + \beta^2 \right] \frac{\alpha_j - \alpha_{j-1}}{\alpha_{j-1}} \right\} > 0$$

that is,

$$2(1 + \gamma) + \left[(1 + \gamma)^2 + \beta^2 \right] \frac{\eta}{1 - \eta} > 0$$

which is equivalent to (5.3.6).

Define the function f by $f(t) = 1 + 2(1 + \gamma)t + \left[(1 + \gamma)^2 + \beta^2 \right] t^2$. f has a minumum at $t_0 = -(1 + \gamma)/[(1 + \gamma)^2 + \beta^2]$. The above inequality is equivalent to $\eta(\gamma^2 + \beta^2) + 2\gamma > \eta - 2$ and is also equivalent to

$$\frac{\eta}{2(1 - \eta)} > -\frac{1 + \gamma}{(1 + \gamma)^2 + \beta^2} = t_0. \tag{5.3.7}$$

Therefore, for those values of η satisfying (5.3.7), f is monotone increasing. Let $\epsilon > 0$ and small. Then, $f(\eta/(1 - \eta) - \epsilon) = f(\eta/(1 - \eta)) - (2\epsilon)g(\epsilon)$, where $g(\epsilon) = 1 + \gamma + \left[(1 + \gamma)^2 + \beta^2 \right] [\eta/(1 - \eta) - \epsilon/2]$. Note that $g(\epsilon) > 0$ for small ϵ, since f is monotone increasing for $t > \eta/[2(1 - \eta)]$. Thus, we shall show that $f(\eta/(1 - \eta)) > 1$. From (5.3.7),

$$\gamma^2 + \beta^2 + \frac{2\gamma}{\eta} > \frac{\eta - 2}{\eta},$$

which is equivalent to

$$\left| \frac{1}{1 - \eta} - \frac{\eta}{\alpha(1 - \eta)} \right| > 1.$$

But $1/(1 - \eta) = 1 + \eta/(1 - \eta)$, so we have $f(\eta/(1 - \eta)) = \left| 1 + (1 - \alpha^{-1}) \eta/(1 - \eta) \right|^2 > 1$. Now choose $\epsilon > 0$ and so small that $f(\eta/(1 - \eta) - \epsilon) = f(\eta/(1 - \eta)) - (2\epsilon)g(\epsilon) = m^2 > 1$. Then, by the definition of η, there exists an N such that $n > N$ implies $(\alpha_{n+1}-\alpha_n)/\alpha_n > \eta/(1-\eta)-\epsilon$, so that $f((\alpha_{n+1} - \alpha_n)/\alpha_n) > f(\eta/(1 - \eta) - \epsilon) = m^2$. Using (5.3.1),

$$\frac{|b_{nk}|}{|b_{n+1,k}|} = \frac{\dfrac{\alpha_k - \alpha_{k-1}}{\alpha_{k-1}}|\alpha|^2 \prod_{j=k}^{n}\left|1 + \left(1 - \dfrac{1}{\alpha}\right)\dfrac{\alpha_j - \alpha_{j-1}}{\alpha_{j-1}}\right|}{\dfrac{\alpha_k - \alpha_{k-1}}{\alpha_{k-1}}|\alpha|^2 \prod_{j=k}^{n+1}\left|1 + \left(1 - \dfrac{1}{\alpha}\right)\dfrac{\alpha_j - \alpha_{j-1}}{\alpha_{j-1}}\right|}$$

$$= \left|1 + \left(1 - \frac{1}{\alpha}\right)\frac{\alpha_{n+1} - \alpha_n}{\alpha_n}\right|$$

$$= f\left(\frac{\alpha_{n+1} - \alpha_n}{\alpha_n}\right) > m^2 > 1$$

for all $n \geq N$. Therefore, $(|b_{nk}|)$ is monotone decreasing in n for each $k, n > N$, so that B has bounded columns. It remains to show that B has finite norm.

For the ϵ being used, from (5.3.5), we can enlarge N, if necessary, to ensure that $(\alpha_n - \alpha_{n-1})/\alpha_{n-1} < \mu/(1 - \mu) + 1$ for $n \geq N$. From (5.3.1),

$$\sum_{k=N}^{n-1} |b_{nk}| = \sum_{k=N}^{n-1} \frac{\alpha_k - \alpha_{k-1}}{\alpha_{k-1}|\alpha|^2 \prod_{j=k}^{n}\left|1 + \left(1 - \dfrac{1}{\alpha}\right)\dfrac{\alpha_j - \alpha_{j-1}}{\alpha_{j-1}}\right|}$$

$$\leq \frac{1}{|\alpha|^2}\left(\frac{\mu}{1 - \mu} + 1\right) \sum_{k=N}^{n-1} \frac{1}{\prod_{j=k}^{n}\left|1 + \left(1 - \dfrac{1}{\alpha}\right)\dfrac{\alpha_j - \alpha_{j-1}}{\alpha_{j-1}}\right|}$$

$$\leq \frac{1}{|\alpha|^2}\left(\frac{\mu}{1 - \mu} + 1\right) \sum_{k=N}^{n-1} m^{-n+k-1} < H,$$

where H is a constant independent of n. Further

$$|b_{nn}| = \frac{\alpha_n}{|\alpha|\left|\alpha_n - \dfrac{\alpha_n - \alpha_{n-1}}{\alpha}\right|}$$

$$= \frac{\alpha_n}{|\alpha|\left|\alpha_{n-1} + \left(1 - \dfrac{1}{\alpha}\right)(\alpha_n - \alpha_{n-1})\right|}$$

$$= \frac{\dfrac{\alpha_n}{\alpha_{n-1}}}{|\alpha|\left|1 + \left(1 - \dfrac{1}{\alpha}\right)\dfrac{\alpha_n - \alpha_{n-1}}{\alpha_{n-1}}\right|}$$

$$= \frac{1 + \dfrac{\alpha_n - \alpha_{n-1}}{\alpha_{n-1}}}{|\alpha|\left|1 + \left(1 - \dfrac{1}{\alpha}\right)\dfrac{\alpha_n - \alpha_{n-1}}{\alpha_{n-1}}\right|}$$

$$< \frac{1 + \dfrac{\mu}{1 - \mu} + 1}{|\alpha|m}.$$

Hence, B has a finite norm. $\qquad\square$

Corollary 5.3.8. [232, Corollary 9] *Let $c_j \to \delta$, as $j \to \infty$. Then,*

$$\sigma(\Lambda, c_0) = \left\{\alpha \in \mathbb{C} : \left|\alpha - \frac{1}{2-\delta}\right| \le \frac{1-\delta}{2-\delta}\right\} \cup S.$$

If $T \in B(c_0)$ with the matrix A, then it is known that the adjoint operator $T^* : c_0^* \to c_0^*$ is defined by the transpose A^t of the matrix A. It should be noted that the dual space c_0^* of c_0 is isometrically isomorphic to the Banach space ℓ_1 of absolutely summable sequences normed by $\|x\| = \sum_k |x_k|$.

Theorem 5.3.9. $\sigma_p(\Lambda, c_0) = \emptyset$.

Proof. Suppose that $\Lambda x = \alpha x$ for $x \neq \theta = (0, 0, 0, \ldots)$ in c_0. Then, by solving the system of linear equations

$$\left.\begin{array}{l} (1-\alpha)x_0 = 0 \\[2mm] x_1 = \dfrac{\alpha_0}{\alpha_1\left(\alpha - \dfrac{\alpha_1 - \alpha_0}{\alpha_1}\right)} x_0 \\[5mm] x_2 = \dfrac{\alpha_0}{\alpha_2\left(\alpha - \dfrac{\alpha_2 - \alpha_1}{\alpha_2}\right)\left(\alpha - \dfrac{\alpha_1 - \alpha_0}{\alpha_1}\right)} x_0 \\[5mm] \vdots \\[2mm] x_n = \dfrac{\alpha_0}{\alpha_n\left(\alpha - \dfrac{\alpha_n - \alpha_{n-1}}{\alpha_n}\right)\cdots\left(\alpha - \dfrac{\alpha_2 - \alpha_1}{\alpha_2}\right)\left(\alpha - \dfrac{\alpha_1 - \alpha_0}{\alpha_1}\right)} x_0 \\[5mm] \vdots \end{array}\right\} \quad (5.3.8)$$

one can obtain by the assumption $(1 - \alpha)x_0 = 0$ with $x_0 = 0$ that $x = \theta = (0, 0, 0, \ldots)$, a contradiction. So, we must have $x_0 \neq 0$ and $\alpha = 1$. But, if $\alpha = 1$, we have $x = (x_0, x_0, x_0, \ldots)$ which contradicts the fact that $x \in c_0$.

This completes the proof. $\qquad\square$

Theorem 5.3.10. [232, Theorem 10] *Let δ be defined as in Corollary 5.3.8. Then,*

$$\sigma_p(\Lambda^*, c_0^*) = \left\{\alpha \in \mathbb{C} : \left|\alpha - \frac{1}{2-\delta}\right| < \frac{1-\delta}{2-\delta}\right\} \cup S.$$

Proof. Suppose that $\Lambda^* x = \alpha x$ for $x \neq \theta$ in $c_0^* \cong \ell_1$. Then, by solving the system of linear equations

$$
\begin{aligned}
x_1 &= \frac{\alpha_1 - \alpha_0}{\alpha_0}\left(1 - \frac{1}{\alpha}\right) x_0, \\
x_2 &= \frac{\alpha_2 - \alpha_1}{\alpha_0}\left(1 - \frac{\alpha_1 - \alpha_0}{\alpha \alpha_1}\right)\left(1 - \frac{1}{\alpha}\right) x_0 \\
&\;\;\vdots \\
x_n &= \frac{\alpha_n - \alpha_{n-1}}{\alpha_0}\left(1 - \frac{1}{\alpha}\right) x_0 \prod_{j=1}^{n-1}\left(1 - \frac{\alpha_j - \alpha_{j-1}}{\alpha \alpha_j}\right) \\
&\;\;\vdots
\end{aligned}
$$

we can write

$$
x_n = \frac{\alpha_n - \alpha_{n-1}}{\alpha_{n-1}}\left(1 - \frac{1}{\alpha}\right) x_0 \prod_{j=1}^{n-1}\left[1 + \left(1 - \frac{1}{\alpha}\right)\frac{\alpha_j - \alpha_{j-1}}{\alpha_{j-1}}\right].
$$

Let $|\alpha - 1/(2 - \delta)| < (1 - \delta)/(2 - \delta)$ or $\alpha \in S$ and $u_n = \prod_{j=1}^{n-1}\left[1 + \left(1 - \alpha^{-1}\right)\left(\alpha_j - \alpha_{j-1}\right)/\alpha_{j-1}\right]$. One can see that

$$
\left|1 + \left(1 - \alpha^{-1}\right)\left(\alpha_j - \alpha_{j-1}\right)/\alpha_{j-1}\right| < 1
$$

for all sufficiently large j if and only if

$$
\left[1 + (1 + \gamma)\frac{\alpha_j - \alpha_{j-1}}{\alpha_{j-1}}\right]^2 + \left(\beta\frac{\alpha_j - \alpha_{j-1}}{\alpha_{j-1}}\right)^2 < 1, \quad \text{where} \quad -\frac{1}{\alpha} = \gamma + i\beta.
$$

Then, we have from the discussion in Theorem 5.3.6 and the hypothesis on α,

$$
\left|\frac{u_{n+1}}{u_n}\right| = \left|1 + \left(1 - \frac{1}{\alpha}\right)\frac{\alpha_n - \alpha_{n-1}}{\alpha_{n-1}}\right| < 1
$$

for all sufficiently large n, so $\sum_n |u_n|$ is convergent. Since $\left|(\alpha_n - \alpha_{n-1})\left(1 - \alpha^{-1}\right) x_0/\alpha_{n-1}\right|$ is bounded, it follows that $\sum_n |x_n|$ is convergent, so that $\Lambda^* x = \alpha x$ has non-zero solutions.

Therefore, the proof is completed. $\qquad \square$

Theorem 5.3.11. [232, Theorem 11] *Let δ be defined as in Corollary 5.3.8. Then, $\sigma_p(\Lambda, c_0) = \left\{\alpha = c_n \in \mathbb{C} : 0 \leq \alpha \leq \dfrac{\delta}{2 - \delta}\right\} \cup \{1\}$.*

Proof. Let c_k be any diagonal entry satisfying $0 < c_k \leq \delta/(2-\delta)$. Let j be the smallest positive integer such that $c_j = c_k$. By setting $x_n = 0$ for $n > j + 1$, $x_0 = 0$, the system $(\Lambda^* - c_j I)x = \theta$ reduces to a homogeneous linear system of j equations in $j + 1$ unknowns, so that non-trivial solutions exist. Therefore, $\Lambda - c_j I \in 3$.

$\Lambda - \alpha I$ is not one to one for $\alpha = 0, 1$ and so $\Lambda - \alpha I \in 3$.

This step concludes the proof. $\qquad \square$

Theorem 5.3.12. [232, Theorem 13] $\sigma_r(\Lambda, c_0) = \sigma_p(\Lambda^*, c_0^*) \setminus \sigma_p(\Lambda, c_0)$.

Proof. For $\alpha \in \sigma_p(\Lambda^*, c_0^*) \setminus \sigma_p(\Lambda, c_0)$, the operator $\Lambda - \alpha I$ is triangle, so has an inverse. But $\Lambda^* - \alpha I$ is not one to one by Theorem 5.3.10. Therefore, by Lemma 5.2.10, $\overline{R(\Lambda - \alpha I)} \neq c_0$ which completes the proof. □

Theorem 5.3.13. [232, Theorem 14] *Let δ be defined as in Corollary 5.3.8 and $c_n \geq \delta$ for all sufficiently large n. Then,*

$$\sigma_c(\Lambda, c_0) = \left\{ \alpha \in \mathbb{C} : \left| \alpha - \frac{1}{2 - \delta} \right| = \frac{1 - \delta}{2 - \delta}, \alpha \neq 1, \frac{\delta}{2 - \delta} \right\}.$$

Proof. Fix $\alpha \neq 1, \delta/(2 - \delta)$ and satisfy $|\alpha - 1/(2 - \delta)| = (1 - \delta)/(2 - \delta)$. Since the operator $\Lambda - \alpha I$ is a triangle, it has an inverse. Consider the adjoint operator $\Lambda^* - \alpha I$. As in Theorem 5.3.10, x_0 is arbitrary and

$$x_n = \frac{\alpha_n - \alpha_{n-1}}{\alpha_{n-1}} \left(1 - \frac{1}{\alpha} \right) x_0 \prod_{j=k}^{n} \left[1 + \left(1 - \frac{1}{\alpha} \right) \frac{\alpha_j - \alpha_{j-1}}{\alpha_{j-1}} \right]$$

for all $n \in \mathbb{N}_1$. From the hypothesis, there exists a positive integer N such that $n \geq N$ implies $c_n \geq \delta$. This fact, together with the condition on α, implies that $\left| 1 + \left(1 - \alpha^{-1} \right) (\alpha_n - \alpha_{n-1})/\alpha_{n-1} \right| \geq 1$ for $n \geq N$. Thus, $|x_n| = C(\alpha_n - \alpha_{n-1})/\alpha_{n-1}$ for $n \geq N$, where C is a positive constant independent of n. We can write

$$\frac{\alpha_n - \alpha_{n-1}}{\alpha_{n-1}} = c_n \left(1 + \frac{\alpha_n - \alpha_{n-1}}{\alpha_{n-1}} \right) \geq c_n.$$

Therefore $(x_n) \in \ell_1 \Leftrightarrow x_0 = 0$, that is, $\Lambda^* - \alpha I$ is one to one. From Lemma 5.2.10, the range of $\Lambda - \alpha I$ is dense in c_0.
This completes the proof. □

Theorem 5.3.14. [232, Theorem 16] *Let δ be defined as in Corollary 5.3.8 and is less than 1. If α satisfies $\left| \alpha - (2 - \delta)^{-1} \right| < (1 - \delta)/(2 - \delta)$ and $\alpha \notin S$, then $\alpha \in III_1\sigma(\Lambda, c_0)$.*

Proof. First of all $\Lambda - \alpha I$ is a triangle, hence one to one. Therefore, $\Lambda - \alpha I \in 1 \cup 2$. To verify that $\Lambda - \alpha I \in III_1\sigma(\Lambda, c_0)$ it is sufficient to show that $\Lambda^* - \alpha I$ is an onto operator by Lemma 5.2.11.
Suppose $y = (\Lambda^* - \alpha I)x$, where $x = (x_k), y = (y_k) \in \ell_1$. Then,

$$x_0 = \frac{1}{1 - \alpha} y_0 - \frac{\alpha_0}{(\alpha_1 - \alpha_0)(1 - \alpha)} y_1$$

and

$$(c_n - \alpha) x_n + (\alpha_n - \alpha_{n-1}) \sum_{k=n+1}^{\infty} \frac{x_k}{\alpha_k} = y_n \qquad (5.3.9)$$

for all $n \in \mathbb{N}_1$. Choose $x_1 = 0$ and solve (5.3.9) for x in terms of y to get

$$(\alpha_1 - \alpha_0) \sum_{k=2}^{\infty} \frac{x_k}{\alpha_k} = y_1 \tag{5.3.10}$$

$$(c_n - \alpha) x_n = y_n - (\alpha_n - \alpha_{n-1}) \sum_{k=n+1}^{\infty} \frac{x_k}{\alpha_k}. \tag{5.3.11}$$

For example, substituting (5.3.10) into (5.3.11), with $n = 2$, yields

$$(c_2 - \alpha) x_2 = y_2 - (\alpha_2 - \alpha_1) \sum_{k=3}^{\infty} \frac{x_k}{\alpha_k},$$

so that $x_2 = (\alpha_2 - \alpha_1)/[\alpha(\alpha_1 - \alpha_0)]y_1 - (1/\alpha)y_2$. For $n = 3$,

$$x_3 = -\left(\frac{\alpha_3 - \alpha_2}{\alpha_1 - \alpha_0}\right)\left(\frac{\alpha_2 - \alpha_1}{\alpha_2} - \alpha\right)\frac{1}{\alpha^2}y_1 + \left(\frac{\alpha_3 - \alpha_2}{\alpha_2}\right)\frac{1}{\alpha^2}y_2 - \frac{1}{\alpha}y_3.$$

Continuing this process, the elements of the matrix $B = (b_{nk})$ such that $By = x$ are calculated as

$$b_{00} = \frac{1}{1 - \alpha},$$

$$b_{01} = -\frac{\alpha_0}{(\alpha_1 - \alpha_0)(1 - \alpha)}$$

$$b_{21} = \frac{\alpha_2 - \alpha_1}{(\alpha_1 - \alpha_0)\alpha},$$

$$b_{nn} = -\frac{1}{\alpha}, \quad n > 1,$$

$$b_{n,n-1} = \frac{\alpha_n - \alpha_{n-1}}{\alpha_{n-1}\alpha^2}, \quad n > 2,$$

$$b_{n1} = \frac{\alpha_n - \alpha_{n-1}}{(\alpha_1 - \alpha_0)\alpha} \prod_{j=2}^{n-1}\left(1 - \frac{\alpha_j - \alpha_{j-1}}{\alpha\alpha_j}\right), \quad n > 2,$$

$$b_{nk} = \frac{\alpha_n - \alpha_{n-1}}{\alpha_k\alpha^2} \prod_{j=k+1}^{n-1}\left(1 - \frac{\alpha_j - \alpha_{j-1}}{\alpha\alpha_j}\right), \quad 1 < k < n - 1,$$

and $b_{nk} = 0$, otherwise.

To show that $B \in B(\ell_1)$, it is sufficient to establish that $\sum_n |b_{nk}|$ is finite independent of k. $\sum_n |b_{n0}| = \frac{1}{|1 - \alpha|}$. We can write

$$1 - \frac{c_j}{\alpha} = \frac{\alpha_{j-1}}{\alpha_j}\left[1 + \left(1 - \frac{1}{\alpha}\right)\frac{\alpha_j - \alpha_{j-1}}{\alpha_{j-1}}\right].$$

Also, $\sup_{n \in \mathbb{N}} |(\alpha_n - \alpha_{n-1})/\alpha_{n-1}| \le M < \infty$. Therefore,

$$\sum_n |b_{n1}| \le \frac{1}{|\alpha|} \left[M + M \sum_{n=3}^{\infty} \prod_{j=2}^{n-1} \left| 1 + \left(1 - \frac{1}{\alpha} \right) \frac{\alpha_j - \alpha_{j-1}}{\alpha_{j-1}} \right| \right] \qquad (5.3.12)$$

and, for $k > 1$,

$$\sum_n |b_{nk}| \le \frac{1}{|\alpha|} + \frac{M}{|\alpha|^2} + \frac{M}{|\alpha|^2} \sum_{n=k+2}^{\infty} \prod_{j=k+1}^{n-1} \left| 1 + \left(1 - \frac{1}{\alpha} \right) \frac{\alpha_j - \alpha_{j-1}}{\alpha_{j-1}} \right|. \qquad (5.3.13)$$

Since $k > 1$, the series in the inequality (5.3.13) is absolutely convergent from Theorem 5.3.6. Therefore, $\|B\|_{(\ell_1 : \ell_1)} < \infty$.

Since $(\Lambda - \alpha I)^{-1}$ is bounded, it is continuous, and $\alpha \in III_1\sigma(\Lambda, c_0)$. This completes the proof. $\qquad\qquad\qquad\qquad\qquad\qquad\qquad\qquad\qquad\qquad\qquad\qquad\square$

Theorem 5.3.15. [232, Theorem 17] *Let δ be defined as in Corollary 5.3.8 and $\delta < 1$. If $\alpha = \delta$ or $\alpha = (\alpha_n - \alpha_{n-1})/\alpha_n$ for all $n \in \mathbb{N}_0$ and $\delta/(2 - \delta) < \alpha < 1$, then $\alpha \in III_1\sigma(\Lambda, c_0)$.*

Proof. First assume that Λ has distinct diagonal elements, and fix $j \ge 1$. Then, the system $(\Lambda - c_j I)x = \theta$ implies that $x_n = 0$ for $n = 0, 1, \ldots, j - 1$, and, for $n \ge j$,

$$(c_j - c_n)x_n - \sum_{k=0}^{n-1} \alpha_{nk}x_k = 0. \qquad (5.3.14)$$

The system (5.3.14) yields the following recursion relation

$$x_{n+1} = \frac{\alpha_n c_j x_n}{\alpha_{n+1}(c_j - c_{n+1})},$$

which can be solved for x_n to obtain

$$
\begin{aligned}
x_{j+m} &= \frac{\alpha_j x_j c_j^m}{\alpha_{j+m} \prod_{i=0}^{m}(c_j - c_{j+i})} = x_j \prod_{i=0}^{m} \frac{\alpha_{j+i-1}}{\alpha_{j+i}\left(1 - \frac{c_{j+i}}{c_j} \right)} \\
&= x_j \left\{ \prod_{i=0}^{m} \frac{\alpha_{j+i}}{\alpha_{j+i-1}} \left[1 - \frac{\alpha_j(\alpha_{j+i} - \alpha_{j+i-1})}{\alpha_{j+i}(\alpha_j - \alpha_{j-1})} \right] \right\}^{-1} \\
&= x_j \left\{ \prod_{i=0}^{m} \left[\frac{\alpha_{j+i}}{\alpha_{j+i-1}} - \frac{\alpha_j \cdot (\alpha_{j+i} - \alpha_{j+i-1})}{\alpha_{j+i-1}(\alpha_j - \alpha_{j-1})} \right] \right\}^{-1} \\
&= x_j \left\{ \prod_{i=0}^{m} \left[\frac{\alpha_{j+i}}{\alpha_{j+i-1}} - \frac{\alpha_{j+i} - \alpha_{j+i-1}}{\alpha_{j+i-1}} + \left(1 - \frac{1}{c_j} \right) \frac{\alpha_{j+i} - \alpha_{j+i-1}}{\alpha_{j+i-1}} \right] \right\}^{-1} \\
&= x_j \left\{ \prod_{i=0}^{m} \left[1 + \left(1 - \frac{1}{c_j} \right) \frac{\alpha_{j+i} - \alpha_{j+i-1}}{\alpha_{j+i-1}} \right] \right\}^{-1}. \qquad (5.3.15)
\end{aligned}
$$

Since $0 < c_j < 1$, the argument of Theorem 5.3.6 applies and (5.3.2) is true. Therefore, $x \in c_0$ implies $x = \theta$ and $\Lambda - c_j I$ is injective so that $\Lambda - c_j I \in 1 \cup 2$.

Clearly, $\Lambda - c_j I \in C$. It remains to show that $\Lambda^* - c_j I$ is onto.

Suppose that $(\Lambda^* - c_j I)x = y$ with $x = (x_k), y = (y_k) \in \ell_1$. By choosing $x_{j+1} = 0$, we can solve for $x_0, x_1, \ldots, x_j$ in terms of $y_0, y_1, \ldots, y_{j+1}$. As in Theorem 5.3.14, the remaining equations can be written in the form $x = By$, where the non-zero elements of B are

$$b_{j+m,j+m} = -\frac{1}{c_j}$$

$$b_{j+2,j+1} = \frac{\alpha_{j+2} - \alpha_{j+1}}{c_j(\alpha_{j+1} - \alpha_j)};$$

$$b_{j+m,j+m-1} = \frac{\alpha_{j+m} - \alpha_{j+m-1}}{c_j^2 \alpha_{j+m-1}}, \quad m > 2; \tag{5.3.16}$$

$$b_{j+m,j+k} = \frac{\alpha_{j+m} - \alpha_{j+m-1}}{c_j^2 \alpha_{j+k}} \prod_{i=j+k+1}^{j+m-1} \left(1 - \frac{c_i}{c_j}\right), \quad 1 < k < m-1, \quad m > 3;$$

$$b_{j+m,j+1} = \frac{\alpha_{j+m} - \alpha_{j+m-1}}{c_j(\alpha_{j+1} - \alpha_j)} \prod_{i=j+2}^{j+m-1} \left(1 - \frac{c_i}{c_j}\right), \quad m > 2.$$

From (5.3.16),

$$\sum_{n=j+1}^{\infty} |b_{n,j+1}| = \frac{\alpha_{j+2} - \alpha_{j+1}}{c_j(\alpha_{j+1} - \alpha_j)} + \frac{1}{c_j(\alpha_{j+1} - \alpha_j)} \sum_{n=j+3}^{\infty} (\alpha_n - \alpha_{n-1})$$

$$\times \prod_{i=j+2}^{n-1} \left|1 - \frac{c_i}{c_j}\right| \tag{5.3.17}$$

For $m > 1$,

$$\sum_{n=m+j}^{\infty} |b_{n,m+j}| = \frac{1}{c_j} + \frac{\alpha_{j+m+1} - \alpha_{j+m}}{c_j^2 \alpha_{j+m}} + \frac{1}{c_j^2} \sum_{n=j+m+2}^{\infty} \frac{\alpha_n - \alpha_{n-1}}{\alpha_{j+m}}$$

$$\times \prod_{i=j+m+1}^{n-1} \left|1 - \frac{c_i}{c_j}\right| \tag{5.3.18}$$

Using the relation

$$1 - \frac{c_j}{\alpha} = \frac{\alpha_{j-1}}{\alpha_j}\left[1 + \left(1 - \frac{1}{\alpha}\right)\frac{\alpha_j - \alpha_{j-1}}{\alpha_{j-1}}\right],$$

one can convert (5.3.17) and (5.3.18) into the expressions similar to (5.3.12) and (5.3.13), and therefore $\|B\|_{(\ell_1:\ell_1)} < \infty$.

Suppose that Λ does not have distinct diagonal elements. The restriction on α guarantees that no zero diagonal elements are being considered. Let $c_j \neq 0$ be any diagonal element which occurs more than once, and let k, r denote, respectively, the smallest and largest integers for which $c_j = c_k = c_r$.

From (5.3.15) it follows that $x_n = 0$ for $n \geq r$. Also, $x_n = 0$ for $0 \leq n < k$. Therefore, the system $(\Lambda - c_j I)x = \theta$ becomes

$$(c_j - c_n)x_n - \sum_{i=j}^{n-1} \alpha_{ni} x_i = 0, \quad k < n \leq r. \tag{5.3.19}$$

Case I. Let $r = k + 1$. Then, (5.3.19) reduces to the single equation

$$(c_j - c_{k+1})x_{j+1} - \frac{\alpha_k - \alpha_{k-1}}{\alpha_{k+1}} x_k = 0$$

which implies that $x_k = 0$, since $c_j = c_r = c_{k+1}$, and $c_j \neq 0$. Therefore, $x = \theta$.

Case II. Let $r > k + 1$. From (5.3.19), one obtains the recursion formula $x_n = \alpha_{n+1}(c_j - c_{n+1})x_{n+1}/(\alpha_n c_j)$ with $k < n < r$. Since $x_r = 0$, it then follows that $x_n = 0$ for $k < n < r$. Using (5.3.19) with $n = k + 1$ yields $x_k = 0$, and so, again we get $x = \theta$.

To show that $\Lambda^* - c_j I$ is onto, suppose $(\Lambda^* - c_j I)x = y$ with $x = (x_k)$, $y = (y_k) \in \ell_1$. By choosing $x_{j+1} = 0$, we can solve for $x_0, x_1, \ldots, x_j$ in terms of $y_0, y_1, \ldots, y_{j+1}$. As in Theorem 5.3.14, the remaining equations can be written in the form $x = By$, where the non-zero elements of B are as in (5.3.16) with the other elements of B clearly zero.

Since $k \leq j \leq r$, there are two cases to consider.

Case I. Let $j = r$. Then, the proof proceeds exactly as in the argument following (5.3.16).

Case II. Let $j < r$. Then, from (5.3.16), $b_{j+m,j+k} = b_{j+m,j+1} = 0$ at least for $m \geq r - j + 2$. If there are other values of $n \in \mathbb{N}_0$, $j < n < r$ for which $c_n - c_j$, then the additional elements of B will be zero. These zero elements do not affect the validity of the argument showing that (5.3.17) converges.

If $\delta = 0$, then 0 does not lie inside the disc, and so it is not considered in this theorem.

Let $\alpha = \delta > 0$. If $\alpha_{nn} \leq \delta$ for each $n \geq 1$, all i sufficiently large, then the argument of Theorem 5.3.14 applies and $\Lambda - \delta I \in III_1$. If $\alpha_{nn} = \delta$ for some $n \in \mathbb{N}_0$, then the proof of Theorem 5.3.15 applies, with c_j replaced by δ, and, again, $\Lambda - \delta I \in III_1$.

Therefore, in all cases, $\Lambda - c_j I \in 1 \cup 2$. $\qquad\square$

Theorem 5.3.16. *Let η be defined as in Theorem 5.3.7. If there exists the values of n such that $0 \leq c_n \leq \eta/(2-\eta)$, then $\alpha = c_n$ implies $\alpha \in III_3\sigma(\Lambda, c_0)$. Also, $1 \in III_3\sigma(\Lambda, c_0)$.*

Proof. Let c_k be any diagonal element satisfying $0 < c_k \leq \eta/(2 - \eta)$. Let j be the smallest integer such that $c_j = c_k$. By setting $x_n = 0$ for $n > j + 1$, $x_0 = 0$, the system $(\Lambda^* - c_j I)x = \theta$ reduces to a homogeneous linear system of j equations in $j + 1$ unknowns, so that non-trivial solutions exist. Therefore, $\Lambda - c_j I \in 3$.

If $c_j = \eta/(2-\eta)$, then clearly $\Lambda - c_j I \in 3$. Assume that $0 < c_j \leq \eta/(2-\eta)$ and let r denote the largest integer such that $c_r = c_k$. Solving $(\Lambda - c_j I)x = \theta$

leads to (5.3.15) with $j = r$. For $m \geq n$, from (5.3.15),

$$
\begin{aligned}
\frac{|x_{j+m+1}|}{|x_{j+m}|} &= \frac{\alpha_{j+m} c_j}{\alpha_{j+m+1} |c_{j+m+1} - c_j|} \\
&= \frac{1 - c_{j+m+1}}{\frac{c_{j+m+1}}{c_j} - 1} \\
&< \frac{1 - \eta}{\frac{\eta}{c_j} - 1} < 1,
\end{aligned}
$$

since $0 < c_j \leq \eta/(2-\eta)$. Consequently, $(x_n) \in \ell_1$, hence $(x_n) \in c_0$, and $\Lambda - c_j I$ is not injective.

     Suppose that Λ has a zero on the main diagonal and $\eta > 0$. Let j denote the smallest positive integer for which $c_j = 0$. Let $e^{(j)}$ denote the coordinate sequence with a 1 in the jth position and all other elements zero. Then, $A e^{(j)} = 0$ and $\Lambda - c_j I = -\Lambda$ is not one to one. By setting $x_0 = 0$, $x_n = 0$ for $n > j+1$, the system $(\Lambda^* - c_j I) x = \theta$ reduces to a homogeneous linear system of j equations in $j+1$ unknowns.

     When the diagonal elements of Λ do not converge, it was shown in [54] that, even for weighted mean methods, the spectrum need no longer be a disc.

     For $\alpha = 1$, since $\Lambda - \alpha I$ is not one to one by Theorem 5.3.9 and hence $\Lambda - \alpha I \in 3$. Also, since $\Lambda^* - \alpha I$ is not one to one by Theorem 5.3.10, $\overline{R(\Lambda - \alpha I)} \neq c_0$ by Lemma 5.2.10.

     This step concludes the proof.   □

Theorem 5.3.17. [232, Theorem 19] $I_3\sigma(\Lambda, c_0) = III_2\sigma(\Lambda, c_0) = \emptyset$.

Proof. Let δ be defined as in Corollary 5.3.8 and $c_n \geq \delta$ for all sufficiently large n. Then, $I_3\sigma(\Lambda, c_0) = \emptyset$ and $III_2\sigma(\Lambda, c_0) = \emptyset$ follow from Corollary 5.3.8, Theorem 5.3.13 and Theorems 5.3.14-5.3.16.   □

     Let $E = \left\{ \overline{\frac{\alpha_j - \alpha_{j-1}}{\alpha_j}} : \frac{\alpha_j - \alpha_{j-1}}{\alpha_j} < \eta/(2-\eta) \right\}$, where η is as in Theorem 5.3.7.

     We shall consider $\delta = \eta$, i.e., for which the main diagonal elements converge, where δ is as in Corollary 5.3.8.

Theorem 5.3.18. [232, Theorem 20] *The following statements hold:*

(a) $\sigma_{ap}(\Lambda, c_0) = \left\{ \alpha \in \mathbb{C} : \left| \alpha - (2-\delta)^{-1} \right| = (1-\delta)/(2-\delta) \right\} \cup E$.

(b) $\sigma_\delta(\Lambda, c_0) = \sigma(\Lambda, c_0)$.

(c) $\sigma_{co}(\Lambda, c_0) = \left\{ \alpha \in \mathbb{C} : \left| \alpha - (2-\delta)^{-1} \right| < (1-\delta)/(2-\delta) \right\} \cup S$.

Proof. (a) Since the equality

$$III_1\sigma(\Lambda, c_0) = \left[\left\{\alpha \in \mathbb{C} : \left|\alpha - \frac{1}{2-\delta}\right| < \frac{1-\delta}{2-\delta}\right\} \backslash S\right] \bigcup \left\{\alpha = \alpha_{nn} : \frac{\delta}{2-\delta} < \alpha < 1\right\}$$

holds by Theorems 5.3.14-5.3.15 and from Table 8.2, $\sigma_{ap}(\Lambda, c_0) = \sigma(\Lambda, c_0)\backslash III_1\sigma(\Lambda, c_0)$. Therefore, we have

$$\sigma_{ap}(\Lambda, c_0) = \left\{\alpha \in \mathbb{C} : \left|\alpha - \frac{1}{2-\delta}\right| = \frac{1-\delta}{2-\delta}\right\} \cup E.$$

(b) Since $\sigma_\delta(\Lambda, c_0) = \sigma(\Lambda, c_0)\backslash I_3\sigma(\Lambda, c_0)$ from Table 8.2 and $I_3\sigma(\Lambda, c_0) = \emptyset$ by Theorem 5.3.17, we have $\sigma_\delta(\Lambda, c_0) = \sigma(\Lambda, c_0)$.

(c) Since the equality $\sigma_{co}(\Lambda, c_0) = III_1\sigma(\Lambda, c_0) \cup III_2\sigma(\Lambda, c_0) \cup III_3\sigma(\Lambda, c_0)$ holds from Table 8.2, we have

$$\sigma_{co}(\Lambda, c_0) = \left\{\alpha \in \mathbb{C} : \left|\alpha - \frac{1}{2-\delta}\right| < \frac{1-\delta}{2-\delta}\right\} \cup S$$

by Theorems 5.3.14-5.3.17. □

The next corollary can be obtained from Proposition 5.2.9.

Corollary 5.3.19. [232, Corollary 21] *The following results hold:*

(a) $\sigma_{ap}(\Lambda^*, \ell_1) = \sigma(\Lambda, c_0)$.

(b) $\sigma_\delta(\Lambda^*, \ell_1) = \{\alpha \in \mathbb{C} : |\alpha - (2-\delta)^{-1}| = (1-\delta)/(2-\delta)\} \cup E$.

(c) $\sigma_p(\Lambda^*, \ell_1) = \{\alpha \in \mathbb{C} : |\alpha - (2-\delta)^{-1}| < (1-\delta)/(2-\delta)\} \cup S$.

5.3.2 The Fine Spectrum of the Operator Λ on the Sequence Space c

In this subsection, we investigate the fine spectrum of the operator Λ over the sequence space c.

Theorem 5.3.20. [232, Theorem 22] $\sigma(\Lambda, c) \subseteq \{\alpha \in \mathbb{C} : |2\alpha - 1| \leq 1\}$.

Proof. This is obtained in the similar way used in the proof of Theorem 5.3.5. □

Theorem 5.3.21. [232, Theorem 23] *Suppose that μ and η are defined as in Theorems 5.3.6 and 5.3.7, respectively. Then,*

$$\left\{\alpha \in \mathbb{C} : \left|\alpha - \frac{1}{2-\mu}\right| \leq \frac{1-\mu}{2-\mu}\right\} \cup S \subseteq \sigma(\Lambda, c) \subseteq \left\{\alpha \in \mathbb{C} : \left|\alpha - \frac{1}{2-\eta}\right| \leq \frac{1-\eta}{2-\eta}\right\} \cup S,$$

where $S = \left\{\dfrac{\alpha_j - \alpha_{j-1}}{\alpha_j}, j \geq 0\right\}$.

Proof. This is similar to the proof of Theorem 5.3.6 and Theorem 5.3.7. To avoid the repetition of similar statements, we omit details. □

Corollary 5.3.22. [232, Corollary 24] *Let δ be defined as in Corollary 5.3.8. Then,* $\sigma(\Lambda, c) = \left\{ \alpha \in \mathbb{C} : \left| \alpha - \dfrac{1}{2 - \delta} \right| \leq \dfrac{1 - \delta}{2 - \delta} \right\} \cup S.$

If $T : c \to c$ is a bounded linear operator with the matrix A, then $T^* :$ $c^* \to c^*$ acting on $\mathbb{C} \oplus \ell_1$ has a matrix representation of the form $\begin{bmatrix} \chi & 0 \\ b & A^t \end{bmatrix}$, where χ is the limit of the sequence of row sums of A minus the sum of the limit of the columns of A, and b is the column vector whose k^{th} element is the limit of the kth column of A for each $k \in \mathbb{N}_0$. For $\Lambda : c \to c$, the matrix $\Lambda^* \in B(\ell_1)$ is of the form $\Lambda^* = \begin{bmatrix} 1 & 0 \\ 0 & \Lambda^t \end{bmatrix}$.

Theorem 5.3.23. [232, Theorem 25] *Let δ be defined as in Corollary 5.3.8. Then,* $\sigma_p(\Lambda^*, c^*) = \left\{ \alpha \in \mathbb{C} : \left| \alpha - \dfrac{1}{2 - \delta} \right| < \dfrac{1 - \delta}{2 - \delta} \right\} \cup S.$

Proof. Suppose that $\Lambda^* x = \alpha x$ for $x \neq \theta$ in $c^* \cong \ell_1$. Then, by solving the system of linear equations

$$
\left.
\begin{aligned}
(1 - \alpha)x_0 &= 0, \\
x_2 &= \frac{\alpha_1 - \alpha_0}{\alpha_0} \left(1 - \frac{1}{\alpha} \right) x_1, \\
x_3 &= \frac{\alpha_2 - \alpha_1}{\alpha_0} \left(1 - \frac{c_1}{\alpha} \right) \left(1 - \frac{1}{\alpha} \right) x_1 \\
&\vdots \\
x_n &= \frac{\alpha_{n-1} - \alpha_{n-2}}{\alpha_0} \left(1 - \frac{1}{\alpha} \right) x_1 \prod_{j=1}^{n-1} \left(1 - \frac{c_j}{\alpha} \right) \\
&\vdots
\end{aligned}
\right\},
$$

we get by assumption $(1 - \alpha)x_0 = 0$ with $\alpha = 1$ that $x = (x_0, x_1, 0, 0, \ldots) \in c$. If $\alpha \neq 1$, then we have $x_0 = 0$ and $(x_n) \in \ell_1$ if and only if $\left| 1 + (1 - \alpha^{-1})(\alpha_j - \alpha_{j-1})/\alpha_{j-1} \right| < 1$, by Theorem 5.3.10.
This completes the proof. □

Theorem 5.3.24. [232, Theorem 26] *Let δ be defined as in Corollary 5.3.8. Then,*

$$
\sigma_p(\Lambda, c) = \left\{ \alpha = c_n \in \mathbb{C} : 0 \leq \alpha \leq \frac{\delta}{2 - \delta} \right\} \cup \{1\}.
$$

Proof. The proof is identical to the proof of Theorem 5.3.11. □

Theorem 5.3.25. [232, Theorem 27] $\sigma_r(\Lambda, c) = \sigma_p(\Lambda^*, c^*) \backslash \sigma_p(\Lambda, c)$.

Proof. For $\alpha \in \sigma_p(\Lambda^*, c^*) \backslash \sigma_p(\Lambda, c)$, the operator $\Lambda - \alpha I$ is a triangle, so it has an inverse. But $\Lambda^* - \alpha I$ is not one to one by Theorem 5.3.23. Therefore, by Lemma 5.2.10, $\overline{R(\Lambda - \alpha I)} \neq c_0$ and this step concludes the proof. □

Theorem 5.3.26. [232, Theorem 28] *Let δ be defined as in Corollary 5.3.8 and $c_n \geq \delta$ for all sufficiently large n. Then,*

$$\sigma_c(\Lambda, c) = \left\{ \alpha \in \mathbb{C} : \left| \alpha - \frac{1}{2 - \delta} \right| = \frac{1 - \delta}{2 - \delta}, \alpha \neq 1, \frac{\delta}{2 - \delta} \right\}.$$

Proof. This is obtained in the similar way used in the proof of Theorem 5.3.13. □

Theorem 5.3.27. [232, Theorem 29] *Let δ be defined as in Corollary 5.3.8 and is less than 1. If α satisfies $|\alpha - 1/(2 - \delta)| < (1 - \delta)/(2 - \delta)$ and $\alpha \notin S$, then $\alpha \in III_1\sigma(\Lambda, c)$.*

Proof. The proof is identical to that of Theorem 5.3.14. □

Theorem 5.3.28. [232, Theorem 30] *Let δ be defined as in Corollary 5.3.8 and $\delta < 1$. If $\alpha = \delta$ or $\alpha = c_n$ for all $n \in \mathbb{N}_0$ and $\delta/(2 - \delta) < \alpha < 1$, then $\alpha \in III_1\sigma(\Lambda, c)$.*

Proof. The proof is identical to that of Theorem 5.3.15. □

Theorem 5.3.29. [232, Theorem 31] *Let η be defined as in Theorem 5.3.7. If there exists the values of n such that $0 \leq c_n \leq \eta/(2 - \eta)$, then $\alpha = c_n$ implies $\alpha \in III_3\sigma(\Lambda, c)$. Also, $1 \in III_3\sigma(\Lambda, c)$.*

Proof. The proof is identical to that of Theorem 5.3.16. □

Theorem 5.3.30. [232, Theorem 32] *The following statement holds: $I_3\sigma(\Lambda, c) = III_2\sigma(\Lambda, c) = \emptyset$.*

Proof. Let δ be defined as in Corollary 5.3.8 and $(\alpha_n - \alpha_{n-1})/\alpha_n \geq \delta$ for all sufficiently large n, then $I_3\sigma(\Lambda, c) = \emptyset$ and $III_2\sigma(\Lambda, c) = \emptyset$ are obtained from Corollary 5.3.22, Theorems 5.3.26-5.3.29. □

Theorem 5.3.31. [232, Theorem 33] *The following statements hold:*

(a) $\sigma_{ap}(\Lambda, c) = \left\{ \alpha \in \mathbb{C} : \left| \alpha - (2 - \delta)^{-1} \right| = (1 - \delta)/(2 - \delta) \right\} \cup E$.

(b) $\sigma_\delta(\Lambda, c) = \sigma(\Lambda, c)$.

(c) $\sigma_{co}(\Lambda, c) = \left\{ \alpha \in \mathbb{C} : \left| \alpha - (2 - \delta)^{-1} \right| < (1 - \delta)/(2 - \delta) \right\} \cup S$.

Proof. (a) Since the relation

$$III_1\sigma(\Lambda, c) = \left[\left\{ \alpha \in \mathbb{C} : \left| \alpha - \frac{1}{2 - \delta} \right| < \frac{1 - \delta}{2 - \delta} \right\} \backslash S \right] \cup \left\{ \alpha = \alpha_{nn} : \frac{\delta}{2 - \delta} < \alpha < 1 \right\}$$

holds by Theorems 5.3.27–5.3.28 and from Table 8.2, $\sigma_{ap}(\Lambda, c) = \sigma(\Lambda, c) \backslash III_1\sigma(\Lambda, c)$. Therefore, we have

$$\sigma_{ap}(\Lambda, c) = \left\{\alpha \in \mathbb{C} : \left|\alpha - \frac{1}{2-\delta}\right| = \frac{1-\delta}{2-\delta}\right\} \cup E.$$

(b) Since $\sigma_\delta(\Lambda, c) = \sigma(\Lambda, c) \backslash I_3\sigma(\Lambda, c)$ from Table 8.2 and $I_3\sigma(\Lambda, c) = \emptyset$ by Theorem 5.3.30, it is immediate that $\sigma_\delta(\Lambda, c) = \sigma(\Lambda, c)$.

(c) Since the equality $\sigma_{co}(\Lambda, c) = III_1\sigma(\Lambda, c) \cup III_2\sigma(\Lambda, c) \cup III_3\sigma(\Lambda, c)$ holds from Table 8.2, we have

$$\sigma_{co}(\Lambda, c) = \left\{\alpha \in \mathbb{C} : \left|\alpha - \frac{1}{2-\delta}\right| < \frac{1-\delta}{2-\delta}\right\} \cup S$$

by Theorems 5.3.27–5.3.30.   □

The next corollary can be obtained from Proposition 5.2.9.

Corollary 5.3.32. [232, Corollary 34] *The following statements hold:*

(a) $\sigma_{ap}(\Lambda^*, \ell_1) = \sigma(\Lambda, c)$.

(b) $\sigma_\delta(\Lambda^*, \ell_1) = \left\{\alpha \in \mathbb{C} : \left|\alpha - (2-\delta)^{-1}\right| = (1-\delta)/(2-\delta)\right\} \cup E$.

(c) $\sigma_p(\Lambda^*, \ell_1) = \left\{\alpha \in \mathbb{C} : \left|\alpha - (2-\delta)^{-1}\right| < (1-\delta)/(2-\delta)\right\} \cup S$.

Let A be an infinite matrix and the set c_A denotes the *convergence domain of a matrix* A. A theorem which proves that $c_A = c$ is called a *Mercerian theorem*, after Mercer, who proved a significant theorem of this type [145, p. 186].

Now, we can give our final theorem of this section.

Theorem 5.3.33. [232, Theorem 35] *Suppose that* $|\alpha + 1| > |\alpha - 1|$. *Then, the convergence field of* $A = \alpha I + (1 - \alpha)\Lambda$ *is* c.

Proof. By Theorem 5.3.20, $\Lambda - [\alpha/(\alpha - 1)]I$ has an inverse in $B(c)$. That is to say that

$$A^{-1} = \frac{1}{1-\alpha}\left(\Lambda - \frac{\alpha}{\alpha-1}I\right)^{-1} \in B(c).$$

Since A is a triangle and is in $B(c)$, A^{-1} is also conservative, which implies that $c_A = c$; [228, p. 12].   □

5.4   On the Fine Spectrum of the Upper Triangle Double Band Matrix Δ^+ on the Sequence Space c_0

In this section, following Dündar and Başar [75], we determine the fine spectrum of the matrix operator Δ^+ defined by an *upper triangle double band matrix* acting on the space c_0 of null sequences, with respect to the Goldberg's classification. Also, we give the approximate point spectrum, defect spectrum and compression spectrum of the matrix operator Δ^+ on c_0.

Now, we may quote the following lemmas which are needed in proving the next theorems:

Lemma 5.4.1. [121, Theorem 2.6] $\sigma_p(\Delta, c_0) = \emptyset$.

Lemma 5.4.2. [8, Theorem 2.6] $\sigma_c(\Delta, c_0) = \{\alpha \in \mathbb{C} : |\alpha - 1| = 1\}$.

5.4.1 The Spectrum and the Fine Spectrum of the Upper Triangle Double Band Matrix Δ^+ on the Sequence Space c_0

In this subsection, we study spectrum and fine spectrum of the operator represented by the *upper triangle double band matrix* Δ^+ on the sequence space c_0. The operator Δ^+ is represented by the matrix $\Delta^+ = (d_{nk})$, as follows;

$$d_{nk} := \begin{cases} 1 & , \quad k = n, \\ -1 & , \quad k = n+1, \\ 0 & , \quad \text{otherwise} \end{cases}$$

for all $k, n \in \mathbb{N}_0$. First, we give the following theorem which presents the null space $N(\Delta^+ - \alpha I)$ of the operator $\Delta^+ - \alpha I$ on the sequence space c_0.

Theorem 5.4.3. [75, Theorem 1] *Let* $|\alpha - 1| \geq 1$. *Then, the null space* $N(\Delta^+ - \alpha I)$ *of the operator* $\Delta^+ - \alpha I$ *on the sequence space* c_0 *is* $\{\theta\}$.

Proof. By solving the system of linear equations

$$\left. \begin{aligned} x_0 - x_1 &= \alpha x_0 \\ x_1 - x_2 &= \alpha x_1 \\ x_2 - x_3 &= \alpha x_2 \\ &\vdots \end{aligned} \right\}$$

we obtain that $x_n = (1 - \alpha)^{n-n_0} x_{n_0}$ for $n > n_0$ with $x_{n_0} \neq 0$, which leads us to the fact that

$$N(\Delta^+ - \alpha I) := \{x = (x_n) \in c_0 : x_n = (1 - \alpha)^{n-n_0} x_{n_0}, (n > n_0)\}.$$

Therefore, it is clear that the null space $N(\Delta^+ - \alpha I)$ of the operator $\Delta^+ - \alpha I$ consists of zero vector θ for $|\alpha - 1| \geq 1$, as asserted. $\qquad \square$

Theorem 5.4.4. [75, Theorem 2] $\sigma(\Delta^+, c_0) := \{\alpha \in \mathbb{C} : |\alpha - 1| \leq 1\}$.

Proof. Define $D := \{\alpha \in \mathbb{C} : |\alpha - 1| > 1\}$. It is enough to prove that $(\Delta^+ - \alpha I)^{-1}$ exists and is in $(c_0 : c_0)$ for $\alpha \in D$ and $(\Delta^+ - \alpha I)^{-1} \notin (c_0 : c_0)$ for

$\alpha \notin D$. Let $\alpha \in D$. Since $\Delta^+ - \alpha I$ is triangle, so $(\Delta^+ - \alpha I)^{-1}$ exists and solving $(\Delta^+ - \alpha I)x = y$ for x in terms of y gives the matrix $B = (b_{nk})$ of the equation $x = By$. Therefore, the matrix $B = (b_{nk})$ is given by

$$b_{nk} := \begin{cases} (1-\alpha)^{n-k-1} & , \quad k \geq n, \\ 0 & , \quad k < n \end{cases}$$

for all $k, n \in \mathbb{N}_0$. Then, we have

$$\|(\Delta^+ - \alpha I)^{-1}\|_{(c_0:c_0)} = \sup_{n\in\mathbb{N}_0} \sum_{k=n}^{\infty} \frac{|1-\alpha|^n}{|1-\alpha|^{k+1}}$$

$$= \sup_{n\in\mathbb{N}_0} \sum_{k} \frac{1}{|1-\alpha|^{k+1}} < \infty. \qquad (5.4.1)$$

This shows that $(\Delta^+ - \alpha I)^{-1} \in (c_0 : c_0)$. Also from (5.4.1), we can see that

$$\|(\Delta^+ - \alpha I)^{-1}\|_{(c_0:c_0)} = \infty$$

for $\alpha \notin D$.

This completes the proof of the theorem. $\qquad\square$

Theorem 5.4.5. [75, Theorem 3] $\sigma_p(\Delta^+, c_0) = \{\alpha \in \mathbb{C} : |\alpha - 1| < 1\}$.

Proof. Suppose $\Delta^+ x = \alpha x$ for $x \neq \theta = (0, 0, 0, \ldots)$ in c_0. Then, by solving the system of linear equations

$$\left. \begin{array}{rcl} x_0 - x_1 & = & \alpha x_0 \\ x_1 - x_2 & = & \alpha x_1 \\ x_2 - x_3 & = & \alpha x_2 \\ & \vdots & \end{array} \right\},$$

we can find that, if x_{n_0} is the first non-zero element of the sequence $x = (x_n)$, then we have

$$x_n = (1-\alpha)^{n-n_0} x_{n_0} \quad \text{for all} \quad n > n_0.$$

Hence, we show that $|1 - \alpha| < 1$ if and only if $x \in c_0$. $\qquad\square$

Since the adjoint operator of matrix transformation on c_0 is the transpose of the matrix, then by Lemma 5.5.1, we have the following result:

Corollary 5.4.6. [75, Corollary 1] $\sigma_p((\Delta^+)^*, c_0^*) = \emptyset$.

Theorem 5.4.7. [75, Theorem 4] $\sigma_c(\Delta^+, c_0) = \{\alpha \in \mathbb{C} : |\alpha - 1| = 1\}$.

Proof. From Theorem 6.8.25, $\alpha \notin \sigma_p(\Delta^+, c_0)$. Then $\alpha I - \Delta^+$ is one to one, and hence has an inverse. Also, by Corollary 5.4.6 $\alpha I - (\Delta^+)^*$ is one to one and by Lemma 5.2.10, we have $\overline{R(\alpha I - \Delta^+)} = c_0$.

This completes the proof of the theorem. $\qquad\square$

Theorem 5.4.8. [75, Theorem 5] $\sigma_r(\Delta^+, c_0) = \emptyset$.

Proof. Since the set of the spectrum is union of the point spectrum, the continuous spectrum and the residual spectrum, then from Theorems 5.4.4, 6.8.25 and 5.4.7 we observe that $\sigma_r(\Delta^+, c_0) = \emptyset$.

This completes the proof of the theorem. $\qquad\qquad\square$

Theorem 5.4.9. [75, Theorem 6] *If* $\alpha \neq 1$ *and* $\alpha \in \sigma_p(\Delta^+, c_0)$, *then* $\alpha \in II_3\sigma(\Delta^+, c_0)$.

Proof. By Theorem 6.8.25, we have $\alpha I - \Delta^+ \in II_3 \cup I_3$. Now, we show that the operator $\alpha I - \Delta^+$ is not onto. For the sequence $y = \{(1 - \alpha)^n\} \in c_0$, since $x \in c_0$ is not present that supply the equality $(\alpha I - \Delta^+)x = y$, so the transformation $\alpha I - \Delta^+$ is not onto which is what we wished to prove. $\quad\square$

Theorem 5.4.10. [75, Theorem 7] $1 \in I_3\sigma(\Delta^+, c_0)$.

Proof. For $\alpha = 1$, the matrix $\alpha I - \Delta^+ = I - \Delta^+$ is

$$
\alpha I - \Delta^+ = \begin{bmatrix} 0 & 1 & 0 & 0 & \cdots \\ 0 & 0 & 1 & 0 & \cdots \\ 0 & 0 & 0 & 1 & \cdots \\ 0 & 0 & 0 & 0 & \cdots \\ \vdots & \vdots & \vdots & \vdots & \ddots \end{bmatrix}.
$$

It is clear by Theorem 6.8.25 that $I - \Delta^+ \in 3$.

To show $I - \Delta^+ \in I$, we prove that the transformation $I - \Delta^+$ is onto. From the equation $(I - \Delta^+)x = y$, we have for $x, y \in c_0$ that

$$
\left.\begin{array}{rcl} x_1 & = & y_0 \\ x_2 & = & y_1 \\ x_3 & = & y_2 \\ & \vdots & \end{array}\right\},
$$

such that for every $y \in c_0$ there is a sequence $x \in c_0$. Then, this show that the transformation $I - \Delta^+$ is onto.

This step concludes the proof. $\qquad\qquad\square$

Theorem 5.4.11. [75, Theorem 8] *If* $\alpha \in \sigma_c(\Delta^+, c_0)$, *then* $\alpha I - \Delta^+ \in II_2\sigma(\Delta^+, c_0)$.

Proof. Let $|\alpha - 1| = 1$. By Theorem 6.8.25 the transformation $\alpha I - \Delta^+$ has an inverse, from Theorem 5.4.4 the transformation $(\alpha I - \Delta^+)^{-1}$ is discontinuous and by Theorem 5.4.2 since $\overline{R(\alpha I - \Delta^+)} = c_0$, we have

$$
\alpha I - \Delta^+ \in I_2 \cup II_2\sigma(\Delta^+, c_0).
$$

Also, we can obtain from Theorem 5.4.9 that the transformation $\alpha I - \Delta^+$ is not onto.

This step completes the proof. $\qquad\qquad\square$

Theorem 5.4.12. [75, Theorem 9] *The following statements hold:*

(a) $\sigma_{ap}\left(\Delta^+, c_0\right) = \{\alpha \in \mathbb{C} : |\alpha - 1| \leq 1\}$.

(b) $\sigma_\delta\left(\Delta^+, c_0\right) = \{\alpha \in \mathbb{C} : |\alpha - 1| \leq 1\} \setminus \{1\}$.

(c) $\sigma_{co}\left(\Delta^+, c_0\right) = \emptyset$.

Proof. Since the following equality

$$\sigma\left(\Delta^+, c_0\right) = I_3\sigma\left(\Delta^+, c_0\right) \cup II_2\sigma\left(\Delta^+, c_0\right) \cup II_3\sigma\left(\Delta^+, c_0\right)$$

holds by Theorems 5.4.4–5.4.10, and the subdivisions in Goldberg's classification are disjoint, then we must have

$$III_1\sigma\left(\Delta^+, c_0\right) = III_2\sigma\left(\Delta^+, c_0\right) = III_3\sigma\left(\Delta^+, c_0\right) = \emptyset.$$

(a) $\sigma_{ap}\left(\Delta^+, c_0\right) = \sigma\left(\Delta^+, c_0\right) \setminus III_1\sigma\left(\Delta^+, c_0\right)$ is obtained from Table 8.2. Then, one can easily see that

$$\sigma_{ap}\left(\Delta^+, c_0\right) = \sigma\left(\Delta^+, c_0\right) = \{\alpha \in \mathbb{C} : |\alpha - 1| \leq 1\},$$

as desired.

(b) Since the following equality

$$\sigma_\delta\left(\Delta^+, c_0\right) = \sigma\left(\Delta^+, c_0\right) \setminus I_3\sigma\left(\Delta^+, c_0\right)$$

holds from Table 8.2, we derive that $\sigma_\delta\left(\Delta^+, c_0\right) = \{\alpha \in \mathbb{C} : |\alpha - 1| \leq 1\} \setminus \{1\}$.

(c) From Table 8.2,

$$\sigma_{co}\left(\Delta^+, c_0\right) = III_1\sigma\left(\Delta^+, c_0\right) \cup III_2\sigma\left(\Delta^+, c_0\right) \cup III_3\sigma\left(\Delta^+, c_0\right),$$

then we have $\sigma_{co}\left(\Delta^+, c_0\right) = \emptyset$. □

The following corollary can be obtained by Proposition 5.2.9.

Corollary 5.4.13. [75, Corollary 2] *The following statements hold:*

(a) $\sigma_{ap}\left((\Delta^+)^* = \Delta, c_0^* \cong \ell_1\right) = \{\alpha \in \mathbb{C} : |\alpha - 1| \leq 1\} \setminus \{1\}$.

(b) $\sigma_\delta\left((\Delta^+)^* = \Delta, c_0^* \cong \ell_1\right) = \{\alpha \in \mathbb{C} : |\alpha - 1| \leq 1\}$.

(c) $\sigma_p\left((\Delta^+)^* = \Delta, c_0^* \cong \ell_1\right) = \emptyset$, [121, Theorem 2.6].

5.5 On the Fine Spectrum of the Generalized Difference Operator Defined by a Double Sequential Band Matrix over the Sequence Space ℓ_p, $(1 < p < \infty)$

In this section, following Karaisa and Başar [116], we investigate the fine spectrum with respect to the Goldberg's classification of the operator $B(\tilde{r}, \tilde{s})$

defined by a double sequential band matrix over the sequence space ℓ_p, where $1 < p < \infty$. Since $B(\widetilde{r}, \widetilde{s})$ is reduced in the special case $r_k = r \neq 0$ and $s_k = s \neq 0$ for all $k \in \mathbb{N}_0$ to $B(r, s)$, these results are much more general than the spectrum of the generalized difference operator $B(r, s)$ over ℓ_p obtained by Bilgiç and Furkan [50]. We quote some lemmas which are needed in proving the theorems given in the present section.

Lemma 5.5.1. [174, p. 253, Theorem 34.16] *The matrix $A = (a_{nk})$ gives rise to a bounded linear operator $T \in B(\ell_1)$ from ℓ_1 into itself if and only if the supremum of ℓ_1 norms of the columns of A is bounded.*

Lemma 5.5.2. [174, p. 245, Theorem 34.3] *The matrix $A = (a_{nk})$ gives rise to a bounded linear operator $T \in B(\ell_\infty)$ from ℓ_∞ into itself if and only if the supremum of ℓ_1 norms of the rows of A is bounded.*

Lemma 5.5.3. [174, p. 254, Theorem 34.18] *Let $1 < p < \infty$ and $A \in (\ell_\infty : \ell_\infty) \cap (\ell_1 : \ell_1)$. Then, $A \in (\ell_p : \ell_p)$.*

Let $\widetilde{r} = (r_k)$ and $\widetilde{s} = (s_k)$ be two sequences whose elements are either constants or distinct non-zero real numbers such that:

$$\lim_{k \to \infty} r_k = r,$$
$$\lim_{k \to \infty} s_k = s \neq 0$$
$$|r_k - r| \neq |s|.$$

Then, we define the sequential generalized difference matrix $B(\widetilde{r}, \widetilde{s})$ by

$$B(\widetilde{r}, \widetilde{s}) = \begin{bmatrix} r_0 & 0 & 0 & 0 & \cdots \\ s_0 & r_1 & 0 & 0 & \cdots \\ 0 & s_1 & r_2 & 0 & \cdots \\ 0 & 0 & s_2 & r_3 & \cdots \\ \vdots & \vdots & \vdots & \vdots & \ddots \end{bmatrix}.$$

Therefore, we introduce the operator $B(\widetilde{r}, \widetilde{s})$ from ℓ_p into itself by

$$B(\widetilde{r}, \widetilde{s})x = (r_k x_k + s_{k-1} x_{k-1})_{k=0}^{\infty} \text{ with } s_{-1} = x_{-1} = 0, \text{ where } x = (x_k) \in \ell_p.$$

5.5.1   The Fine Spectrum of the Operator $B(\widetilde{r}, \widetilde{s})$ on the Sequence Space ℓ_p

Theorem 5.5.4. [38, 3.1. Theorem] *The operator $B(\widetilde{r}, \widetilde{s}) : \ell_p \to \ell_p$ is a bounded linear operator and*

$$(|r_k|^p + |s_k|^p)^{1/p} \leq \|B(\widetilde{r}, \widetilde{s})\|_p \leq \|\widetilde{s}\|_\infty + \|\widetilde{r}\|_\infty. \tag{5.5.1}$$

Proof. Since the linearity of the operator $\widetilde{B}(\tilde{r}, \tilde{s})$ is easy to show, we omit details.

Now, we prove that (5.5.1) holds for the operator $B(\tilde{r}, \tilde{s})$ on the space ℓ_p. It is trivial that $B(\tilde{r}, \tilde{s})e^{(k)} = (0, 0, \ldots, r_k, s_k, 0, 0, \ldots)$ for $e^{(k)} \in \ell_p$. Therefore, we have

$$\frac{\|B(\tilde{r}, \tilde{s})e^{(k)}\|_p}{\|e^{(k)}\|_p} = (|r_k|^p + |s_k|^p)^{1/p},$$

which implies that

$$(|r_k|^p + |s_k|^p)^{1/p} \leq \|B(\tilde{r}, \tilde{s})\|_p. \tag{5.5.2}$$

Let $x = (x_k) \in \ell_p$, where $p > 1$. Then, since $(s_{k-1}x_{k-1})_{k\in\mathbb{N}_1}, (r_k x_k)_{k\in\mathbb{N}_0} \in \ell_p$ it is easy to see by Minkowski's inequality that

$$
\begin{aligned}
\|B(\tilde{r}, \tilde{s})x\|_p &= \left(\sum_k |s_{k-1}x_{k-1} + r_k x_k|^p \right)^{1/p} \\
&\leq \left(\sum_{k=1}^{\infty} |s_{k-1}x_{k-1}|^p \right)^{1/p} + \left(\sum_k |r_k x_k|^p \right)^{1/p} \\
&\leq (\|\tilde{s}\|_\infty + \|\tilde{r}\|_\infty)\|x\|_p,
\end{aligned}
$$

which leads us to the result that

$$\|B(\tilde{r}, \tilde{s})\|_p \leq \|\tilde{s}\|_\infty + \|\tilde{r}\|_\infty. \tag{5.5.3}$$

Therefore, by combining the inequalities in (5.5.2) and (5.5.3) we have (5.5.1), as desired. $\qquad\square$

Theorem 5.5.5. [38, 3.2. Theorem] *Let* $\mathcal{A} = \{\alpha \in \mathbb{C} : |r - \alpha| \leq |s|\}$ *and* $\mathcal{B} = \{r_k : k \in \mathbb{N}_0, |r - r_k| > |s|\}$. *Then, the set* $\mathcal{B}$ *is finite and* $\sigma[B(\tilde{r}, \tilde{s}), \ell_p] = \mathcal{A} \cup \mathcal{B}$.

Proof. We firstly prove that $\sigma[B(\tilde{r}, \tilde{s}), \ell_p] \subseteq \mathcal{A} \cup \mathcal{B}$ which is equivalent to show that $\alpha \in \mathbb{C}$ such that $|r - \alpha| > |s|$ and $\alpha \neq r_k$ for all $k \in \mathbb{N}_0$ implies $\alpha \notin \sigma[B(\tilde{r}, \tilde{s}), \ell_p]$. It is easy to see that $\mathcal{B}$ is finite and $\{r_k \in \mathbb{C} : k \in \mathbb{N}_0\} \subseteq \mathcal{A} \cup \mathcal{B}$. So, we omit details.

It is immediate that $B(\tilde{r}, \tilde{s}) - \alpha I$ is a triangle and so has an inverse. Let $y = (y_k) \in \ell_1$. Then, by solving the equation

$$
[B(\tilde{r}, \tilde{s}) - \alpha I]x =
\begin{bmatrix}
r_0 - \alpha & 0 & 0 & \cdots \\
s_0 & r_1 - \alpha & 0 & \cdots \\
0 & s_1 & r_2 - \alpha & \cdots \\
\vdots & \vdots & \vdots & \ddots
\end{bmatrix}
\begin{bmatrix}
x_0 \\ x_1 \\ x_2 \\ \vdots
\end{bmatrix}
$$

$$
= \begin{bmatrix}
(r_0 - \alpha)x_0 \\
s_0 x_0 + (r_1 - \alpha)x_1 \\
s_1 x_1 + (r_2 - \alpha)x_2 \\
\vdots
\end{bmatrix}
= \begin{bmatrix}
y_0 \\ y_1 \\ y_2 \\ \vdots
\end{bmatrix}
$$

for $x = (x_k)$ in terms of y, we obtain

$$
\begin{aligned}
x_0 &= \frac{y_0}{r_0 - \alpha} \\
x_1 &= \frac{y_1}{r_1 - \alpha} + \frac{-s_0 y_0}{(r_1 - \alpha)(r_0 - \alpha)} \\
x_2 &= \frac{y_2}{r_2 - \alpha} + \frac{-s_1 y_1}{(r_2 - \alpha)(r_1 - \alpha)} + \frac{s_0 s_1 y_0}{(r_2 - \alpha)(r_1 - \alpha)(r_0 - \alpha)} \\
&\vdots \\
x_k &= \frac{(-1)^k s_0 s_1 s_2 \cdots s_{k-1} y_0}{(r_0 - \alpha)(r_1 - \alpha)(r_2 - \alpha) \cdots (r_k - \alpha)} + \cdots - \frac{s_{k-1} y_{k-1}}{(r_k - \alpha)(r_{k-1} - \alpha)} + \frac{y_k}{r_k - \alpha} \\
&\vdots
\end{aligned}
$$

Therefore, we obtain $B = [B(\widetilde{r}, \widetilde{s}) - \alpha I]^{-1}$, as follows:

$$
B = \begin{bmatrix}
\dfrac{1}{r_0 - \alpha} & 0 & 0 & \cdots \\
\dfrac{-s_0}{(r_1 - \alpha)(r_0 - \alpha)} & \dfrac{1}{r_1 - \alpha} & 0 & \cdots \\
\dfrac{s_0 s_1}{(r_0 - \alpha)(r_1 - \alpha)(r_2 - \alpha)} & \dfrac{-s_1}{(r_2 - \alpha)(r_1 - \alpha)} & \dfrac{1}{r_2 - \alpha} & \cdots \\
\vdots & \vdots & \vdots & \ddots
\end{bmatrix}.
$$

Then, $\sum_k |x_k| \leq \sum_k S^k |y_k|$, where

$$
S^k = \left| \frac{1}{r_k - \alpha} \right| + \left| \frac{s_k}{(r_k - \alpha)(r_{k+1} - \alpha)} \right| + \left| \frac{s_k s_{k+1}}{(r_k - \alpha)(r_{k+1} - \alpha)(r_{k+2} - \alpha)} \right| + \cdots .
$$

Define S_n^k by

$$
\begin{aligned}
S_n^k ={}& \left| \frac{1}{r_k - \alpha} \right| + \left| \frac{s_k}{(r_k - \alpha)(r_{k+1} - \alpha)} \right| + \left| \frac{s_k s_{k+1}}{(r_k - \alpha)(r_{k+1} - \alpha)(r_{k+2} - \alpha)} \right| + \cdots + \\
&+ \left| \frac{s_k s_{k+1} \cdots s_{n+k}}{(r_k - \alpha)(r_{k+1} - \alpha)(r_{k+2} - \alpha) \cdots (r_{k+n+1} - \alpha)} \right|
\end{aligned}
$$

for all $k, n \in \mathbb{N}_0$. Then, since

$$
S_n = \lim_{k \to \infty} S_n^k = \left| \frac{1}{r - \alpha} \right| + \left| \frac{s}{(r - \alpha)^2} \right| + \left| \frac{s^2}{(r - \alpha)^3} \right| + \cdots + \left| \frac{s^{n+1}}{(r - \alpha)^{n+2}} \right|,
$$

we have

$$
S = \lim_{n \to \infty} S_n = \left| \frac{1}{r - \alpha} \right| \left(1 + \left| \frac{s}{r - \alpha} \right| + \left| \frac{s}{r - \alpha} \right|^2 + \cdots \right) < \infty, \qquad (5.5.4)
$$

since $|r - \alpha| > |s|$. Then, we have

$$
\lim_{n \to \infty} \lim_{k \to \infty} S_n^k = \lim_{k \to \infty} \lim_{n \to \infty} S_n^k = S
$$

and $(S^k)_{k \in \mathbb{N}_0} \in c$. Thus,

$$\sum_k |x_k| \le \sum_k S^k |y_k| \le \|(S^k)\|_\infty \sum_k |y_k| < \infty,$$

since $y \in \ell_1$. This shows that $[B(\tilde{r}, \tilde{s}) - \alpha I]^{-1} \in (\ell_1 : \ell_1)$.

Suppose that $y = (y_k) \in \ell_\infty$. By solving the equation $[B(\tilde{r}, \tilde{s}) - \alpha I]x = y$ for $x = (x_k)$ in terms of y, we get $|x_k| \le S_k \left(\sup_{k \in \mathbb{N}_0} |y_k| \right)$, where;

$$S_k = \left| \frac{1}{r_k - \alpha} \right| + \left| \frac{s_{k-1}}{(r_{k-1} - \alpha)(r_k - \alpha)} \right| + \left| \frac{s_{k-1} s_{k-2}}{(r_{k-2} - \alpha)(r_{k-1} - \alpha)(r_k - \alpha)} \right| +$$
$$+ \cdots + \left| \frac{s_0 s_1 \cdots s_{k-1}}{(r_0 - \alpha)(r_1 - \alpha) \cdots (r_k - \alpha)} \right|.$$

Now, we prove that $(S_k) \in \ell_\infty$. Since $|s_k/(r_k - \alpha)| \to |s/(r - \alpha)| = p < 1$, as $k \to \infty$, then there exists $k_0 \in \mathbb{N}_0$ such that $|s_k/(r_k - \alpha)| < p_0$ with $p_0 < 1$ for all $k \ge k_0 + 1$,

$$S_k = \frac{1}{|r_k - \alpha|} \left[1 + \left| \frac{s_{k-1}}{r_{k-1} - \alpha} \right| + \left| \frac{s_{k-1} s_{k-2}}{(r_{k-1} - \alpha)(r_{k-2} - \alpha)} \right| + \right.$$
$$\left. + \cdots + \left| \frac{s_{k-1} s_{k-2} \cdots s_{k_0+1} s_{k_0} \cdots s_0}{(r_{k-1} - \alpha)(r_{k-2} - \alpha) \cdots (r_{k_0+1} - \alpha)(r_{k_0} - \alpha) \cdots (r_0 - \alpha)} \right| \right]$$
$$\le \frac{1}{|r_k - \alpha|} \left[1 + p_0 + p_0^2 + \cdots + p_0^{k-k_0} + p_0^{k-k_0} \frac{|s_{k_0-1}|}{|r_{k_0-1} - \alpha|} + \right.$$
$$\left. + \cdots + p_0^{k-k_0} \left| \frac{s_{k_0-1} s_{k_0-2} \cdots s_0}{(r_{k_0-1} - \alpha)(r_{k_0-2} - \alpha) \cdots (r_0 - \alpha)} \right| \right].$$

Therefore;

$$S_k \le \frac{1}{|r_k - \alpha|} \left(1 + p_0 + p_0^2 + \cdots p_0^{k-k_0} + p_0^{k-k_0} M k_0 \right),$$

where

$$M k_0 = 1 + \left| \frac{s_{k_0-1}}{r_{k_0-1} - \alpha} \right| + \cdots + \left| \frac{s_{k_0-1} s_{k_0-2} \cdots s_0}{(r_{k_0-1} - \alpha)(r_{k_0-2} - \alpha) \cdots (r_0 - \alpha)} \right|.$$

Then, $M k_0 \ge 1$ and so,

$$S_k \le \frac{M k_0}{|r_k - \alpha|} \left(1 + p_0 + p_0^2 + \cdots + p_0^{k-k_0} \right).$$

But there exists $k_1 \in \mathbb{N}_0$ and a real number p_1 such that $1/|r_k - \alpha| < p_1$ for all $k \ge k_1$. Then, $S_k \le (M p_1 k_0)/(1 - p_0)$ for all $k > \max\{k_0, k_1\}$. Hence, $\sup_{k \in \mathbb{N}_0} S_k < \infty$. This shows that $\|x\|_\infty \le \|(S_k)\|_\infty \|y\|_\infty < \infty$, which means $[B(\tilde{r}, \tilde{s}) - \alpha I]^{-1} \in (\ell_\infty : \ell_\infty)$. By Lemma 5.5.3, we have

$$[B(\tilde{r}, \tilde{s}) - \alpha I]^{-1} \in (\ell_p : \ell_p) \text{ for } \alpha \in \mathbb{C} \text{ with } |r - \alpha| > |s| \text{ and } \alpha \ne r_k. \quad (5.5.5)$$

Hence,

$$\sigma[B(\tilde{r},\tilde{s}),\ell_p] \subseteq \mathcal{A} \cup \mathcal{B}. \qquad (5.5.6)$$

Now, we show that $\mathcal{A} \cup \mathcal{B} \subseteq \sigma[B(\tilde{r},\tilde{s}),\ell_p]$. We assume that $\alpha \neq r_k$ for all $k \in \mathbb{N}_0$ and $\alpha \in \mathbb{C}$ with $|r - \alpha| \leq |s|$. Clearly, $B(\tilde{r},\tilde{s}) - \alpha I$ is a triangle and so, $[B(\tilde{r},\tilde{s}) - \alpha I]^{-1}$ exists. For $e^{(0)} = (1,0,0,\dots) \in \ell_p$, $[B(\tilde{r},\tilde{s}) - \alpha I]^{-1}e^{(0)} = S^0 \notin \ell_p$, and so $[B(\tilde{r},\tilde{s}) - \alpha I]^{-1} \notin B(\ell_p)$. Then, $\alpha \in \sigma[B(\tilde{r},\tilde{s}),\ell_p]$. In the case of $r_k = \alpha$ for some $k \in \mathbb{N}_0$, we then have either $\alpha = r$ or $\alpha = r_k \neq r$ for some $k \in \mathbb{N}_0$. Therefore, we get that

$$[B(\tilde{r},\tilde{s}) - r_k I]x = \begin{bmatrix} r_0 - r_k & 0 & 0 & \cdots \\ s_0 & r_1 - r_k & 0 & \cdots \\ 0 & s_1 & r_2 - r_k & \cdots \\ \vdots & \vdots & \vdots & \ddots \end{bmatrix} \begin{bmatrix} x_0 \\ x_1 \\ x_2 \\ \vdots \end{bmatrix}$$

$$= \begin{bmatrix} (r_0 - r_k)x_0 \\ s_0 x_0 + (r_1 - r_k)x_1 \\ s_1 x_1 + (r_2 - r_k)x_2 \\ \vdots \\ s_{k-2}x_{k-2} + (r_{k-1} - r_k)x_{k-1} \\ s_{k-1}x_{k-1} + (r_k - r_k)x_k \\ s_k x_k + (r_{k+1} - r_k)x_{k+1} \\ \vdots \end{bmatrix}.$$

Let $\alpha = r_k = r$ for all $k \in \mathbb{N}_0$ and solving the equation $[B(\tilde{r},\tilde{s}) - \alpha I]x = \theta$, we obtain $x_0 = x_1 = x_2 = \cdots = 0$, which shows that $B(\tilde{r},\tilde{s}) - \alpha I$ is one to one but its range $R[B(\tilde{r},\tilde{s}) - \alpha I] = \{y = (y_k) \in \omega : y \in \ell_p, \; y_1 = 0\}$ is not dense in ℓ_p and $\alpha = r \in \sigma[B(\tilde{r},\tilde{s}),\ell_p]$. Now, let $\alpha = r_k$ for some $k \in \mathbb{N}_0$. Then, the equation $[B(\tilde{r},\tilde{s}) - \alpha I]x = \theta$ yields

$$x_0 = x_1 = x_2 = \cdots = x_{k-1} = 0 \quad \text{and} \quad x_n = \frac{s_{n-1}}{r_k - r_n}x_{n-1} \quad \text{for all} \ \ n \geq k+1.$$

This shows that $B(\tilde{r},\tilde{s}) - \alpha I$ is not injective for $\alpha = r_k$, such that $|\alpha - r| > |s|$. Therefore, $[B(\tilde{r},\tilde{s}) - \alpha I]^{-1}$ does not exist. So, $r_k \in \sigma[B(\tilde{r},\tilde{s}),\ell_p]$ for all $k \in \mathbb{N}_0$. Thus,

$$\mathcal{A} \cup \mathcal{B} \subseteq \sigma[B(\tilde{r},\tilde{s}),\ell_p]. \qquad (5.5.7)$$

Combining the inclusions (5.5.6) and (5.5.7), we get $\sigma[B(\tilde{r},\tilde{s}),\ell_p] = \mathcal{A} \cup \mathcal{B}$. This completes the proof. $\qquad\qquad\qquad\qquad\qquad\qquad\qquad\qquad\qquad\qquad\quad\square$

Here and after, by $\mathcal{C}$ and $\mathcal{SD}$ we denote the set of constant sequences and the set of sequences of distinct non-zero real numbers, respectively.

Theorem 5.5.6. [38, 3.3. Theorem] $\sigma_p[B(\tilde{r}, \tilde{s}), \ell_p] = \begin{cases} \emptyset & , \quad \tilde{r}, \tilde{s} \in \mathcal{C}, \\ \mathcal{B} & , \quad \tilde{r}, \tilde{s} \in \mathcal{SD}. \end{cases}$

Proof. We prove the theorem by dividing into two parts.

Part I. Assume that $\tilde{r}, \tilde{s} \in \mathcal{C}$. Consider $B(\tilde{r}, \tilde{s})x = \alpha x$ for $x \neq \theta = (0, 0, 0, \dots)$ in ℓ_p. That is to say that we should solve the system of linear equations

$$\left. \begin{aligned} rx_0 &= \alpha x_0 \\ sx_0 + rx_1 &= \alpha x_1 \\ sx_1 + rx_2 &= \alpha x_2 \\ &\vdots \\ sx_{k-1} + rx_k &= \alpha x_k \\ &\vdots \end{aligned} \right\} .$$

Case $\alpha = r$. Suppose that x_{n_0} is the first non zero element of the sequence $x = (x_n)$ and $\alpha = r$, then we get $sx_{n_0} + rx_{n_0+1} = \alpha x_{n_0+1}$ which implies $x_{n_0} = 0$ which contradicts the assumption $x_{n_0} \neq 0$. Hence, the equation $B(\tilde{r}, \tilde{s})x = \alpha x$ has no solution $x \neq \theta$.

Part II. Assume that $\tilde{r}, \tilde{s} \in \mathcal{SD}$. Then, by solving the equation $B(\tilde{r}, \tilde{s})x = \alpha x$ for $x \neq \theta = (0, 0, 0, \dots)$ in ℓ_p we obtain $(r_0 - \alpha)x_0 = 0$ and $(r_{k+1} - \alpha)x_{k+1} + s_k x_k = 0$ for all $k \in \mathbb{N}_0$. Hence, for all $\alpha \notin \{r_k : k \in \mathbb{N}_0\}$, we have $x_k = 0$ for all $k \in \mathbb{N}_0$ which contradicts our assumption. So, $\alpha \notin \sigma_p[B(\tilde{r}, \tilde{s}), \ell_p]$. This shows that $\sigma_p[B(\tilde{r}, \tilde{s}), \ell_p] \subseteq \{r_k : k \in \mathbb{N}_0\}\backslash\{r\}$. Now, we prove that

$$\alpha \in \sigma_p[B(\tilde{r}, \tilde{s}), \ell_p] \text{ if and only if } \alpha \in \mathcal{B}.$$

Let $\alpha \in \sigma_p[B(\tilde{r}, \tilde{s}), \ell_p]$. We consider the case $\alpha = r_0$ and $\alpha = r_k$ for some $k \in \mathbb{N}_1$. Then, by solving the equation $B(\tilde{r}, \tilde{s})x = \alpha x$ for $x \neq \theta = (0, 0, 0, \dots)$ in ℓ_p with $\alpha = r_0$ we derive that

$$x_k = \frac{s_0 s_1 s_2 \dots s_{k-1}}{(r_0 - r_k)(r_0 - r_{k-1})(r_0 - r_{k-2}) \cdots (r_0 - r_1)} x_0 \text{ for all } k \in \mathbb{N}_1,$$

which can be expressed by the recursion relation

$$x_k = \frac{s_{k-1}}{r_0 - r_k} x_{k-1} \text{ for all } k \in \mathbb{N}_1.$$

Therefore, we have

$$\lim_{k \to \infty} \left| \frac{x_k}{x_{k-1}} \right|^p = \lim_{k \to \infty} \left| \frac{s_{k-1}}{r_k - r_0} \right|^p = \left| \frac{s}{r - r_0} \right|^p \leq 1.$$

But $|s/(r - r_0)|^p \neq 1$. Then, $\alpha = r_0 \in \{r_k : k \in \mathbb{N}_0, |r_k - r| > |s|\} = \mathcal{B}$. If we choose $\alpha = r_k \neq r$ for all $k \in \mathbb{N}_1$, then we get $x_0 = x_1 = x_2 = \cdots = x_{k-1} = 0$ and

$$x_{n+1} = \frac{s_n s_{n-1} s_{n-2} \dots s_k}{(r_k - r_{n+1})(r_k - r_n)(r_k - r_{n-1}) \cdots (r_k - r_{k+1})} x_k \text{ for all } n \in \mathbb{N}_k,$$

which can also be expressed by the recursion relation

$$x_{n+1} = \frac{s_n}{r_k - r_{n+1}} x_n \quad \text{for all } n \in \mathbb{N}_k.$$

Therefore, we have

$$\lim_{n \to \infty} \left| \frac{x_{n+1}}{x_n} \right|^p = \lim_{n \to \infty} \left| \frac{s_n}{r_{n+1} - r_k} \right|^p = \left| \frac{s}{r - r_k} \right|^p \leq 1.$$

But $|s/(r - r_k)| \neq 1$. Then, $\alpha = r_k \in \{r_k : k \in \mathbb{N}_0, |r_k - r| > |s|\} = \mathcal{B}$. Thus, $\sigma_p[B(\tilde{r}, \tilde{s}), \ell_p] \subseteq \mathcal{B}$.

Conversely, let $\alpha \in \mathcal{B}$. Then, there exists $k \in \mathbb{N}_0$ with $\alpha = r_k \neq r$ and

$$\lim_{n \to \infty} \left| \frac{s_n}{r_{n+1} - r_k} \right| = \left| \frac{s}{r - r_k} \right| < 1,$$

so we have $x \in \ell_p$. Thus, $\mathcal{B} \subseteq \sigma_p[B(\tilde{r}, \tilde{s}), \ell_p]$.

This completes the proof. $\qquad \square$

Theorem 5.5.7. [38, 3.4. Theorem] *The point spectrum of $B(\tilde{r}, \tilde{s})$ on the space ℓ_p is the set* $\begin{cases} \{\alpha \in \mathbb{C} : |r - \alpha| < |s|\} & , \; \tilde{r}, \tilde{s} \in \mathcal{C}, \\ \{\alpha \in \mathbb{C} : |r - \alpha| \leq |s|\} \cup \mathcal{B} & , \; \tilde{r}, \tilde{s} \in SD. \end{cases}$

Proof. By solving the equation $B(\tilde{r}, \tilde{s})^* f = \alpha f$ for $\theta \neq f \in \ell_p^* \cong \ell_q$, that is, the system of linear equations

$$\left. \begin{aligned} r_0 f_0 + s_0 f_1 &= \alpha f_0 \\ r_1 f_1 + s_1 f_2 &= \alpha f_1 \\ r_2 f_2 + s_2 f_3 &= \alpha f_2 \\ &\;\;\vdots \\ r_{k-1} f_{k-1} + s_{k-1} f_k &= \alpha f_{k-1} \\ &\;\;\vdots \end{aligned} \right\},$$

we derive that $f_k = (\alpha - r_{k-1}) f_{k-1}/s_{k-1}$ for all $k \in \mathbb{N}_1$. Therefore, we have

$$|f_k| = \left| \frac{\alpha - r_{k-1}}{s_{k-1}} \right| |f_{k-1}| \quad \text{for all } k \in \mathbb{N}_1. \tag{5.5.8}$$

We also prove this theorem by dividing into two parts.

Part I. Assume that $\tilde{r}, \tilde{s} \in \mathcal{C}$ with $r_k = r$ and $s_k = s$ for all $k \in \mathbb{N}_0$. Using (5.5.8), we get

$$f_k = \left(\frac{\alpha - r}{s} \right)^k f_0 \quad \text{for all } k \in \mathbb{N}_1.$$

Then, since

$$\lim_{k \to \infty} \left| \frac{f_{k+1}}{f_k} \right|^q = \left| \frac{\alpha - r}{s} \right|^q < 1 \quad \text{provided} \quad \left| \frac{r - \alpha}{s} \right| < 1,$$

the series $\sum_{k=1}^{\infty} |f_k|^q = \sum_{k=1}^{\infty} |(\alpha - r)/s|^{q(k-1)} |f_0|$ converges by the ratio test, i.e., $f \in \ell_q$.

If $\alpha \in \mathbb{C}$ with $|\alpha - r| = |s|$, then the ratio test fails. But, since $|f_k| \to |f_0| \neq 0$, as $k \to \infty$, the series $\sum_k |f_k|^q$ is divergent. This means that $f \in \ell_q$ if and only if $f_0 \neq 0$ and $|r - \alpha| < |s|$. Hence, $\sigma_p[B(\widetilde{r}, \widetilde{s})^*, \ell_p^*] = \{\alpha \in \mathbb{C} : |r - \alpha| < |s|\}$.

Part II. Let $\widetilde{r}, \widetilde{s} \in \mathcal{SD}$. It is clear that for all $k \in \mathbb{N}_0$, the vector $f = (f_0, f_1, \ldots, f_k, 0, 0, \ldots)$ is an eigenvector of the operator $B(\widetilde{r}, \widetilde{s})^*$ corresponding to the eigenvalue $\alpha = r_k$, where $f_0 \neq 0$ and $f_n = (\alpha - r_{n-1})f_{n-1}/s_{n-1}$ for all $k \in \{1, 2, 3, \ldots, n\}$. Thus, $\mathcal{B} \subseteq \sigma_p[B(\widetilde{r}, \widetilde{s})^*, \ell_p^*]$. If $|r - \alpha| < |s|$ and $\alpha = r_k$, by taking into account (5.5.8), since

$$\lim_{k \to \infty} \left| \frac{f_k}{f_{k-1}} \right|^q = \lim_{k \to \infty} \left| \frac{\alpha - r_{k-1}}{s_{k-1}} \right|^q = \left| \frac{r - \alpha}{s} \right|^q < 1,$$

the ratio test gives that $f \in \ell_q$. If $\alpha \in \mathbb{C}$ with $|r - \alpha| = |s|$, the ratio test fails. But one can easily find a decreasing sequence of positive real numbers $f = (f_k) \in \ell_q$ such that $(|f_k/f_{k-1}|) \to 1$, as $k \to \infty$, for example $f = (f_k) = (1/k^2)$. Hence, $|r - \alpha| \leq s$ implies $f \in \ell_q$.

Conversely, we have to show that $f \in \ell_q$ implies $|r - \alpha| \leq s$. If the condition $|r - \alpha| \leq |s|$ does not hold, then $|r - \alpha| > |s|$ which implies that $\sum_k |f_k|^q$ is divergent. This means that $f \in \ell_q$ if and only if $f_0 \neq 0$ and $|r - \alpha| \leq |s|$. Hence,

$$\sigma_p[B(\widetilde{r}, \widetilde{s})^*, \ell_p^*] = \{\alpha \in \mathbb{C} : |r - \alpha| \leq |s|\} \cup \mathcal{B}.$$

This completes the proof. $\qquad \square$

Theorem 5.5.8. [38, 3.7. Theorem] *The residual spectrum of $B(\widetilde{r}, \widetilde{s})$ on the space ℓ_p is the set* $\begin{cases} \{\alpha \in \mathbb{C} : |r - \alpha| < |s|\}, & \widetilde{r}, \widetilde{s} \in \mathcal{C}, \\ \{\alpha \in \mathbb{C} : |r - \alpha| \leq |s|\}, & \widetilde{r}, \widetilde{s} \in \mathcal{SD}. \end{cases}$

Proof. We prove the theorem by dividing into two parts.

Part I. Let $\widetilde{r}, \widetilde{s} \in \mathcal{C}$. We show that the operator $B(\widetilde{r}, \widetilde{s}) - \alpha I$ has an inverse and $\overline{R(B(\widetilde{r}, \widetilde{s}) - \alpha I)} \neq \ell_p$ for α satisfying $|r - \alpha| < |s|$. For $\alpha \neq r$, $B(\widetilde{r}, \widetilde{s}) - \alpha I$ is triangle and so, has an inverse. For $\alpha = r$, the operator $B(\widetilde{r}, \widetilde{s}) - \alpha I$ is one to one by Theorem 5.5.6. So, it has an inverse. By Theorem 5.5.7, the operator $[B(\widetilde{r}, \widetilde{s}) - \alpha I)]^* = B(\widetilde{r}, \widetilde{s})^* - \alpha I$ is not one to one for $\alpha \in \mathbb{C}$ such that $|r - \alpha| < |s|$. Hence, the range of the operator $B(\widetilde{r}, \widetilde{s}) - \alpha I$ is not dense in ℓ_p by Lemma 5.2.10. So, $\sigma_r[B(\widetilde{r}, \widetilde{s}), \ell_p] = \{\alpha \in \mathbb{C} : |r - \alpha| < |s|\}$.

Part II. Let $\widetilde{r}, \widetilde{s} \in \mathcal{SD}$ with $r_k \to r$ and $s_k \to s$, as $k \to \infty$, for $\alpha \in \mathbb{C}$ such that $|r - \alpha| \leq |s|$. Then, the operator $B(\widetilde{r}, \widetilde{s}) - \alpha I$ is triangle with $\alpha \neq r_k$ for all $k \in \mathbb{N}_0$. So, the operator $B(\widetilde{r}, \widetilde{s}) - \alpha I$ has an inverse. By Theorem 5.5.6, the operator $B(\widetilde{r}, \widetilde{s}) - \alpha I$ is one to one for $\alpha = r_k$ for all $k \in \mathbb{N}_0$. Thus, $[B(\widetilde{r}, \widetilde{s}) - \alpha I]^{-1}$ exists. But by Theorem 5.5.7, $[B(\widetilde{r}, \widetilde{s}) - \alpha I]^* = B(\widetilde{r}, \widetilde{s})^* - \alpha I$ is not one to one with $\alpha \in \mathbb{C}$ such that $|r - \alpha| \leq |s|$. Hence, the range of the operator $B(\widetilde{r}, \widetilde{s}) - \alpha I$ is not dense in ℓ_p, by Lemma 5.2.10. So, $\sigma_r[B(\widetilde{r}, \widetilde{s}), \ell_p] = \{\alpha \in \mathbb{C} : |r - \alpha| \leq |s|\}$.

This completes the proof. $\qquad \square$

Theorem 5.5.9. [38, 3.8. Theorem] *The continuous spectrum of $B(\widetilde{r}, \widetilde{s})$ on the space ℓ_p is the set* $\begin{cases} \{\alpha \in \mathbb{C} : |r - \alpha| = |s|\}, & \widetilde{r}, \widetilde{s} \in \mathcal{C}, \\ \emptyset & , \quad \widetilde{r}, \widetilde{s} \in \mathcal{SD}. \end{cases}$

Proof. We prove the theorem by dividing into two parts.

Part I. Let $\widetilde{r}, \widetilde{s} \in \mathcal{C}$ for $\alpha \in \mathbb{C}$ such that $|r - \alpha| = |s|$. Since $\sigma[B(\widetilde{r}, \widetilde{s}), \ell_p]$ is the disjoint union of the parts $\sigma_p[B(\widetilde{r}, \widetilde{s}), \ell_p]$, $\sigma_r[B(\widetilde{r}, \widetilde{s}), \ell_p]$ and $\sigma_c[B(\widetilde{r}, \widetilde{s}), \ell_p]$, we must have $\sigma_c[B(\widetilde{r}, \widetilde{s}), \ell_p] = \{\alpha \in \mathbb{C} : |r - \alpha| = |s|\}$.

Part II. Let $\widetilde{r}, \widetilde{s} \in \mathcal{SD}$. It is known that $\sigma_p[B(\widetilde{r}, \widetilde{s}), \ell_p]$, $\sigma_r[B(\widetilde{r}, \widetilde{s}), \ell_p]$ and $\sigma_c[B(\widetilde{r}, \widetilde{s}), \ell_p]$ are mutually disjoint sets and their union is $\sigma[B(\widetilde{r}, \widetilde{s}), \ell_p]$. Therefore, it is immediate from Theorems 5.5.5, 5.5.6 and 5.5.8 that $\sigma[B(\widetilde{r}, \widetilde{s}), \ell_p] = \sigma_p[B(\widetilde{r}, \widetilde{s}), \ell_p] \cup \sigma_r[B(\widetilde{r}, \widetilde{s}), \ell_p]$ and hence $\sigma_c[B(\widetilde{r}, \widetilde{s}), \ell_p] = \emptyset$.

This completes the proof. $\qquad \square$

Theorem 5.5.10. [38, 3.9. Theorem] *When $|r - \alpha| > |s|$ for $\alpha \neq r_k$, $[B(\widetilde{r}, \widetilde{s}) - \alpha I] \in I_1$.*

Proof. We show that the operator $B(\widetilde{r}, \widetilde{s}) - \alpha I$ is bijective and has a continuous inverse for $\alpha \in \mathbb{C}$ such that $|r - \alpha| > |s|$. Since $\alpha \neq r_k$, then $B(\widetilde{r}, \widetilde{s}) - \alpha I$ is a triangle. So, it has an inverse. The inverse of the operator $B(\widetilde{r}, \widetilde{s}) - \alpha I$ is continuous for $\alpha \in \mathbb{C}$ such that $|r - \alpha| > |s|$, by the relation (5.5.5). Thus, for every $y \in \ell_p$, we can find that $x \in \ell_p$ such that

$$[B(\widetilde{r}, \widetilde{s}) - \alpha I]x = y, \quad \text{since} \quad [B(\widetilde{r}, \widetilde{s}) - \alpha I]^{-1} \in (\ell_p : \ell_p).$$

This shows that the operator $B(\widetilde{r}, \widetilde{s}) - \alpha I$ is onto and so, $B(\widetilde{r}, \widetilde{s}) - \alpha I \in I_1$. $\quad \square$

Theorem 5.5.11. [38, 3.10. Theorem] *Let $\widetilde{r}, \widetilde{s} \in \mathcal{C}$ with $r_k = r$ and $s_k = s$ for all $k \in \mathbb{N}_0$. Then, $r \in \sigma[B(\widetilde{r}, \widetilde{s}), \ell_p] III_1$.*

Proof. We have $\sigma_r[B(\widetilde{r}, \widetilde{s}), \ell_p] = \{\alpha \in \mathbb{C} : |r - \alpha| < |s|\}$, by Theorem 5.5.8. Clearly, $r \in \sigma_r[B(\widetilde{r}, \widetilde{s}), \ell_p]$. It is sufficient to show that the operator $[B(\widetilde{r}, \widetilde{s}) - rI]^{-1}$ is continuous. By Lemma 5.2.11, it is enough to show that $[B(\widetilde{r}, \widetilde{s}) - Ir]^*$ is an onto operator and for given $y = (y_k) \in \ell_p^* = \ell_q$, we have to find $x = (x_k) \in \ell_q$ such that $[B(\widetilde{r}, \widetilde{s}) - Ir]^* x = y$. Solving the system of linear equations

$$\left. \begin{array}{rcl} s_0 x_1 & = & y_0 \\ s_1 x_2 & = & y_1 \\ s_2 x_3 & = & y_2 \\ & \vdots & \\ s_{k-1} x_k & = & y_{k-1} \\ & \vdots & \end{array} \right\}$$

one can easily observe that $sx_k = y_{k-1}$ for all $k \in \mathbb{N}_1$ which implies that $(x_k) \in \ell_q$, since $y = (y_k) \in \ell_q$. This shows that $[B(\widetilde{r}, \widetilde{s}) - Ir]^*$ is onto. Hence, $r \in \sigma[B(\widetilde{r}, \widetilde{s}), \ell_p]III_1$. $\square$

Theorem 5.5.12. [38, 3.11. Theorem] *Let $\widetilde{r}, \widetilde{s} \in \mathcal{C}$ with $r_k = r$ and $s_k = s$ for all $k \in \mathbb{N}_0$ and $\alpha \in \sigma_r[B(\widetilde{r}, \widetilde{s}), \ell_p]$ for all $r \neq \alpha$. Then, $\alpha \in \sigma[B(\widetilde{r}, \widetilde{s}), \ell_p]III_2$.*

Proof. It is sufficient to show that the operator $[B(\widetilde{r}, \widetilde{s}) - I\alpha]^{-1}$ is discontinuous for $r \neq \alpha$ and $\alpha \in \sigma_r[B(\widetilde{r}, \widetilde{s}), \ell_p]$. It is obvious that the operator $[B(\widetilde{r}, \widetilde{s}) - I\alpha]^{-1}$ is discontinuous for $r \neq \alpha$ and $\alpha \in \mathbb{C}$ such that $|r - \alpha| < |s|$ with $r_k \neq \alpha$, by (5.5.5). $\square$

Theorem 5.5.13. [38, 3.12. Theorem] *If $\widetilde{r}, \widetilde{s} \in \mathcal{SD}$ and $\alpha \in \sigma_r[B(\widetilde{r}, \widetilde{s}), \ell_p]$, then $\alpha \in \sigma[B(\widetilde{r}, \widetilde{s}), \ell_p]III_2$.*

Proof. It is sufficient to show that the operator $[B(\widetilde{r}, \widetilde{s}) - I\alpha]^{-1}$ is discontinuous for $\alpha \in \sigma_r[B(\widetilde{r}, \widetilde{s}), \ell_p]$. By (5.5.4), the operator $[B(\widetilde{r}, \widetilde{s}) - I\alpha]^{-1}$ is discontinuous for $r_k \neq \alpha$ and $\alpha \in \mathbb{C}$ with $|r - \alpha| \leq |s|$. $\square$

Theorem 5.5.14. [38, 3.13. Theorem] *Let $\widetilde{r}, \widetilde{s} \in \mathcal{C}$ with $r_k = r$, $s_k = s$ for all $k \in \mathbb{N}_0$. Then, the following statements hold:*

(i) $\sigma_{ap}[B(\widetilde{r}, \widetilde{s}), \ell_p] = \mathcal{A} \setminus \{r\}$.

(ii) $\sigma_\delta[B(\widetilde{r}, \widetilde{s}), \ell_p] = \mathcal{A}$.

(iii) $\sigma_{co}[B(\widetilde{r}, \widetilde{s}), \ell_p] = \mathcal{A}^\circ$.

Proof. (i) From Table 8.2, we get

$$\sigma_{ap}[B(\widetilde{r}, \widetilde{s}), \ell_p] = \sigma[B(\widetilde{r}, \widetilde{s}), \ell_p] \setminus \sigma[B(\widetilde{r}, \widetilde{s}), \ell_p]III_1.$$

We have by Theorem 5.5.11 and Theorem 5.5.5 that

$$\sigma_{ap}[B(\widetilde{r}, \widetilde{s}), \ell_p] = (\mathcal{A} \cup \mathcal{B}) \setminus \{r\} = \mathcal{A} \setminus \{r\}.$$

(ii) Since the following equality

$$\sigma_\delta[B(\widetilde{r}, \widetilde{s}), \ell_p] = \sigma[B(\widetilde{r}, \widetilde{s}), \ell_p] \setminus \sigma[B(\widetilde{r}, \widetilde{s}), \ell_p]I_3$$

holds from Table 8.2, we derive by Theorem 5.5.5 and Theorem 5.5.6 that $\sigma_\delta[B(\widetilde{r}, \widetilde{s}), \ell_p] = \mathcal{A}$.

(iii) From Table 8.2, we have

$$\sigma_\delta[B(\widetilde{r}, \widetilde{s}), \ell_p] = \sigma[B(\widetilde{r}, \widetilde{s}), \ell_p]III_1 \cup \sigma[B(\widetilde{r}, \widetilde{s}), \ell_p]III_2 \cup \sigma[B(\widetilde{r}, \widetilde{s}), \ell_p]III_3$$

and since $\sigma[B(\widetilde{r}, \widetilde{s}), \ell_p]III_3 = \emptyset$ by Theorem 5.5.6, it is immediate that $\sigma_{co}[B(\widetilde{r}, \widetilde{s}), \ell_p] = \sigma_r[B(\widetilde{r}, \widetilde{s}), \ell_p]$. Therefore, we obtain by Theorem 5.5.12 that $\sigma_{co}[B(\widetilde{r}, \widetilde{s}), \ell_p] = \mathcal{A}^\circ$. $\square$

Theorem 5.5.15. [38, 3.14. Theorem] *Let* $\widetilde{r}, \widetilde{s} \in \mathcal{SD}$. *Then,*

$$\sigma_{ap}[B(\widetilde{r}, \widetilde{s}), \ell_p] = \sigma_\delta[B(\widetilde{r}, \widetilde{s}), \ell_p] = \sigma_{co}[B(\widetilde{r}, \widetilde{s}), \ell_p] = \mathcal{A} \cup \mathcal{B}.$$

Proof. We have by Theorem 5.5.7 and Part (e) of Proposition 5.2.9 that

$$\sigma_p[B^*(\widetilde{r}, \widetilde{s}), \ell_p^*] = \sigma_{co}[B(\widetilde{r}, \widetilde{s}), \ell_p] = \{\alpha \in \mathbb{C} \ : \ |r - \alpha| \le |s|\}.$$

Furthermore, because $\sigma_p[B(\widetilde{r}, \widetilde{s}), \ell_p] = \{r_k\}$ by Theorem 5.5.6 and the subdivisions in Goldberg's classification are disjoint, we must have

$$\sigma\left[B(\widetilde{r}, \widetilde{s}), \ell_p\right] I_3 = \sigma\left[B(\widetilde{r}, \widetilde{s}), \ell_p\right] II_3 = \emptyset.$$

Hence, $\sigma\left[B(\widetilde{r}, \widetilde{s}), \ell_p\right] III_3 = \{r_k\}$. Additionally, since $\sigma\left[B(\widetilde{r}, \widetilde{s}), \ell_p\right] III_1 = \emptyset$ by Theorem 5.5.8 and Theorem 5.5.13, we have

$$\sigma\left[B(\widetilde{r}, \widetilde{s}), \ell_p\right] = \sigma\left[B(\widetilde{r}, \widetilde{s}), \ell_p\right] III_2 \cup \sigma\left[B(\widetilde{r}, \widetilde{s}), \ell_p\right] III_3.$$

Therefore, we derive from Table 8.2 that

$$
\begin{aligned}
\sigma_{ap}[B(\widetilde{r}, \widetilde{s}), \ell_p] &= \sigma\left[B(\widetilde{r}, \widetilde{s}), \ell_p\right] \backslash \sigma\left[B(\widetilde{r}, \widetilde{s}), \ell_p\right] III_1 = \sigma\left[B(\widetilde{r}, \widetilde{s}), \ell_p\right] \\
\sigma_\delta[B(\widetilde{r}, \widetilde{s}), \ell_p] &= \sigma\left[B(\widetilde{r}, \widetilde{s}), \ell_p\right] \backslash \sigma\left[B(\widetilde{r}, \widetilde{s}), \ell_p\right] I_3 = \sigma\left[B(\widetilde{r}, \widetilde{s}), \ell_p\right] \\
\sigma_\delta[B(\widetilde{r}, \widetilde{s}), \ell_p] &= \sigma\left[B(\widetilde{r}, \widetilde{s}), \ell_p\right] III_2 \cup \sigma\left[B(\widetilde{r}, \widetilde{s}), \ell_p\right] III_3 = \sigma\left[B(\widetilde{r}, \widetilde{s}), \ell_p\right].
\end{aligned}
$$

$\square$

5.6 Fine Spectrum of the Generalized Difference Operator Δ_{uv} on the Sequence Space ℓ_1

In this section, following Srivastava and Kumar [205], we give the main results on the spectrum and point spectrum of the operator Δ_{uv} on the space ℓ_1 of absolutely convergent series. The operator Δ_{uv} is defined on ℓ_1 by (5.6.1), below, under certain conditions on the sequences $u = (u_n)$ and $v = (v_n)$. Further, the results related to the continuous spectrum, residual spectrum and fine spectrum of the operator Δ_{uv} on the sequence space ℓ_1 are also derived.

5.6.1 Introduction

Let $u = (u_k)$ and $v = (v_k)$ be two sequences satisfying the following conditions:

(i) u is either a constant sequence or sequence of distinct real numbers with $u_k \to U$, as $k \to \infty$.

(ii) v is a sequence of non-zero real numbers with $v_k \to V$, as $k \to \infty$.

(iii) $|U - u_k| < |V|$ for each $k \in \mathbb{N}_0$.

We define the operator Δ_{uv} on the sequence space ℓ_1, as follows:

$$\Delta_{uv}x = (u_n x_n + v_{n-1}x_{n-1})_{n \in \mathbb{N}_0}, \quad \text{where} \quad x = (x_n) \in \ell_1. \qquad (5.6.1)$$

It is easy to verify that the operator Δ_{uv} can be represented by the matrix

$$\Delta_{uv} = \begin{pmatrix} u_0 & 0 & 0 & 0 & \cdots \\ v_0 & u_1 & 0 & 0 & \cdots \\ 0 & v_1 & u_2 & 0 & \cdots \\ 0 & 0 & v_2 & u_3 & \cdots \\ \vdots & \vdots & \vdots & \vdots & \ddots \end{pmatrix}.$$

It is known by Theorem 3.1 of Srivastava and Kumar [205] that the operator $\Delta_{uv} : \ell_1 \to \ell_1$ is a bounded linear operator and

$$\|\Delta_{uv}\|_{B(\ell_1)} = \sup_{k \in \mathbb{N}_0} \{|u_k|, |v_k|\}.$$

5.6.2 Spectrum and Point Spectrum of the Operator Δ_{uv} on the Sequence Space ℓ_1

In this subsection, we determine the spectrum and point spectrum of the operator Δ_{uv} on ℓ_1.

Theorem 5.6.1. [205, Theorem 3.2] *Spectrum of the operator Δ_{uv} on the sequence space ℓ_1 is given by $\sigma(\Delta_{uv}, \ell_1) = \{\alpha \in \mathbb{C} : |U - \alpha| \leq |V|\}$.*

Proof. Proof of this theorem is divided into two parts. In the first part, we show that $\sigma(\Delta_{uv}, \ell_1) \subseteq \{\alpha \in \mathbb{C} : |U - \alpha| \leq |V|\}$ which is equivalent to

$\alpha \in \mathbb{C}$ with $|U - \alpha| \leq |V|$ implies $\sigma(\Delta_{uv}, \ell_1)$; i.e., $\alpha \in \rho(\Delta_{uv}, \ell_1)$.

In the second part, we establish the reverse inclusion $\{\alpha \in \mathbb{C} : |U - \alpha| \leq |V|\} \subseteq \sigma(\Delta_{uv}, \ell_1)$.

Part I. Let $\alpha \in \mathbb{C}$ with $|U - \alpha| > |V|$. Clearly, $\alpha \neq U$ and $\alpha \neq u_k$ for each $k \in \mathbb{N}_0$ as it does not satisfy this condition. Further, $\Delta_{uv} - \alpha I = (a_{nk})$ reduces to a triangle and hence has an inverse $(\Delta_{uv} - \alpha I)^{-1} = B$, where

$$B = \begin{pmatrix} \dfrac{1}{u_0 - \alpha} & 0 & 0 & \cdots \\[2ex] \dfrac{-v_0}{(u_0 - \alpha)(u_1 - \alpha)} & \dfrac{1}{u_1 - \alpha} & 0 & \cdots \\[2ex] \dfrac{v_0 v_1}{(u_0 - \alpha)(u_1 - \alpha)(u_2 - \alpha)} & \dfrac{-v_1}{(u_1 - \alpha)(u_2 - \alpha)} & \dfrac{1}{u_2 - \alpha} & \cdots \\[2ex] \dfrac{v_0 v_1 v_2}{(u_0 - \alpha)(u_1 - \alpha)(u_2 - \alpha)(u_3 - \alpha)} & \dfrac{v_1 v_2}{(u_1 - \alpha)(u_2 - \alpha)(u_3 - \alpha)} & \dfrac{-v_2}{(u_2 - \alpha)(u_3 - \alpha)} & \cdots \\[2ex] \vdots & \vdots & \vdots & \ddots \end{pmatrix}$$

By Lemma 5.5.1, the operator $(\Delta_{uv} - \alpha I)^{-1} \in B(\ell_1)$ if the supremum of ℓ_1 norms of the columns of (b_{nk}) is bounded, i.e., $\sup_{k \in \mathbb{N}_0} \sum_n |b_{nk}| < \infty$. In order to show $\sup_{k \in \mathbb{N}_0} \sum_n |b_{nk}| < \infty$, first we prove that the series $\sum_n |b_{nk}|$ is convergent for each $k \in \mathbb{N}_0$. Let $S_k = \sum_n |b_{nk}|$. Then, since

$$\lim_{n \to \infty} \left| \frac{b_{n+1,0}}{b_{n0}} \right| = \lim_{n \to \infty} \left| \frac{v_n}{u_{n+1} - \alpha} \right| = \left| \frac{V}{U - \alpha} \right| < 1 \text{ for } \alpha \in \mathbb{C} \text{ with } |U - \alpha| > |V|,$$

the series

$$S_0 = \sum_n |b_{n0}| = \left| \frac{1}{u_0 - \alpha} \right| + \sum_{n=1}^{\infty} \left| \frac{v_0 v_1 \cdots v_{n-1}}{(u_0 - \alpha)(u_1 - \alpha)(u_2 - \alpha) \cdots (u_n - \alpha)} \right| \quad (5.6.2)$$

is convergent. Similarly, one can show the convergence of the series $S_k = \sum_n |b_{nk}|$ for each $k \in \mathbb{N}_1$.

Now, we claim that $\sup_{k \in \mathbb{N}_0} S_k$ is finite. Let $|v_k/(u_{k+1} - \alpha)| \to \beta$, as $k \to \infty$. Since modulus function is continuous, we have

$$\beta = \left| \frac{V}{U - \alpha} \right|. \quad (5.6.3)$$

Clearly, $0 < \beta < 1$ and

$$\frac{\beta}{|V|} = \left| \frac{1}{U - \alpha} \right|. \quad (5.6.4)$$

Therefore, we have

$$S_k = \left| \frac{1}{u_k - \alpha} \right| + \left| \frac{v_k}{(u_k - \alpha)(u_{k+1} - \alpha)} \right| + \left| \frac{v_k v_{k+1}}{(u_k - \alpha)(u_{k+1} - \alpha)(u_{k+2} - \alpha)} \right| + \cdots \quad (5.6.5)$$

Taking limit on both sides of the equality (5.6.5) and using the equalities (5.6.3) and (5.6.4), we get

$$\lim_{k \to \infty} S_k = \frac{\beta}{|V|(1 - \beta)} < \infty.$$

Since (S_k) is a sequence of positive real numbers and $\lim_{k \to \infty} S_k < \infty$, $\sup_{k \in \mathbb{N}_0} S_k < \infty$. Thus,

$$(\Delta_{uv} - \alpha I)^{-1} \in B(\ell_1) \text{ for } \alpha \in \mathbb{C} \text{ with } |U - \alpha| > |V|. \quad (5.6.6)$$

Next, we show that domain of the operator $(\Delta_{uv} - \alpha I)^{-1}$ is dense in ℓ_1 which is equivalent to say that range of the operator $\Delta_{uv} - \alpha I$ is dense in ℓ_1. Then, it follows immediately that $(\Delta_{uv} - \alpha I)^{-1} \in B(\ell_1)$. Hence, we have

$$\sigma(\Delta_{uv}, \ell_1) \subseteq \{\alpha \in \mathbb{C} : |U - \alpha| \leq |V|\}. \quad (5.6.7)$$

Part II. We now prove the reverse inclusion. First, we prove the inclusion

$$\{\alpha \in \mathbb{C} : |U - \alpha| \leq |V|\} \subseteq \sigma(\Delta_{uv}, \ell_1) \quad (5.6.8)$$

under the assumption that $\alpha \neq U$ and $\alpha \neq u_k$ for each $k \in \mathbb{N}_0$, i.e., we want to show that one of the conditions of Definition 5.2.1 fails. Let $\alpha \in \mathbb{C}$ with $|U - \alpha| \leq |V|$. Clearly, $\Delta_{uv} - \alpha I$ is a triangle and hence $(\Delta_{uv} - \alpha I)^{-1}$ exists. So, the condition (R1) is fulfilled but the condition (R2) fails as can be seen, below:

Suppose that $\alpha \in \mathbb{C}$ with $|U - \alpha| < |V|$. Since

$$\lim_{n \to \infty} \left| \frac{b_{n+1,0}}{b_{n0}} \right| = \lim_{n \to \infty} \left| \frac{v_n}{u_{n+1} - \alpha} \right| = \left| \frac{V}{U - \alpha} \right| > 1,$$

the series on the right side of (5.6.2) is divergent. So, (S_k) is unbounded. Hence,

$$(\Delta_{uv} - \alpha I)^{-1} \notin B(\ell_1) \quad \text{for} \quad \alpha \in \mathbb{C} \quad \text{with} \quad |U - \alpha| < |V|. \tag{5.6.9}$$

Next, we consider $\alpha \in \mathbb{C}$ with $|U - \alpha| = |V|$. Taking limit on both sides of equality (5.6.5), we get

$$\lim_{k \to \infty} S_k = \left| \frac{1}{u - \alpha} \right| + \left| \frac{V}{(u - \alpha)^2} \right| + \left| \frac{V^2}{(u - \alpha)^3} \right| + \cdots = \frac{1}{|V|} + \frac{1}{|V|} + \frac{1}{|V|} + \cdots = \infty.$$

This shows that (S_k) is unbounded. Hence,

$$(\Delta_{uv} - \alpha I)^{-1} \notin B(\ell_1) \quad \text{for} \quad \alpha \in \mathbb{C} \quad \text{with} \quad |U - \alpha| = |V|. \tag{5.6.10}$$

Finally, we prove the inclusion (5.6.8) under the assumption that $\alpha = U$ and $\alpha = u_k$ for each $k \in \mathbb{N}_0$.

Case (i). If (u_k) is a constant sequence, say $u_k = U$ for all $k \in \mathbb{N}_0$, then

$$(\Delta_{uv} - U I)x = \begin{pmatrix} 0 \\ v_0 x_0 \\ v_1 x_1 \\ v_2 x_2 \\ \vdots \end{pmatrix} = \theta \text{ implies } x_k = 0 \text{ for each } k \in \mathbb{N}_0.$$

This shows that the operator $\Delta_{uv} - UI$ is one to one, but $R(\Delta_{uv} - UI)$ is not dense in ℓ_1. That is, the condition (R3) fails. Hence, $U \in \sigma(\Delta_{uv}, \ell_1)$.

Case (ii). If (u_k) is a sequence of distinct real numbers, then the series on the right side of (5.6.2) is divergent for each $\alpha = u_k$ and consequently, (S_k) is unbounded. Hence,

$$(\Delta_{uv} - \alpha I)^{-1} \notin B(\ell_1) \quad \text{for} \quad \alpha = u_k. \tag{5.6.11}$$

So, the condition (R2) fails. Hence, $u_k \in \sigma(\Delta_{uv}, \ell_1)$ for all $k \in \mathbb{N}_0$.

Again, taking limit on the equality (5.6.5), we get that $S_k \to \infty$, as $k \to \infty$, for $\alpha = U$ which means that (S_k) is unbounded. Hence,

$$(\Delta_{uv} - \alpha I)^{-1} \notin B(\ell_1) \quad \text{for} \quad \alpha = U. \tag{5.6.12}$$

So, $U \in \sigma(\Delta_{uv}, \ell_1)$. Thus, in this case also $u_k \in \sigma(\Delta_{uv}, \ell_1)$ for all $k \in \mathbb{N}_0$ and $U \in \sigma(\Delta_{uv}, \ell_1)$. Therefore, we have

$$\{\alpha \in \mathbb{C} : |U - \alpha| \leq |V|\} \subseteq \sigma(\Delta_{uv}, \ell_1) \tag{5.6.13}$$

Therefore, by the inclusions (5.6.7) and (5.6.13), we get $\sigma(\Delta_{uv}, \ell_1) = \{\alpha \in \mathbb{C} : |U - \alpha| \leq |V|\}$.

This completes the proof. $\qquad\square$

Theorem 5.6.2. [205, Theorem 3.3] *The point spectrum of the operator Δ_{uv} on the sequence space ℓ_1 is an empty set.*

Proof. For the point spectrum of the operator Δ_{uv}, we find those α in $\mathbb{C}$ such that the matrix equation $\Delta_{uv}x = \alpha x$ is satisfied for non-zero vector $x = (x_k)$ in ℓ_1.

Consider $\Delta_{uv}x = \alpha x$ for $x \neq \theta = (0, 0, 0, \ldots)$ in ℓ_1, which gives

$$\left.\begin{array}{rcl}
u_0 x_0 &=& \alpha x_0 \\
v_0 x_0 + u_1 x_1 &=& \alpha x_1 \\
v_0 x_0 + u_1 x_1 &=& \alpha x_2 \\
&\vdots& \\
v_{k-1} x_{k-1} + u_k x_k &=& \alpha x_k \\
&\vdots&
\end{array}\right\}.$$

We prove this theorem by dividing the proof into two cases.

Case (i). Suppose that (u_k) is a constant sequence, say $u_k = U$ for all $k \in \mathbb{N}_0$. Let x_t be the first non-zero element of the sequence $x = (x_n)$. So, the equation $v_{t-1}x_{t-1} + Ux_t = \alpha x_t$ implies $\alpha = U$ and from the equation $v_t x_t + Ux_{t+1} = \alpha x_{t+1}$, we get $x_t = 0$ which is a contradiction our assumption. Hence, $\sigma_p(\Delta_{uv}, \ell_1) = \emptyset$.

Case (ii). Suppose that (u_k) is a sequence of distinct real numbers. Clearly, $x_k = v_{k-1}x_{k-1}/(\alpha - u_k)$ for all $k \in \mathbb{N}_1$. If $\alpha = u_0$, then

$$\lim_{k\to\infty} \left| \frac{x_k}{x_{k-1}} \right| = \left| \frac{V}{u_0 - U} \right| > 1,$$

because $|U - u_0| < |V|$. So, $x \notin \ell_1$ for $x_0 \neq 0$.

Similarly, if $\alpha = u_k$ for all $k \in \mathbb{N}_1$, then $x_{k-1} = x_{k-2} = \cdots = x_0 = 0$ and $x_{n+1} = v_n x_n/(u_k - u_{n+1})$ for all $n \in \mathbb{N}_k$. This implies that

$$\lim_{n\to\infty} \left| \frac{x_{n+1}}{x_n} \right| = \left| \frac{V}{u_k - U} \right| > 1,$$

because $|U - u_k| < |V|$ for all $k \in \mathbb{N}_1$. So, $x \notin \ell_1$ for $x_0 \neq 0$.

If $x_0 = 0$, then $x_k = 0$ for all $k \in \mathbb{N}_1$. Only possibility is $x = \theta = (0, 0, \ldots)$. Hence, $\sigma_p(\Delta_{uv}, \ell_1) = \emptyset$. $\qquad\square$

5.6.3 Residual and Continuous Spectrum of the Operator Δ_{uv} on the Sequence Space ℓ_1

We need the result for the point spectrum of Δ_{uv}^* on the sequence space ℓ_1^* for obtaining residual and continuous spectrum. So, we first determine the point spectrum of the operator Δ_{uv}^* on ℓ_1^*.

Theorem 5.6.3. [205, Theorem 4.1] *The point spectrum $\sigma_p(\Delta_{uv}^*, \ell_1^*)$ of the adjoint operator Δ_{uv}^* on ℓ_1^* is the set $\{\alpha \in \mathbb{C} : |U - \alpha| \leq |V|\}$.*

Proof. Suppose $\Delta_{uv}^* f = \alpha f$ for $\theta \neq f \in \ell_1^* \cong \ell_\infty$, where

$$
\Delta_{uv}^* = \begin{pmatrix} u_0 & v_0 & 0 & 0 & \cdots \\ 0 & u_1 & v_1 & 0 & \cdots \\ 0 & 0 & u_2 & v_2 & \cdots \\ 0 & 0 & 0 & u_3 & \cdots \\ \vdots & \vdots & \vdots & \vdots & \ddots \end{pmatrix} \quad \text{and} \quad f = \begin{pmatrix} f_0 \\ f_1 \\ f_2 \\ f_3 \\ \vdots \end{pmatrix}.
$$

This gives that

$$
f_k = \frac{\alpha - u_{k-1}}{v_{k-1}} f_{k-1} = \frac{(\alpha - u_{k-1})(\alpha - u_{k-2}) \cdots (\alpha - u_0)}{v_{k-1} v_{k-2} \cdots v_0} f_0 \tag{5.6.14}
$$

for all $k \in \mathbb{N}_1$. Hence,

$$
|f_k| = \left| \frac{\alpha - u_{k-1}}{v_{k-1}} f_{k-1} \right| = \left| \frac{(\alpha - u_{k-1})(\alpha - u_{k-2}) \cdots (\alpha - u_0)}{v_{k-1} v_{k-2} \cdots v_0} \right| |f_0|
$$

$$
\text{for all } k \in \mathbb{N}_1. \tag{5.6.15}
$$

Using equality (5.6.14), we get

$$
\lim_{k \to \infty} \left| \frac{f_k}{f_{k-1}} \right| = \lim_{k \to \infty} \left| \frac{\alpha - u_{k-1}}{v_{k-1}} \right| = \left| \frac{\alpha - U}{V} \right| < 1 \quad \text{provided } |U - \alpha| < |V|.
$$

So, $f \in \ell_1$ and consequently, $f \in \ell_\infty$. Again, for $\alpha \in \mathbb{C}$ with $|U - \alpha| = |V|$,

$$
\lim_{k \to \infty} \left| \frac{f_k}{f_{k-1}} \right| = \lim_{k \to \infty} \left| \frac{\alpha - u_{k-1}}{v_{k-1}} \right| = \left| \frac{\alpha - U}{V} \right| = 1.
$$

Thus, the ratio test fails. In this case, we take $f = (f_k)$ in such a way it is a decreasing sequence of positive real numbers and $(f_k/f_{k-1}) \to 1$, as $k \to \infty$. Clearly, $f = (f_k)$ is a convergent sequence and consequently, $f \in \ell_\infty$. Hence, $|U - \alpha| \leq |V|$, which implies that $\sup_{k \in \mathbb{N}_0} |f_k| < \infty$.

Conversely, suppose that $\sup_{k \in \mathbb{N}_0} |f_k| < \infty$. From the equality (5.6.15), it follows that $|(\alpha - u_{k-1})/v_{k-1}| \leq 1$ for all $k \in \mathbb{N}_m$; where m is a positive integer. So, $\lim_{k \to \infty} |(\alpha - u_{k-1})/v_{k-1}| \leq 1$, i.e., $|U - \alpha| \leq |V|$. Hence, $\sup_{k \in \mathbb{N}_0} |f_k| < \infty$ which implies that $|U - \alpha| \leq |V|$. This means that $f \in \ell_1^*$ if and only if $f_0 \neq 0$, $f = (f_k)$ is a decreasing sequence of positive real numbers such that $(f_k/f_{k-1}) \to 1$, as $k \to \infty$, and $|U - \alpha| \leq |V|$. Thus, $\sigma_p(\Delta_{uv}^*, \ell_1^*) = \{\alpha \in \mathbb{C} : |U - \alpha| \leq |V|\}$. $\qquad\square$

Theorem 5.6.4. [205, Theorem 4.2] *The residual spectrum* $\sigma_r(\Delta_{uv}^*, \ell_1^*)$ *of the operator* Δ_{uv}^* *on* ℓ_1^* *is the set*

$$\{\alpha \in \mathbb{C} : |U - \alpha| \leq |V|\}.$$

Proof. The proof of this theorem is divided into the following two cases.

Case (i). Suppose that (u_k) is a constant sequence, say $u_k = U$ for all $k \in \mathbb{N}_0$. For $\alpha \in \mathbb{C}$ with $|U - \alpha| \leq |V|$, the operator $\Delta_{uv} - \alpha I$ is a triangle except $\alpha = U$ and consequently, the operator $\Delta_{uv} - \alpha I$ has an inverse. Further by Theorem 5.6.2, the operator $\Delta_{uv} - \alpha I$ is one to one for $\alpha = U$ and hence, has an inverse. But by Theorem 5.6.3, the operator $(\Delta_{uv} - \alpha I)^*$ is not injective for $\alpha \in \mathbb{C}$ with $|U - \alpha| \leq |V|$. Hence, by Lemma 5.2.10, the range of the operator $\Delta_{uv} - \alpha I$ is not dense in ℓ_1. Thus, $\sigma_r(\Delta_{uv}^*, \ell_1^*) = \{\alpha \in \mathbb{C} : |U - \alpha| \leq |V|\}$.

Case (ii). Suppose that (u_k) is a sequence of distinct real numbers. For $\alpha \in \mathbb{C}$ with $|U - \alpha| \leq |V|$, the operator $\Delta_{uv} - \alpha I$ is a triangle except $\alpha = u_k$ for all $k \in \mathbb{N}_0$ and consequently, the operator $\Delta_{uv} - \alpha I$ has an inverse. Further by Theorem 5.6.2, the operator $\Delta_{uv} - \alpha I$ is one to one and hence, $(\Delta_{uv} - \alpha I)^{-1}$ exists for all $k \in \mathbb{N}_0$.

On the basis of the argument given in Case (i), it is easy to verify that the range of the operator $\Delta_{uv} - \alpha I$ is not dense in ℓ_1. Thus, $\sigma_r(\Delta_{uv}^*, \ell_1^*) = \{\alpha \in \mathbb{C} : |U - \alpha| \leq |V|\}$. $\qquad\square$

Theorem 5.6.5. [205, Theorem 4.3] *The continuous spectrum* $\sigma_c(\Delta_{uv}, \ell_1)$ *of the operator* Δ_{uv} *on* ℓ_1 *is an empty set.*

Proof. It is known that $\sigma_p(\Delta_{uv}, \ell_1)$, $\sigma_r(\Delta_{uv}, \ell_1)$ and $\sigma_c(\Delta_{uv}, \ell_1)$ are pairwise disjoint sets and union of these sets is $\sigma(\Delta_{uv}, \ell_1)$. But, by Theorems 5.6.1 and 5.6.4, we get $\sigma(\Delta_{uv}, \ell_1) = \sigma_r(\Delta_{uv}, \ell_1)$ which shows that the continuous spectrum $\sigma_c(\Delta_{uv}, \ell_1)$ is empty set, as desired. $\qquad\square$

5.6.4   Fine Spectrum of the Operator Δ_{uv} on the Sequence Space ℓ_1

Theorem 5.6.6. [205, Theorem 5.1] *If* α *satisfies* $|U - \alpha| > |V|$, *then* $\Delta_{uv} - \alpha I \in I_1$.

Proof. It is required to show that the operator $\Delta_{uv} - \alpha I$ is bijective and has a continuous inverse for $\alpha \in \mathbb{C}$ with $|U - \alpha| \leq |V|$. Since $\alpha \neq U$ and $\alpha \neq u_k$ for each $k \in \mathbb{N}_0$, therefore the operator $\Delta_{uv} - \alpha I$ is a triangle. Hence, it has an inverse. The operator $(\Delta_{uv} - \alpha I)^{-1}$ is continuous for $\alpha \in \mathbb{C}$ with $|U - \alpha| \leq |V|$ by the statement (5.6.6). Also, the equation $(\Delta_{uv} - \alpha I)x = y$ gives $x = (\Delta_{uv} - \alpha I)^{-1}y$, i.e., $x_{n} = \{(\Delta_{uv} - \alpha I)^{-1}y\}_n$ for all $n \in \mathbb{N}_0$. Thus, for every $y \in \ell_1$, we can find $x \in \ell_1$ such that $(\Delta_{uv} - \alpha I)x = y$, since $(\Delta_{uv} - \alpha I)^{-1} \in B(\ell_1)$. This shows that $\Delta_{uv} - \alpha I$ is an onto operator and hence, $\Delta_{uv} - \alpha I \in I_1$. $\qquad\square$

Theorem 5.6.7. [205, Theorem 5.2] *Let u be a constant sequence, say $u_k = U$ for all $k \in \mathbb{N}_0$. Then, $U \in III_1\sigma(\Delta_{uv}, \ell_1)$.*

Proof. We have $\sigma_r(\Delta_{uv}^*, \ell_1^*) = \{\alpha \in \mathbb{C} : |U - \alpha| \leq |V|\}$. Clearly, $U \in III_1\sigma(\Delta_{uv}, \ell_1)$. It is sufficient to show that the operator $(\Delta_{uv} - UI)^{-1}$ is continuous. By Lemma 5.2.11, it is enough to show that $(\Delta_{uv} - UI)^*$ is onto, i.e., for given $y = (y_n) \in \ell_1^*$, we have to find $x = (x_n) \in \ell_1^*$ such that $(\Delta_{uv} - UI)^* x = y$. Now, $(\Delta_{uv} - UI)^* x = y$, i.e.,

$$\left. \begin{array}{rcl} v_0 x_1 & = & y_0 \\ v_1 x_2 & = & y_1 \\ v_2 x_3 & = & y_2 \\ & \vdots & \\ v_{k-1} x_k & = & y_{k-1} \\ & \vdots & \end{array} \right\} .$$

Thus, $v_{n-1} x_n = y_{n-1}$ for all $n \in \mathbb{N}_1$ which implies that $\sup_{n \in \mathbb{N}_0} |x_n| < 1$, since $y \in \ell_\infty$ and $v = (v_k)$ is a convergent sequence. This shows that the operator $(\Delta_{uv} - UI)^*$ is onto and hence, $U \in III_1\sigma(\Delta_{uv}, \ell_1)$. $\qquad \square$

Theorem 5.6.8. [205, Theorem 5.3] *Let u be a constant sequence, say $u_k = U$ for all $k \in \mathbb{N}_0$ and $\alpha \neq U$, $\alpha \in \sigma_r(\Delta_{uv}, \ell_1)$. Then, $\alpha \in III_2\sigma(\Delta_{uv}, \ell_1)$.*

Proof. It is sufficient to show that the operator $(\Delta_{uv} - UI)^{-1}$ is discontinuous for $\alpha \neq U$, $\alpha \in \sigma_r(\Delta_{uv}, \ell_1)$. The operator $(\Delta_{uv} - UI)^{-1}$ is discontinuous by the statements (5.6.9) and (5.6.10) for $U \neq \alpha \in \mathbb{C}$ with $|U - \alpha| \leq |V|$. $\qquad \square$

Theorem 5.6.9. [205, Theorem 5.4] *Let u be a sequence of distinct real numbers and $\alpha \in \sigma_r(\Delta_{uv}, \ell_1)$. Then, $\alpha \in III_2\sigma(\Delta_{uv}, \ell_1)$.*

Proof. It is sufficient to show that the operator $(\Delta_{uv} - UI)^{-1}$ is discontinuous for $\alpha \in \sigma_r(\Delta_{uv}, \ell_1)$. The operator $(\Delta_{uv} - UI)^{-1}$ is discontinuous by the statements (5.6.9)-(5.6.12) for $U \neq \alpha \in \mathbb{C}$ with $|U - \alpha| \leq |V|$. $\qquad \square$

5.6.5   Conclusion

Although the matrix Λ is used for obtaining some new sequence spaces by its domain from the classical sequence spaces, it is not considered for determining the spectrum or fine spectrum acting as a linear operator on any of the classical sequence spaces c_0, c or ℓ_p. Following Altay and Başar [8], Karakaya and Altun [118], the fine spectrum with respect to Goldberg's classification of the operator defined by the triangle matrix Λ is determined over the sequence spaces c_0 and c. Additionally; the approximate point spectrum, defect spectrum and compression spectrum of the matrix operator Λ are given over the spaces c_0 and c. Since Section 5.3 is devoted to the fine spectrum of the operator defined by the lambda matrix over the sequence spaces c_0 and c

with new subdivision of spectrum, this makes it significant. We should note that the main results of Section 5.3 are given as an extended abstract, without proof, by Yeşilkayagil and Başar [231].

One can determine the fine spectrum of the matrix operator Λ on the spaces ℓ_p and bv_p, in the cases $0 < p < 1$ and $1 \leq p < \infty$, where bv_p denotes the space of all sequences whose Δ-transforms are in the space ℓ_p and is recently studied in the case $0 < p < 1$ by Altay and Başar [13] and in the case $1 \leq p \leq \infty$ by Başar and Altay [35].

Many researchers determined the spectrum and fine spectrum of a matrix operator in some sequence spaces. In addition to this, we give the related results for the matrix operator Δ^+ on the space c_0 with respect to the Goldberg's classification together with the approximate point spectrum, defect spectrum and compression spectrum. We should note here that although the continuous dual of the sequence spaces c_0 and c is the space ℓ_1 the adjoint of the matrix operator Δ^+ on the spaces c_0 ve c are, respectively, defined by

$$(\Delta^+)^* = \Delta \quad \text{and} \quad (\Delta^+)^* = \begin{bmatrix} 0 & 0 \\ 0 & \Delta \end{bmatrix}.$$

So, although the corresponding results of the present study coincide with the results of Başar et al. [37], the adjoint operators are different.

As a natural continuation of Akhmedov and El-Shabrawy [4] and, Srivastava and Kumar [205], the spectrum and the fine spectrum of the double sequential band matrix $B(\tilde{r}, \tilde{s})$ have been determined on the space ℓ_p. Many researchers determine the spectrum and fine spectrum of a matrix operator in some sequence spaces. In addition to this, the definition of some new divisions of spectrum called approximate point spectrum, defect spectrum and compression spectrum of the matrix operator are given and the related results for the matrix operator $B(\tilde{r}, \tilde{s})$ are presented on the space ℓ_p which is a new development for this type works giving the fine spectrum of a matrix operator on a sequence space with respect to the Goldberg's classification.

Finally, we should note that in the case $r_k = r$ and $s_k = s$ for all $k \in \mathbb{N}_0$ since the operator $B(\tilde{r}, \tilde{s})$ defined by a double sequential band matrix reduces to the operator $B(r, s)$ defined by the generalized difference matrix our results are more general and more comprehensive than the corresponding results obtained by Furkan et al. [87] and Bilgiç and Furkan [51], respectively. We note as a further suggestion that one can devote to the investigation of the fine spectrum of the matrix operator $B(\tilde{r}, \tilde{s})$ on the space bv_p.

Chapter 6

Sets of Fuzzy Valued Sequences and Series

Keywords. *Fuzzy series and sequences, fuzzy power series, alternating and binomial fuzzy series, Fourier series of fuzzy-valued functions, slowly decreasing fuzzy sequences, duals of classical sets of fuzzy sequences, sets of fuzzy sequences, sets of fuzzy sequences defined by a modulus.*

6.1 Introduction

Many authors have extensively developed the theory of the different cases of sequence sets with fuzzy metric. Mursaleen and Başarır [170] have recently introduced some new sets of fuzzy-valued sequences generated by a non-negative regular matrix A some of which reduced to the Maddox's spaces $\ell_\infty(p;F)$, $c(p;F)$, $c_0(p;F)$ and $\ell(p;F)$ of fuzzy valued sequences for the special cases of that matrix A. Altın, Et and Çolak [16] have recently defined the concepts of lacunary statistical convergence and lacunary strongly convergence of generalized difference fuzzy valued sequences. Quite recently; Kadak and Başar [104, 105] have recently studied the power series of fuzzy numbers and examined the alternating and binomial series of fuzzy numbers and some sets of fuzzy-valued functions with the level sets, and gave some properties of the level sets together with some inclusion relations, in [103]. Also, Talo and Başar [213] have extended the main results of Başar and Altay [35] to fuzzy numbers and defined the alpha-, beta- and gamma-duals of a set of fuzzy-valued sequences, and gave the duals of the classical sets of fuzzy-valued sequences together with the characterization of the classes of infinite matrices of fuzzy numbers transforming one of the classical set into another one. Finally, Talo and Başar [215] have introduced the sets $\ell_\infty(F;f)$, $c(F;f)$, $c_0(F;f)$ and $\ell(F;f)$ of fuzzy-valued sequences defined by a modulus function and the classes $\ell_\infty(F)$, $c(F)$, $c_0(F)$ and $\ell_p(F)$ of fuzzy-valued sequences consisting of the bounded, convergent, null and absolutely p-summable fuzzy-valued sequences with the level sets.

6.2   Preliminaries, Background and Notations

We begin with giving some required definitions and statements of theorems, propositions and lemmas. A *fuzzy number* is a fuzzy set on the real axis, i.e., a mapping $u : \mathbb{R} \to [0, 1]$, which satisfies the following four conditions:

(i) u is normal, i.e., there exists an $x_0 \in \mathbb{R}$ such that $u(x_0) = 1$.

(ii) u is fuzzy convex, i.e., $u\big(\lambda x + (1 - \lambda)y\big) \geq \min\{u(x), u(y)\}$ for all $x, y \in \mathbb{R}$ and for all $\lambda \in [0, 1]$.

(iii) u is upper semi-continuous.

(iv) The set $[u]_0 = \overline{\{x \in \mathbb{R} : u(x) > 0\}}$ is compact, (cf. Zadeh [236]), where $\overline{\{x \in \mathbb{R} : u(x) > 0\}}$ denotes the closure of the set $\{x \in \mathbb{R} : u(x) > 0\}$ in the usual topology of $\mathbb{R}$.

We denote the set of all fuzzy numbers on $\mathbb{R}$ by $L(\mathbb{R})$ and called it as *the space of fuzzy numbers*. λ-*level set* $[u]_\lambda$ of $u \in L(\mathbb{R})$ is defined by

$$[u]_\lambda := \begin{cases} \overline{\{t \in \mathbb{R} : u(t) \geq \lambda\}} & , \quad 0 < \lambda \leq 1, \\ \overline{\{t \in \mathbb{R} : u(t) > \lambda\}} & , \quad \lambda = 0. \end{cases}$$

The set $[u]_\lambda$ is closed, bounded and a non-empty interval for each $\lambda \in [0, 1]$, which is defined by $[u]_\lambda := [u^-(\lambda), u^+(\lambda)]$. $\mathbb{R}$ can be embedded in $L(\mathbb{R})$, since each $r \in \mathbb{R}$ can be regarded as a fuzzy number $\overline{r}$ defined by

$$\overline{r}(x) := \begin{cases} 1 & , \quad x = r, \\ 0 & , \quad x \neq r. \end{cases}$$

Theorem 6.2.1. [90, Representation theorem] *Let $[u]_\lambda = [u^-(\lambda), u^+(\lambda)]$ for $u \in L(\mathbb{R})$ and for each $\lambda \in [0, 1]$. Then, the following statements hold:*

(i) $u^-(\lambda)$ is a bounded and non-decreasing left continuous function on $(0, 1]$.

(ii) $u^+(\lambda)$ is a bounded and non-increasing left continuous function on $(0, 1]$.

(iii) The functions $u^-(\lambda)$ and $u^+(\lambda)$ are right continuous at the point $\lambda = 0$.

(iv) $u^-(1) \leq u^+(1)$.

Conversely, if the pair of the functions α and β satisfies the conditions (i)–(iv), then there exists a unique $u \in L(\mathbb{R})$ such that $[u]_\lambda := [\alpha(\lambda), \beta(\lambda)]$ for each $\lambda \in [0, 1]$. The fuzzy number u corresponding to the pair of functions α and β is defined by $u : \mathbb{R} \to [0, 1]$, $u(x) := \sup\{\lambda : \alpha(\lambda) \leq x \leq \beta(\lambda)\}$.

Let $u, v, w \in L(\mathbb{R})$ and $\alpha \in \mathbb{R}$. Then the operations addition, scalar multiplication and product defined on $L(\mathbb{R})$ by

$$u + v = w \quad \Leftrightarrow \quad [w]_\lambda = [u]_\lambda + [v]_\lambda \text{ for all } \lambda \in [0, 1]$$
$$\Leftrightarrow \quad w^-(\lambda) = u^-(\lambda) + v^-(\lambda) \text{ and } w^+(\lambda) = u^+(\lambda) + v^+(\lambda)$$
$$\text{for all } \lambda \in [0, 1],$$
$$[\alpha u]_\lambda \quad = \quad \alpha[u]_\lambda \text{ for all } \lambda \in [0, 1]$$
$$uv \quad = \quad w \Leftrightarrow [w]_\lambda = [u]_\lambda [v]_\lambda \text{ for all } \lambda \in [0, 1],$$

where it is immediate that

$$w^-(\lambda) \quad = \quad \min\{u^-(\lambda)v^-(\lambda), u^-(\lambda)v^+(\lambda), u^+(\lambda)v^-(\lambda), u^+(\lambda)v^+(\lambda)\},$$
$$w^+(\lambda) \quad = \quad \max\{u^-(\lambda)v^-(\lambda), u^-(\lambda)v^+(\lambda), u^+(\lambda)v^-(\lambda), u^+(\lambda)v^+(\lambda)\}$$

for all $\lambda \in [0, 1]$. Let W be the set of all closed bounded intervals A of real numbers with endpoints $\underline{A}$ and $\overline{A}$, i.e., $A = [\underline{A}, \overline{A}]$. Define the relation d on W by $d(A, B) = \max\{|\underline{A} - \underline{B}|, |\overline{A} - \overline{B}|\}$. Then, it can easily be observed that d is a metric on W (cf. Diamond and Kloeden [70]) and (W, d) is a complete metric space, (cf. Nanda [174]). Now, we may define the metric D on $L(\mathbb{R})$ by means of the Hausdorff metric d as

$$D(u, v) = \sup_{\lambda \in [0,1]} d([u]_\lambda, [v]_\lambda) = \sup_{\lambda \in [0,1]} \max\{|u^-(\lambda) - v^-(\lambda)|, |u^+(\lambda) - v^+(\lambda)|\}.$$

One can extend the natural-ordering relation on the real line to intervals, as follows: $A \preceq B$ if and only if $\underline{A} \leq \underline{B}$ and $\overline{A} \leq \overline{B}$. The partial-ordering relation on $L(\mathbb{R})$ is defined, as follows:

$$u \preceq v \Leftrightarrow [u]_\lambda \preceq [v]_\lambda \Leftrightarrow u^-(\lambda) \leq v^-(\lambda) \text{ and } u^+(\lambda) \leq v^+(\lambda) \text{ for all } \lambda \in [0, 1].$$

Lemma 6.2.2 (cf. [138]). *Let $u, v, w \in L(\mathbb{R})$. If u is comparable with $\overline{0}$ and v, w are on the same side of $\overline{0}$, i.e., $v \succeq \overline{0}, w \succeq \overline{0}$ or $v \preceq \overline{0}, w \preceq \overline{0}$, then $u(v + w) = uv + uw$.*

Definition 6.2.3. [214, Definition 2.1] *$u \in L(\mathbb{R})$ is said to be a non-negative fuzzy number if and only if $u(x) = 0$ for all $x < 0$. It is immediate that $u \succeq \overline{0}$ if u is a non-negative fuzzy number. By $L(\mathbb{R})^+$, we denote the set of non-negative fuzzy numbers. Similarly, $u \in L(\mathbb{R})$ is said to be a non-positive fuzzy number if and only if $u(x) = 0$ for all $x > 0$. It is immediate that $u \preceq \overline{0}$ if u is a non-positive fuzzy number. By $L(\mathbb{R})^-$, we denote the set of non-positive fuzzy numbers.*

One can see that

$$D(u, \overline{0}) = \sup_{\lambda \in [0,1]} \max\{|u^-(\lambda)|, |u^+(\lambda)|\} = \max\{|u^-(0)|, |u^+(0)|\}.$$

Then, it is trivial that the following statements hold:

(i) $u \succeq \overline{0} \Leftrightarrow u^-(\lambda) \geq 0$ and $u^+(\lambda) \geq 0$ for all $\lambda \in [0, 1]$.

(ii) $u \preceq \overline{0} \Leftrightarrow u^-(\lambda) \leq 0$ and $u^+(\lambda) \leq 0$ for all $\lambda \in [0,1]$.

(iii) $u \not\succ \overline{0} \Leftrightarrow u^-(\lambda) < 0$ and $u^+(\lambda) > 0$ for some $\lambda \in [0,1]$.

Therefore, one can easily see for the multiplication of two fuzzy numbers that

(i) If $u, v \succeq \overline{0}$ then $(uv)^-(\lambda) = u^-(\lambda)v^-(\lambda)$ and $(uv)^+(\lambda) = u^+(\lambda)v^+(\lambda)$ for all $\lambda \in [0,1]$.

(ii) If $u, v \preceq \overline{0}$ then $(uv)^-(\lambda) = u^+(\lambda)v^+(\lambda)$ and $(uv)^+(\lambda) = u^-(\lambda)v^-(\lambda)$ for all $\lambda \in [0,1]$.

(iii) If $u \succeq \overline{0}$, $v \preceq \overline{0}$ then $(uv)^-(\lambda) = u^+(\lambda)v^-(\lambda)$ and $(uv)^+(\lambda) = u^-(\lambda)v^+(\lambda)$ for all $\lambda \in [0,1]$.

(iv) If $u \preceq \overline{0}$, $v \succeq \overline{0}$ then $(uv)^-(\lambda) = u^-(\lambda)v^+(\lambda)$ and $(uv)^+(\lambda) = u^+(\lambda)v^-(\lambda)$ for all $\lambda \in [0,1]$.

Following Matloka [158], we give some definitions concerning with the sequences of fuzzy numbers below, which are needed in the text.

Definition 6.2.4. [213, Definition 2.7] *A sequence $u = (u_k)$ of fuzzy numbers is a function u from the set $\mathbb{N}_0$ into the set $L(\mathbb{R})$ of fuzzy numbers. The fuzzy number u_k denotes the value of the function at $k \in \mathbb{N}_0$ and is called as the general term of the sequence. By $\omega(F)$, we denote the set of all fuzzy-valued sequences.*

Definition 6.2.5. [213, Definition 2.11] *A sequence $(u_k) \in \omega(F)$ is called bounded if and only if the set of fuzzy numbers consisting of the terms of the sequence (u_k) is a bounded set. That is to say that a sequence $(u_k) \in \omega(F)$ is said to be bounded if and only if there exist two fuzzy numbers m and M such that $m \preceq u_k \preceq M$ for all $k \in \mathbb{N}_0$. This means that $m^-(\lambda) \preceq u_k^-(\lambda) \preceq M^-(\lambda)$ and $m^+(\lambda) \preceq u_k^+(\lambda) \preceq M^+(\lambda)$ for all $\lambda \in [0,1]$. By $\ell_\infty(F)$, we denote the set of all fuzzy-valued bounded sequences.*

The boundedness of the sequence $(u_k) \in \omega(F)$ is equivalent to the uniform boundedness of the functions u_k^- and u_k^+ on $[0,1]$. Therefore, one can see by using the relation (2.1.1) that the boundedness of the sequence $(u_k) \in \omega(F)$ is equivalent to the fact that

$$\sup_{k \in \mathbb{N}_0} D(u_k, \overline{0}) = \sup_{k \in \mathbb{N}_0} \sup_{\lambda \in [0,1]} \max \left\{ \left| u_k^-(\lambda) \right|, \left| u_k^+(\lambda) \right| \right\} < \infty.$$

Definition 6.2.6. [104, Definition 4.4] *A sequence $(u_k) \in \omega(F)$ is called convergent with limit $u \in L(\mathbb{R})$ if and only if for every $\varepsilon > 0$ there exists an $n_0 = n_0(\varepsilon) \in \mathbb{N}_0$ such that $D(u_k, u) < \varepsilon$ for all $k \geq n_0$. A sequence of fuzzy mappings is a function whose domain is the set of positive integers and its range is a set of fuzzy numbers. We denote any fuzzy-valued sequence of functions by $\{u_n(x)\}$. By $c(F)$, we denote the set of all fuzzy-valued convergent sequences.*

Now, we can give the following lemma:

Lemma 6.2.7. [158, Theorem 4.1] *Let* $(u_k), (v_k) \in c(F)$ *with* $u_k \to a$, $v_k \to b$, *as* $k \to \infty$. *Then, the following statements hold:*

(i) $u_k + v_k \to a + b$, *as* $k \to \infty$.

(ii) $u_k - v_k \to a - b$, *as* $k \to \infty$.

(iii) $u_k v_k \to ab$, *as* $k \to \infty$.

(iv) $u_k/v_k \to a/b$, *as* $k \to \infty$, *where* $0 \in [u_k]_0$ *for all* $k \in \mathbb{N}$ *and* $0 \in [u]_0$.

A sequence (u_k) of fuzzy numbers is said to be Cauchy if for every $\varepsilon > 0$ there exists a positive integer n_0 such that $D(u_k, u_m) < \varepsilon$ for all $k, m > n_0$. By $C(F)$, we denote the set of all fuzzy-valued Cauchy sequences.

Definition 6.2.8. [213, Definition 5] *Let* $(u_k) \in \omega(F)$. *Then, the expression* $\sum_k u_k$ *is called a series of fuzzy numbers. If the sequence* (s_n) *converges to a fuzzy number* u, *then we say that the series* $\sum_k u_k$ *of fuzzy numbers converges to* u *and write* $\sum_k u_k = u$ *which implies by letting* $n \to \infty$ *that*

$$\sum_{k=0}^{n} u_k^-(\lambda) \to u^-(\lambda) \quad and \quad \sum_{k=0}^{n} u_k^+(\lambda) \to u^+(\lambda),$$

uniformly in $\lambda \in [0,1]$, *where* $s_n = \sum_{k=0}^{n} u_k$ *for all* $n \in \mathbb{N}_0$. *Conversely, if the fuzzy numbers* $u_k = \{(u_k^-(\lambda), u_k^+(\lambda)) : \lambda \in [0,1]\}$, $\sum_k u_k^-(\lambda) = u^-(\lambda)$ *and* $\sum_k u_k^+(\lambda) = u^+(\lambda)$ *converge uniformly in* λ, *then* $u = \{(u^-(\lambda), u^+(\lambda)) : \lambda \in [0,1]\}$ *defines a fuzzy number such that* $u = \sum_k u_k$.

We say otherwise the series of fuzzy numbers diverges. Additionally, if the sequence (s_k) is bounded then we say that the series $\sum_k u_k$ of fuzzy numbers is bounded. By $cs(F)$ and $bs(F)$, we denote the sets of all convergent and bounded series of fuzzy numbers, respectively.

Definition 6.2.9. [214, Definition 2.3] *An absolute value* $|u|$ *of a fuzzy number* u *is defined by*

$$|u|(t) := \begin{cases} \max\{u(t), u(-t)\} & , \quad t \geq 0, \\ 0 & , \quad t < 0. \end{cases}$$

λ-*level set* $[|u|]_\lambda$ *of the absolute value of* $u \in L(\mathbb{R})$ *is in the form* $[|u|]_\lambda = [|u|^-(\lambda), |u|^+(\lambda)]$, *where*

$$|u|^-(\lambda) := \max\{0, u^-(\lambda), -u^+(\lambda)\},$$
$$|u|^+(\lambda) := \max\{|u^-(\lambda)|, |u^+(\lambda)|\}.$$

Now, we may give:

Proposition 6.2.10. [214, Proposition 2.4] *Let* $u, v, w, z \in L(\mathbb{R})$ *and* $\alpha \in \mathbb{R}$. *Then, the following statements hold:*

(i) $(L(\mathbb{R}), D)$ *is a complete metric space,* (cf. Puri and Ralescu [188]).

(ii) $D(\alpha u, \alpha v) = |\alpha| D(u, v)$.

(iii) $D(u + v, w + v) = D(u, w)$.

(iv) $D(u + v, w + z) \leq D(u, w) + D(v, z)$.

(v) $|D(u, \overline{0}) - D(v, \overline{0})| \leq D(u, v) \leq D(u, \overline{0}) + D(v, \overline{0})$.

Proposition 6.2.11. [92, Proposition 2.4] *Let* $u, v, m \in L(\mathbb{R})$ *with* $m \succeq \overline{0}$ *and* $\alpha \in \mathbb{R}$. *Then, the following statements hold:*

(i) $|u| = \begin{cases} u & , & u \succeq \overline{0}, \\ -u & , & u \prec \overline{0}. \end{cases}$

(ii) $|u + v| \preceq |u| + |v|$.

(iii) $|\alpha u| = |\alpha||u|$.

(iv) $|u| = \overline{0}$ *if and only if* $u = \overline{0}$.

(v) $|u| \preceq m$ *if and only if* $-m \preceq u \preceq m$.

We wish to give the following lemma related with the absolute value of the product of two fuzzy numbers:

Lemma 6.2.12. [214, Lemma 2.5] *The absolute value* $|uv|$ *of the product* uv *of* $u, v \in L(\mathbb{R})$ *satisfies the inequalities*

$$|uv|^-(\lambda) \leq |uv|^+(\lambda) \tag{6.2.1}$$
$$\leq \max\{|u^-(\lambda)||v^-(\lambda)|, |u^-(\lambda)||v^+(\lambda)|, |u^+(\lambda)||v^-(\lambda)|, |u^+(\lambda)||v^+(\lambda)|\}.$$

Proof. For the λ-level set $[|uv|]_\lambda$ of $|uv|$, we have

$$|uv|^-(\lambda) = \max\{0, (uv)^-(\lambda), -(uv)^+(\lambda)\},$$
$$|uv|^+(\lambda) = \max\{|(uv)^-(\lambda)|, |(uv)^+(\lambda)|\}.$$

Therefore, since

$$|(uv)^-(\lambda)| = |\min\{u^-(\lambda)v^-(\lambda), u^-(\lambda)v^+(\lambda), u^+(\lambda)v^-(\lambda), u^+(\lambda)v^+(\lambda)\}|$$
$$\leq \max\{|u^-(\lambda)||v^-(\lambda)|, |u^-(\lambda)||v^+(\lambda)|, |u^+(\lambda)||v^-(\lambda)|, |u^+(\lambda)||v^+(\lambda)|\},$$
$$|(uv)^+(\lambda)| = |\max\{u^-(\lambda)v^-(\lambda), u^-(\lambda)v^+(\lambda), u^+(\lambda)v^-(\lambda), u^+(\lambda)v^+(\lambda)\}|$$
$$\leq \max\{|u^-(\lambda)||v^-(\lambda)|, |u^-(\lambda)||v^+(\lambda)|, |u^+(\lambda)||v^-(\lambda)|, |u^+(\lambda)||v^+(\lambda)|\},$$

one can deduce that (6.2.1) holds. $\qquad\square$

Lemma 6.2.13. [214, Lemma 2.13] *If the fuzzy numbers* $u_k = \{(u_k^-(\lambda), u_k^+(\lambda)) : \lambda \in [0, 1]\}$, $\sum_k u_k^-(\lambda) = u^-(\lambda)$ *and* $\sum_k u_k^+(\lambda) = u^+(\lambda)$ *converge uniformly in* λ, *then* $u = \{(u^-(\lambda), u^+(\lambda)) : \lambda \in [0, 1]\}$ *defines a fuzzy number such that* $u = \sum_k u_k$.

Proof. To prove the lemma, we must show that the pair of functions u^- and u^+ satisfies the conditions of Theorem 6.2.1. For this, we prove that u^- is a bounded, non-decreasing, left continuous function on $(0,1]$ and right continuous at the point $\lambda = 0$. u_k^-'s are the bounded, non-decreasing, left continuous functions on $(0,1]$ and right continuous at the point $\lambda = 0$ for each $k \in \mathbb{N}_0$.

(i) Let $\lambda_1 < \lambda_2$. Then, $u_k^-(\lambda_1) \leq u_k^-(\lambda_2)$ for each $k \in \mathbb{N}_0$. Therefore, we have $\sum_k u_k^-(\lambda_1) \leq \sum_k u_k^-(\lambda_2)$ which yields the fact that $u^-(\lambda_1) \leq u^-(\lambda_2)$. Hence, u^- is non-decreasing.

(ii) By taking into account the uniform convergence in λ of $\lim_{\lambda \to \lambda_0^-} u_k^-(\lambda) = u_k^-(\lambda_0)$, $\sum_k u_k^-(\lambda) = u^-(\lambda)$ for each $k \in \mathbb{N}_0$ we obtain for $\lambda_0 \in (0,1]$ that

$$\lim_{\lambda \to \lambda_0^-} u^-(\lambda) = \lim_{\lambda \to \lambda_0^-} \sum_k u_k^-(\lambda) = \sum_k \lim_{\lambda \to \lambda_0^-} u_k^-(\lambda) = \sum_k u_k^-(\lambda_0) = u^-(\lambda_0),$$

which shows that u^- is a left continuous function on $(0,1]$.

(iii) By using the uniform convergence in λ in the expressions $\lim_{\lambda \to 0^+} u_k^-(\lambda) = u_k^-(0)$ for each $k \in \mathbb{N}_0$ and $\sum_k u_k^-(\lambda) = u^-(\lambda)$, we see that

$$\lim_{\lambda \to 0^+} u^-(\lambda) = \lim_{\lambda \to 0^+} \sum_k u_k^-(\lambda) = \sum_k \lim_{\lambda \to 0^+} u_k^-(\lambda) = \sum_k u_k^-(0) = u^-(0).$$

This means that u^- is a right continuous function at the point $\lambda = 0$.

(iv) There exists $M_k > 0$ such that $|u_k^-(\lambda)| \leq M_k$ for all $\lambda \in [0,1]$ and for all $k \in \mathbb{N}_0$. Since the series $\sum_k u_k^-(\lambda) = u^-(\lambda)$ converges uniformly in λ there exists $n_0 \in \mathbb{N}_0$ for all $\varepsilon > 0$ such that $\left| \sum_{k=n+1}^{\infty} u_k^-(\lambda) \right| < \varepsilon$ for all $n \geq n_0$ and for all $\lambda \in [0,1]$. Therefore, we have

$$|u^-(\lambda)| = \left| \sum_k u_k^-(\lambda) \right| = \left| \sum_{k=0}^{n} u_k^-(\lambda) + \sum_{k=n+1}^{\infty} u_k^-(\lambda) \right|$$

$$\leq \sum_{k=0}^{n} |u_k^-(\lambda)| + \left| \sum_{k=n+1}^{\infty} u_k^-(\lambda) \right|$$

$$\leq \sum_{k=0}^{n} M_k + \varepsilon \leq K_\varepsilon.$$

This leads us to the fact that u^- is a bounded function.

Since one can establish in the similar way that u^+ is a bounded, non-increasing, left continuous function on $(0,1]$, and right continuous at the point $\lambda = 0$, we omit details.

Finally, we show that $u^-(1) \leq u^+(1)$. Since $u_k^-(1) \leq u_k^+(1)$ for each $k \in \mathbb{N}_0$, we derive that $\sum_k u_k^-(1) \leq \sum_k u_k^+(1)$ which shows that $u^-(1) \leq u^+(1)$.

Therefore, it is deduced that $[u]_\lambda = [u_k^-(\lambda), u_k^+(\lambda)]$ defines a fuzzy number. Finally, we show that $\sum_k u_k = u$. Since the series of functions $\sum_k u_k^-(\lambda)$

and $\sum_k u_k^+(\lambda)$ converge uniformly in λ to $u^-(\lambda)$ and $u^+(\lambda)$, respectively, for all $\varepsilon > 0$ there exists $n_0 \in \mathbb{N}_0$ such that

$$D\left(\sum_{k=0}^n u_k, u\right) = \sup_{\lambda \in [0,1]} \max\left\{\left|\sum_{k=0}^n u_k^-(\lambda) - u^-(\lambda)\right|, \left|\sum_{k=0}^n u_k^+(\lambda) - u^+(\lambda)\right|\right\}$$

$$\leq \max\left\{\sup_{\lambda \in [0,1]}\left|\sum_{k=0}^n u_k^-(\lambda) - u^-(\lambda)\right|, \sup_{\lambda \in [0,1]}\left|\sum_{k=0}^n u_k^+(\lambda) - u^+(\lambda)\right|\right\} < \varepsilon$$

for all $n \geq n_0$, the sequence $\left(\sum_{k=0}^n u_k\right)$ converges to the fuzzy number u, i.e., $\sum_k u_k = u$.

This step completes the proof.   □

6.2.1 Generalized Hukuhara Difference

Let $\mathcal{K}$ be the space of non-empty compact and convex sets in the n-dimensional Euclidean space $\mathbb{R}^n$. If $n = 1$, by I we denote the set of closed bounded intervals of the real line. Given two elements $A, B \in \mathcal{K}$ and $\alpha \in \mathbb{R}$, the usual interval arithmetic operations, i.e., Minkowski addition and scalar multiplication, are defined by $A + B = \{a + b : a \in A, b \in B\}$ and $\alpha A = \{\alpha a : a \in A\}$. It is well known that addition is associative and commutative, and with neutral element $\{0\}$. If $\alpha = -1$, scalar multiplication gives the opposite $-A = (-1)A = \{-a : a \in A\}$, but, in general, $A + (-A) \neq 0$, i.e., the opposite of A is not the inverse of A in Minkowski addition unless A is a singleton. A first consequence of this fact is that, in general, additive simplification is not valid.

To partially overcome this situation, the *Hukuhara difference*, H-difference for short, has been introduced as a set C for which $A \ominus B \Leftrightarrow A = B + C$ and an important property of $\ominus$ is that $A \ominus A = \{0\}$ for all $A \in \mathcal{K}$ and $(A + B) \ominus B = A$ for all $A, B \in \mathcal{K}$. The H-difference is unique, but it does not always exist. A necessary condition for $A \ominus B$ to exist is that A contains a translate $\{c\} + B$ of B.

A generalization of the Hukuhara difference proposed in [206] aims to overcome this situation.

Definition 6.2.14. [206, Definition 1] *The generalized Hukuhara difference $A \ominus B$ of two sets $A, B \in \mathcal{K}$ is defined, as follows;*

$$A \ominus B = C \Longleftrightarrow \begin{cases} A = B + C, \\ B = A + (-1)C. \end{cases}$$

Proposition 6.2.15. [206, Proposition 3] *Let $A, B \in \mathcal{K}$ be two compact convex sets. Then, we have:*

(i) *If the H-difference exists, it is unique and is a generalization of the usual Hukuhara difference since $A \ominus B = A - B$, whenever $A \ominus B$ exists.*

(ii) $A + (-A) \neq 0$.

(iii) $(A + B) \ominus B = A$.

(iv) $A \ominus B = B \ominus A = C \Leftrightarrow C = \{0\}$ *and* $A = B$.

Proposition 6.2.16. [206, Proposition 4] *The H-difference of two intervals* $A = [\underline{A}, \overline{A}]$ *and* $B = [\underline{B}, \overline{B}]$ *always exists, and the conditions*

$$[\underline{A}, \overline{A}] \ominus [\underline{B}, \overline{B}] = [\underline{C}, \overline{C}], \text{ where}$$
$$\underline{C} = \min\{\underline{A} - \underline{B}, \overline{A} - \overline{B}\},$$
$$\overline{C} = \max\{\underline{A} - \underline{B}, \overline{A} - \overline{B}\}$$

are satisfied simultaneously if and only if the two intervals have the same length and $\underline{C} = \overline{C}$.

Proposition 6.2.17. [207, Proposition 6] *If* A *and* B *are two compact convex sets, then* $d(A, B) = d(A \ominus B, \{0\})$.

Proof. We observe by Proposition 6.2.16 that $d(A \ominus B, \{0\}) = \max\{|\underline{C}|, |\overline{C}|\}$. Then, we obtain that $d(A \ominus B, \{0\}) = \max\{|\underline{A} - \underline{B}|, |\overline{A} - \overline{B}|\} = d(A, B)$. $\square$

Proposition 6.2.18. [207, Proposition 7] *Let* $u : [a, b] \to I$ *be such that* $u(x) = [u^-(x), u^+(x)]$. *Then, we have*

$$\lim_{x \to x_0} u(x) = l \quad \Leftrightarrow \quad \lim_{x \to x_0} \{u(x) \ominus l\} = \{0\},$$
$$\lim_{x \to x_0} u(x) = u(x_0) \quad \Leftrightarrow \quad \lim_{x \to x_0} \{u(x) \ominus u(x_0)\} = \{0\},$$

where the limits are in the Hausdorff metric d *for intervals.*

6.3 Series and Sequences of Fuzzy Numbers

The main purpose of the present section is to derive some results, by using the concept of the level set, related to the convergence of fuzzy-valued series. As a consequence of this approach, following [219], we give the fuzzy analog of certain convergence tests and results concerning the series of non-negative or arbitrary real terms.

6.3.1 Convergence of the Series of Fuzzy Numbers

In this subsection, we emphasize the convergent series of fuzzy numbers. We begin with three well-known consequences concerning the series of functions.

Lemma 6.3.1. [220, p. 376] *Consider the series of functions* $\sum_k f_k(x)$ *with* $f_k : [a, b] \to \mathbb{R}$. *Then, the following statements hold:*

(i) $f_k(x) \to 0$ *uniformly in* $x \in [a, b]$, *as* $k \to \infty$, *if* $\sum_k f_k(x)$ *converges uniformly in* $x \in [a, b]$.

(ii) $\sum_k f_k(x)$ *converges uniformly in* $x \in [a, b]$ *if and only if* $\sum_{k=n+1}^{\infty} f_k(x) \to 0$ *uniformly in* $x \in [a, b]$, *as* $n \to \infty$.

Lemma 6.3.2. [220, p. 377] *Let for the series of functions* $\sum_k f_k(x)$ *and* $\sum_k g_k(x)$, *there exists an* $n_0 \in \mathbb{N}$, *such that* $|f_k(x)| \leq g_k(x)$ *for all* $k \geq n_0$ *and for all* $x \in [a, b]$ *with* $f_k : [a, b] \to \mathbb{R}$ *and* $g_k : [a, b] \to \mathbb{R}$. *If the series* $\sum_k g_k(x)$ *is convergent uniformly in* $[a, b]$, *then the series* $\sum_k |f_k(x)|$ *and* $\sum_k f_k(x)$ *are uniformly convergent in* $[a, b]$.

Weierstrass M test. [220, p. 377] *Let* $f_k : [a, b] \to \mathbb{R}$ *be given. If there exists an* $M_k \geq 0$ *such that* $|f_k(x)| \leq M_k$ *for all* $k \in \mathbb{N}$ *and the series* $\sum_k M_k$ *converges, then the series* $\sum_k f_k(x)$ *is uniformly and absolutely convergent in* $[a, b]$.

Lemma 6.3.3. [221, Example 4.3.1, p. 201] *The geometric series* $\sum_k x^k$ *converges to the sum* $1/(1 - x)$ *if* $|x| < 1$, *and diverges if* $|x| \geq 1$.

We continue by giving one more result which summarizes some basic results on the set $cs(F)$ of a convergent series of fuzzy numbers.

Theorem 6.3.4. [219, Theorem 3.4] *Let* $(u_k), (v_k) \in \omega(F)$ *and* $\alpha \in \mathbb{R}$. *Then, the following statements hold:*

(i) *If* $(u_k) \in cs(F)$, *then* $u_k \to \bar{0}$, *as* $k \to \infty$.

(ii) $(u_k) \in cs(F)$ *if and only if* $\sum_{k=n+1}^{\infty} u_k \to \bar{0}$, *as* $n \to \infty$.

(iii) *If* $(u_k), (v_k) \in cs(F)$, *then* $\sum_k (u_k + v_k) = \sum_k u_k + \sum_k v_k$.

(iv) $\sum_k \alpha u_k = \alpha \sum_k u_k$.

Proof. (i) Let $(u_k) \in cs(F)$. Then, $\sum_k u_k^-(\lambda)$ and $\sum_k u_k^+(\lambda)$ are uniformly convergent in $\lambda \in [0, 1]$. Therefore, $u_k^-(\lambda) \to 0$ and $u_k^+(\lambda) \to 0$ uniformly in $\lambda \in [0, 1]$, as $k \to \infty$, by Part (i) of Lemma 6.3.1 which means that $u_k \to \bar{0}$, as $k \to \infty$, as desired.

(ii) Let $(u_k) \in cs(F)$. Then, the series of functions $\sum_k u_k^-(\lambda)$ and $\sum_k u_k^+(\lambda)$ are uniformly convergent in $\lambda \in [0, 1]$. Thus, $\sum_{k=n+1}^{\infty} u_k^-(\lambda) \to 0$ and $\sum_{k=n+1}^{\infty} u_k^+(\lambda) \to 0$ uniformly in $\lambda \in [0, 1]$, as $n \to \infty$, by Part (ii) of Lemma 6.3.1 which gives that $\sum_{k=n+1}^{\infty} u_k \to \bar{0}$, as $n \to \infty$.

Conversely, suppose that $\sum_{k=n+1}^{\infty} u_k \to \bar{0}$, as $n \to \infty$. Then, $\sum_{k=n+1}^{\infty} u_k^-(\lambda) \to 0$ and $\sum_{k=n+1}^{\infty} u_k^+(\lambda) \to 0$ uniformly in $\lambda \in [0, 1]$, as $n \to \infty$. Hence, the series $\sum_k u_k^-(\lambda)$ and $\sum_k u_k^+(\lambda)$ converge uniformly in $\lambda \in [0, 1]$ by Part (ii) of Lemma 6.3.1 which gives the convergence of the series $\sum_k u_k$.

(iii) Suppose that $\sum_k u_k = u$ and $\sum_k v_k = v$. We have in this situation that $D(\sum_{k=0}^{n} u_k, u) \to 0$ and $D(\sum_{k=0}^{n} v_k, v) \to 0$, as $n \to \infty$. Then, by using the properties of the metric D given by Proposition 6.2.10, we observe that

$$D\left(\sum_{k=0}^{n}(u_k + v_k), u + v\right) = D\left(\sum_{k=0}^{n} u_k + \sum_{k=0}^{n} v_k, u + v\right) \quad (6.3.1)$$

$$\leq D\left(\sum_{k=0}^{n} u_k, u\right) + D\left(\sum_{k=0}^{n} v_k, v\right).$$

By letting $n \to \infty$ in (6.3.1) one can easily see that $\sum_k(u_k + v_k) = u + v = \sum_k u_k + \sum_k v_k$.

(iv) Suppose that $\sum_k u_k = u$, i.e., $D(\sum_{k=0}^{n} u_k, u) \to 0$, as $n \to \infty$. Again by using the properties of the metric D given by Proposition 6.2.10, we obtain

$$D\left(\sum_{k=0}^{n} \alpha u_k, \alpha u\right) = D\left(\alpha \sum_{k=0}^{n} u_k, \alpha u\right) = |\alpha| D\left(\sum_{k=0}^{n} u_k, u\right). \quad (6.3.2)$$

By letting $n \to \infty$ in (6.3.2) we conclude that $D(\sum_{k=0}^{n} \alpha u_k, \alpha u) \to 0$, as $n \to \infty$, which means that Part (iv) holds. $\qquad\square$

Example 6.3.5. [219, Example 3.5] *The converse of Part (i) of Theorem 6.3.4 does not hold, in general. For this, if we define $u = (u_k)$ by*

$$u_k(t) := \begin{cases} 1 - (k+1)t & , \quad 0 \leq t \leq \dfrac{1}{k+1}, \\ 0 & , \quad otherwise \end{cases}$$

for all $k \in \mathbb{N}_0$ [191], then it is immediate that $u_k \to \overline{0}$, as $k \to \infty$. However, since $u_k^+(\lambda) = (1 - \lambda)/(k+1)$ we must have $\sum_k u_k^+(\lambda) = \sum_k(1 - \lambda)/(k+1)$ which diverges for all $\lambda \in [0, 1)$. Hence, the series $\sum_k u_k$ is also divergent.

Theorem 6.3.6. [219, Theorem 3.6] *Suppose that $\sum_n u_n$ is convergent in $L(\mathbb{R})$ and $v \in L(\mathbb{R})$ is comparable with $\overline{0}$, and the terms of $(u_n) \in \omega(F)$ are on the same side of $\overline{0}$. Then, $\sum_n vu_n$ is convergent and $\sum_n vu_n = v\sum_n u_n$.*

Proof. Let $v \succeq \overline{0}$ and $u_n \succeq \overline{0}$ for all $n \in \mathbb{N}_0$. Then, $(vu_n)^+(\lambda) = v^+(\lambda)u_n^+(\lambda)$ and $(vu_n)^-(\lambda) = v^-(\lambda)u_n^-(\lambda)$. Therefore, the series $\sum_n u_n^-(\lambda)$ and $\sum_n u_n^+(\lambda)$ are uniformly convergent on $[0, 1]$, and the series $\sum_n v^-(\lambda)u_n^-(\lambda)$, $\sum_n v^+(\lambda)u_n^+(\lambda)$ are also uniformly convergent on $[0, 1]$, since the functions v^- and v^+ are bounded on $[0, 1]$. This implies the convergence of the series $\sum_n vu_n$. Similarly, the series $\sum_n vu_n$ is also convergent in the cases $v \succeq \overline{0}$, $u_n \preceq \overline{0}$; $v \preceq \overline{0}$, $u_n \succeq \overline{0}$ and $v \preceq \overline{0}$, $u_n \preceq \overline{0}$. Additionally, one can see from Lemmas 6.2.2 and 6.2.7 that

$$\sum_n vu_n = \lim_{k\to\infty} \sum_{n=0}^{k} vu_n = \lim_{k\to\infty} v \sum_{n=0}^{k} u_n = v \lim_{k\to\infty} \sum_{n=0}^{k} u_n = v \sum_n u_n,$$

as was wished. $\qquad\square$

Definition 6.3.7. *A sequence* $(u_n) \in \omega(F)$ *is non-decreasing if* $u_{n+1} \succeq u_n$ *for all* $n \in \mathbb{N}$, *or non-increasing if* $u_{n+1} \preceq u_n$ *for all* $n \in \mathbb{N}$. *A monotonic sequence is a sequence that is either non-increasing or non-decreasing.*

One can conclude from the definition that the sequences $\{u_n^-(\lambda)\}$ and $\{u_n^+(\lambda)\}$ of functions are monotonic for all $\lambda \in [0,1]$, if the sequence (u_n) is monotonic. It is known that a sequence of real numbers is convergent if it is monotonic and bounded. However, this does not hold for the sequences of fuzzy numbers.

Now, following Mares [157], we define u^n for $u \succ \overline{0}$, below.

Definition 6.3.8. [104, Definition 4.2] *Let* u *be any non-negative fuzzy number. We define* u^n *for non-zero real number* n *by* $u^n(x) :=$
$\begin{cases} u(x^{1/n}) & , \quad x > 0, \\ 0 & , \quad x \leq 0. \end{cases}$ *The* λ*-level set of the fuzzy number* u^n *with* $[u]_\lambda = [u_\lambda^-, u_\lambda^+]$ *is determined as follows: Since* $[u^n]_\lambda = \{x : u^n(x) \geq \lambda\} = \{x : u(x^{1/n}) \geq \lambda\}$, *we have*

$$x \in [u^n]_\lambda \quad \Leftrightarrow \quad u(x^{1/n}) \geq \lambda \Leftrightarrow x^{1/n} \in [u]_\lambda \Leftrightarrow u^-(\lambda) \leq x^{1/n} \leq u^+(\lambda)$$
$$\Leftrightarrow \quad (u^-(\lambda))^n \leq x \leq (u^+(\lambda))^n \Leftrightarrow x \in [(u^-(\lambda))^n, (u^+(\lambda))^n]$$
$$\Leftrightarrow \quad x \in [u]_\lambda^n$$

which leads us to the consequence that $[u^n]_\lambda = [u]_\lambda^n$.
In the case of $n = 0$, *we define* u^0 *by* $u^0(x) := \begin{cases} 1 & , \quad x > 0, \\ 0 & , \quad x \leq 0. \end{cases}$

Now, we can give four different forms of u^n with respect to Definition 6.3.8.

Basic Lemma. *Suppose that* $u \in L(\mathbb{R})$ *and* $n \in \mathbb{N}$. *Then, the following four statements hold for* u^n:

(i) *If* $u_\lambda^- > 0$ *and* $u_\lambda^+ > 0$, *then* $[u]_\lambda^n = [(u_\lambda^-)^n, (u_\lambda^+)^n]$.

(ii) *If* $u_\lambda^- < 0$ *and* $u_\lambda^+ > 0$, *then* $[u]_\lambda^n = [(u_\lambda^+)^{n-1}u_\lambda^-, (u_\lambda^+)^n]$.

(iii) *If* $u_\lambda^- < 0$ *and* $u_\lambda^+ < 0$, *then* $[u]_\lambda^n := \begin{cases} [(u_\lambda^+)^n, (u_\lambda^-)^n] & , \quad n \text{ even}, \\ [(u_\lambda^-)^n, (u_\lambda^+)^n] & , \quad n \text{ odd}. \end{cases}$

(iv) *If* $u_\lambda^- < 0$ *and* $u_\lambda^+ = 0$, *then* $[u]_\lambda^n := \begin{cases} [\overline{0}, (u_\lambda^-)^n] & , \quad n \text{ even}, \\ [(u_\lambda^-)^n, \overline{0}] & , \quad n \text{ odd}. \end{cases}$

Proof. Since the cases (ii)–(iv) can be established by the similar way, we consider only Part (i). We prove Part (i) by mathematical induction.

(a) The statement is true for $n = 1$, since $([u]_\lambda)^1 = [u]_\lambda = [u_\lambda^-, u_\lambda^+]$.

(b) Assume that the statement true for $n = m$, i.e., $([u]_\lambda)^m = [(u_\lambda^-)^m, (u_\lambda^+)^m]$.

(c) For $n = m + 1$, one can see by taking into account the hypothesis in Part (b) that

$$\begin{aligned}
([u]_\lambda)^{m+1} &= ([u]_\lambda)^m [u]_\lambda \\
&= [(u_\lambda^-)^m, (u_\lambda^+)^m][(u_\lambda^-), (u_\lambda^+)] \\
&= [(u_\lambda^-)^{m+1}, (u_\lambda^+)^{m+1}].
\end{aligned}$$

This shows that the statement is also true for $n = m+1$, if it is true for $n = m$ which completes the proof. $\qquad\square$

Now, we give the result on the geometric series of non-negative fuzzy numbers.

Corollary 6.3.9. [105, Corollary 4.2] *For a given geometric series* $\sum_{n=1}^\infty u^n(x),$

$$\sum_{n=1}^\infty [u]_\lambda^n = \left[\frac{u_\lambda^-}{1 - u_\lambda^-}, \frac{u_\lambda^+}{1 - u_\lambda^+}\right] \tag{6.3.3}$$

holds, where $\bar{0} \preceq u \prec \bar{1}$.

Proof. Taking into account Part (i) of Basic Lemma, we define the sequence (s_p) by

$$s_p = \sum_{n=1}^p [u]_\lambda^n = \sum_{n=1}^p \left[(u_\lambda^-)^n, (u_\lambda^+)^n\right] = \left[\sum_{n=1}^p (u_\lambda^-)^n, \sum_{n=1}^p (u_\lambda^+)^n\right]$$

for all $p \in \mathbb{N}_1$. Then, we have

$$s_p = \frac{[u]_\lambda \left\{\bar{1} - ([u]_\lambda)^p\right\}}{\bar{1} - [u]_\lambda} \quad \text{for all } p \in \mathbb{N}_1. \tag{6.3.4}$$

At this stage, since u is a non-negative fuzzy number for every $\varepsilon > 0$ there exists a $n_0 = n_0(\varepsilon) \in \mathbb{N}_1$ such that

$$D(u^p, \bar{0}) = \sup_{\lambda \in [0,1]} \max\left\{\left|(u_\lambda^-)^p\right|, \left|(u_\lambda^+)^p\right|\right\} = \max\left\{\left|(u_0^-)^p\right|, \left|(u_0^+)^p\right|\right\} < \varepsilon.$$

Therefore, taking into account this fact, we obtain by letting $p \to \infty$ in (6.3.4) that $s_p \to [u]_\lambda/(\bar{1} - [u]_\lambda)$, as $p \to \infty$, which means that (6.3.3) holds. $\qquad\square$

Example 6.3.10. [105, Example 4.3] *Consider the geometric series* $\sum_{n=1}^\infty u^n$ *of fuzzy numbers, where*

$$u(x) := \begin{cases} 1 & , \quad \dfrac{1}{3} \le x \le \dfrac{1}{2}, \\ 0 & , \quad x < \dfrac{1}{3} \text{ or } x > \dfrac{1}{2}. \end{cases}$$

Then, since $\bar{0} \preceq u \prec \bar{1}$ *with* $u_\lambda^- = 1/3$ *and* $u_\lambda^+ = 1/2$, *it is immediate by* (6.3.3) *that* $\sum_{n=1}^\infty [u]_\lambda^n = [1/2, 1]$.

Example 6.3.11. [105, Example 4.4] *Consider the geometric series* $\sum_{n=1}^{\infty} u^n$ *of fuzzy numbers, where*

$$
u(x) := \begin{cases} 4x - 1 & , \quad \dfrac{1}{4} \leq x < \dfrac{1}{2}, \\[2mm] -4x + 3 & , \quad \dfrac{1}{2} \leq x \leq \dfrac{3}{4}, \\[2mm] 0 & , \quad x < \dfrac{1}{4} \text{ or } x > \dfrac{3}{4}. \end{cases}
$$

Then, since $\overline{0} \preceq u \prec \overline{1}$ *with* $u_\lambda^- = (\lambda + 1)/4$ *and* $u_\lambda^+ = (3 - \lambda)/4$, *we derive from here that*

$$
\sum_{n=1}^{\infty} [u]_\lambda^n = \left[\frac{(\lambda + 1)/4}{1 - \dfrac{\lambda + 1}{4}}, \frac{(3 - \lambda)/4}{1 - \dfrac{3 - \lambda}{4}} \right] = \left[\frac{\lambda + 1}{3 - \lambda}, \frac{3 - \lambda}{1 + \lambda} \right].
$$

6.3.2 The Convergence Tests for the Series of Fuzzy Numbers with Positive Terms

In this subsection, we give some convergence tests for the series of fuzzy numbers with positive terms.

Lemma 6.3.12. [219, Lemma 4.1] *The series* $\sum_n u_n$ *of fuzzy numbers with positive terms converges if and only if the series* $\sum_n u_n^+(0)$ *converges.*

Proof. Suppose that $\sum_n u_n$ converges. Then, the series $\sum_n u_n^+(\lambda)$ converges for each $\lambda \in [0,1]$. Since, this also holds for $\lambda = 0$, the series $\sum_n u_n^+(0)$ is convergent.

Conversely, let $\sum_n u_n^+(0)$ be convergent. Then, since $u_n^-(\lambda) \leq u_n^+(\lambda) \leq u_n^+(0)$ for all $\lambda \in [0,1]$ Weierstrass's M test yields the uniform convergence of the series $\sum_n u_n^-(\lambda)$ and $\sum_n u_n^+(\lambda)$ for all $\lambda \in [0,1]$. Hence, $(u_n) \in cs(F)$. □

Theorem 6.3.13. [219, Theorem 4.2] *Let* $u \in L(\mathbb{R})^+$ *such that* $D(u, \overline{0}) = u^+(0) < 1$. *Then, the series* $\sum_n u^n$ *is convergent.*

Proof. Let $u \in L(\mathbb{R})^+$. Then, $[u^n]_\lambda = [u^-(\lambda)^n, u^+(\lambda)^n]$ for all $n \in \mathbb{N}_0$. Since $D(u, \overline{0}) < 1$ for all $\lambda \in [0,1]$, $u_n^-(\lambda) \leq u_n^+(\lambda) \leq D(u, \overline{0}) < 1$. The convergence of the series $\sum_n D(u, \overline{0})^n$ implies the uniform convergence of the series of functions $\sum_n u^-(\lambda)^n$ and $\sum_n u^+(\lambda)^n$ on $[0,1]$. Additionally, we have

$$
\sum_n u^-(\lambda)^n = \frac{1}{1 - u^-(\lambda)} \quad \text{and} \quad \sum_n u^+(\lambda)^n = \frac{1}{1 - u^+(\lambda)}.
$$

□

Example 6.3.14. [219, Example 4.3] *Define the fuzzy numbers* u *and* v *by*

$$
u(x) := \begin{cases} 2x & , \quad x \in (0, 1/2], \\ 0 & , \quad otherwise \end{cases} \quad \text{and} \quad v(x) := \begin{cases} 2 - \dfrac{2}{x} & , \quad x \in [1, 2], \\ 0 & , \quad otherwise. \end{cases}
$$

In this situation, since $D(u, \bar{0}) = 1/2 < 1$; we have $\sum_n u^{n-1} = v$.

Theorem 6.3.15. [219, Theorem 4.4] *The absolute convergence of a series of fuzzy numbers implies the convergence of the series.*

Proof. Suppose that the series $\sum_k |u_k|$ is convergent. Then, the series

$$\sum_k |u_k|^+(\lambda) = \sum_k \max\{|u_k^-(\lambda)|, |u_k^+(\lambda)|\}$$

converges uniformly in $\lambda \in [0, 1]$. Since $|u_k^-(\lambda)| \leq \max\{|u_k^-(\lambda)|, |u_k^+(\lambda)|\}$ and $|u_k^+(\lambda)| \leq \max\{|u_k^-(\lambda)|, |u_k^+(\lambda)|\}$, the series of functions $\sum_k u_k^-(\lambda)$ and $\sum_k u_k^+(\lambda)$ converge uniformly in $\lambda \in [0, 1]$, by Lemma 6.3.2. Hence, the series $\sum_k u_k$ converges. $\qquad\square$

Theorem 6.3.16 (Comparison test). [219, Theorem 4.5] *Let (u_k) and (v_k) be two sequences of non-negative fuzzy numbers and $u_k \preceq v_k$ for all $k \in \mathbb{N}_0$. Then, the following statements hold:*

(i) $\sum_k u_k$ is convergent, if $\sum_k v_k$ is convergent.

(ii) $\sum_k v_k$ is divergent, if $\sum_k u_k$ is divergent.

Proof. Let $(u_k), (v_k) \in \omega(F)$ such that $u_k \preceq v_k$ and $u_k, v_k \in L(\mathbb{R})^+$ for all $k \in \mathbb{N}_0$.

(i) Let $\sum_k v_k$ be convergent. Then, $u_k^-(\lambda) \leq v_k^-(\lambda)$ and $u_k^+(\lambda) \leq v_k^+(\lambda)$ for all $\lambda \in [0, 1]$ and for all $k \in \mathbb{N}_0$. In this situation, the series of functions $\sum_k v_k^-(\lambda)$ and $\sum_k v_k^+(\lambda)$ converge uniformly in $\lambda \in [0, 1]$. Therefore, the series of functions $\sum_k u_k^-(\lambda)$ and $\sum_k u_k^+(\lambda)$ converge uniformly in $\lambda \in [0, 1]$, by Lemma 6.3.2. Hence, the series $\sum_k u_k$ converges.

(ii) Suppose that $\sum_k u_k$ is divergent while $\sum_k v_k$ is convergent. This contradicts Part (i). Hence, $\sum_k v_k$ must be divergent. $\qquad\square$

Theorem 6.3.17 (Ratio test). [219, Theorem 4.7] *Let $\sum_n u_n$ be a series with positive terms and $[u_{n+1}^+(0)/u_n^+(0)] \to \alpha$, as $n \to \infty$. Then, the following statements hold:*

(i) $\sum_n u_n$ is convergent, if $\alpha < 1$.

(ii) $\sum_n u_n$ is divergent, if $\alpha > 1$.

(iii) The test is inconclusive, if $\alpha = 1$, that is, $\sum_n u_n$ may be convergent or divergent.

Proof. Suppose that $\sum_n u_n$ is a series of fuzzy numbers with positive terms. If $\alpha < 1$, then the series $\sum_n u_n^+(0)$ is convergent which implies the convergence of the series $\sum_n u_n$ by Lemma 6.3.12. If $\alpha > 1$, since the series $\sum_n u_n^+(0)$ is divergent, then the series $\sum_n u_n$ is also divergent by Lemma 6.3.12. $\qquad\square$

Example 6.3.18. [219, Example 4.6] *Consider the sequences $u = (u_n)$ and $v = (v_n)$ of fuzzy numbers defined by*

$$
u_n(t) := \begin{cases} n(n+1)t - 1 & , \quad \dfrac{1}{n(n+1)} \le t \le \dfrac{2}{n(n+1)}, \\[2mm] 3 - n(n+1)t & , \quad \dfrac{2}{n(n+1)} < t \le \dfrac{3}{n(n+1)}, \\[2mm] 0 & , \quad otherwise \end{cases}
$$

$$
v_n(t) := \begin{cases} (n+1)^2 t - 1 & , \quad \dfrac{1}{(n+1)^2} \le t \le \dfrac{2}{(n+1)^2}, \\[2mm] 3 - (n+1)^2 t & , \quad \dfrac{2}{(n+1)^2} < t \le \dfrac{3}{(n+1)^2}, \\[2mm] 0 & , \quad otherwise \end{cases}
$$

for all $n \in \mathbb{N}_0$. It is trivial that $u_n^-(\lambda) = (\lambda+1)/[n(n+1)]$ and $u_n^+(\lambda) = (3-\lambda)/[n(n+1)]$ for all $\lambda \in [0,1]$, similarly $v_n^-(\lambda) = (\lambda+1)/(n+1)^2$ and $v_n^+(\lambda) = (3-\lambda)/(n+1)^2$. Since $v_n^-(\lambda) \le u_n^-(\lambda)$ and $v_n^+(\lambda) \le u_n^+(\lambda)$ for all $\lambda \in [0,1]$ and for all $n \in \mathbb{N}_0$, then the series of functions $\sum_n u_n^-(\lambda)$ and $\sum_n u_n^+(\lambda)$ converge uniformly in $\lambda \in [0,1]$. Hence, the series $\sum_n v_n$ converges with respect to the comparison test.

Example 6.3.19. [219, Example 4.8] *Define the sequence $(u_n) \in \omega(F)$ by*

$$
u_n(t) := \begin{cases} \dfrac{12^n t - 3^n}{4^n - 3^n} & , \quad \dfrac{1}{4^n} \le t \le \dfrac{1}{3^n}, \\[2mm] \dfrac{3^n - 6^k t}{3^n - 2^k} & , \quad \dfrac{1}{3^n} < t \le \dfrac{1}{2^n}, \\[2mm] 0 & , \quad otherwise. \end{cases}
$$

Then, $u_n^-(\lambda) = 3^{-n}\lambda - 4^{-n}(\lambda - 1)$ and $u_n^+(\lambda) = 3^{-n}\lambda + 2^{-n}(1 - \lambda)$, and so we have

$$
\lim_{n \to \infty} \frac{u_{n+1}^+(0)}{u_n^+(0)} = \lim_{n \to \infty} \frac{2^n}{2^{n+1}} = \frac{1}{2} < 1.
$$

Therefore, the series $\sum_n u_n$ converges with respect to the Ratio test.

Since Theorems 6.3.20, 6.3.22, 6.3.24, 6.3.26 and 6.3.27 can be proved in the similar way used in the real case, we give them with the related examples without proof.

Theorem 6.3.20 (Cauchy's root test). [219, Theorem 4.9] *Let $\sum_n u_n$ be a series with positive terms and $\sqrt[n]{u_n^+(0)} \to \alpha$, as $n \to \infty$. Then, the following statements hold:*

(i) *$\sum_n u_n$ is convergent, if $\alpha < 1$.*

(ii) *$\sum_n u_n$ is divergent, if $\alpha > 1$.*

(iii) *The test is inconclusive if, $\alpha = 1$; that is, $\sum_n u_n$ may be convergent or divergent.*

We should remark that the convergence of the series $\sum_n u_n^+(0) < \infty$ guarantees the convergence of the series $\sum_n u_n$ but is not sufficient to determine the sum of the series, that is, $\sum_n u_n^+(0) = v^+(0)$ does not imply $\sum_n u_n = v$. The next example enables us for observing this:

Example 6.3.21. [219, Example 4.10] *Define the sequence* $\{u_n(x)\} \in \omega(F)$ *and* $u, v \in L(\mathbb{R})$ *by*

$$u_n(x) := \begin{cases} 2^n x & , & x \in (0, 2^{-n}], \\ 0 & , & otherwise, \end{cases}$$

and

$$u(x) := \begin{cases} x & , & x \in (0, 1], \\ 0 & , & otherwise, \end{cases} \quad v(x) := \begin{cases} 1 - x & , & x \in [0, 1], \\ 0 & , & otherwise. \end{cases}$$

Then, $\sum_n u_n = u \neq v$ *while* $u^+(0) = v^+(0) = 1$.

Theorem 6.3.22. (Limit Comparison test). [219, Theorem 4.11] *Let* (u_n) *and* (v_n) *be the sequences of non-negative fuzzy numbers, and there exists an* $l \in L(\mathbb{R})$ *such that* $u_n^+(0)/v_n^+(0) \to l$, *as* $n \to \infty$. *Then, the following statements hold:*

(i) *If* $0 < l < \infty$, *then* $\sum_n u_n$ *is convergent (divergent) if and only if* $\sum_n v_n$ *is convergent (divergent).*

(ii) *If* $l = 0$, $\sum_n v_n$ *is convergent, then* $\sum_n u_n$ *is convergent.*

(iii) *If* $l = \infty$, $\sum_n v_n$ *is divergent, then* $\sum_n u_n$ *is also divergent.*

Example 6.3.23. [219, Example 4.12] *Define the sequences* $(u_n), (v_n) \in \omega(F)$ *by*

$$u_n(x) := \begin{cases} (n-1)/2n & , & x \in (1/2, 1], \\ 1 & , & x \in [0, 1/2], \\ 0 & , & otherwise, \end{cases} \quad v_n(x) := \begin{cases} \dfrac{1}{2} + \dfrac{1}{4n} & , & x \in (1/2, 1], \\ 1 & , & x \in [0, 1/2], \\ 0 & , & otherwise \end{cases}$$

for all $n \in \mathbb{N}_1$. *Then,* $u_n^-(\lambda) = 0$ *and* $v_n^-(\lambda) = 0$, *and*

$$u_n^+(\lambda) := \begin{cases} 1 & , & \lambda \in [0, (n-1)/2n], \\ 1/2 & , & \lambda \in ((n-1)/2n, 1], \end{cases} \quad v_n^+(\lambda) := \begin{cases} 1 & , & \lambda \in [0, 1/2 + 1/4n], \\ 1/2 & , & \lambda \in (1/2 + 1/4n, 1]. \end{cases}$$

$u_n^+(0) = 1$ *and* $v_n^+(0) = 1$. *We can easily see that the sequences diverge and there exists* $u_n^+(0)/v_n^+(0) \to l$, *as* $n \to \infty$. *Using Part (i) of Theorem 6.3.22, since* $l = 1$, *we say that if* $\sum_n u_n$ *converges (diverges), then* $\sum_n v_n$ *converges (diverges). Hence, the series* $\sum_n u_n$ *and* $\sum_n v_n$ *are divergent from the limit comparison test.*

Now, we give the ratio test, from Talo et al. [219], for the series of non-negative fuzzy numbers:

Theorem 6.3.24 (Ratio comparison test). [219, Theorem 4.13] *Let (u_n) and (v_n) be the sequences of non-negative fuzzy numbers, there exists $n_0 \in \mathbb{N}$ for all $n \geq n_0$ such that $\dfrac{u_{n+1}^+(0)}{u_n^+(0)} \leq \dfrac{v_{n+1}^+(0)}{v_n^+(0)}$. Then, the following statements hold:*

(i) $\sum_n u_n$ *is convergent, if* $\sum_n v_n$ *is convergent.*

(ii) $\sum_n v_n$ *is divergent, if* $\sum_n u_n$ *is divergent.*

Example 6.3.25. [219, Example 4.14] *Define the fuzzy sequences* $(u_n), (v_n) \in \omega(F)$ *by*

$$u_n(x) := \begin{cases} (n+1)x - 1 & , \quad x \in [(n+1)^{-1}, 2(n+1)^{-1}], \\ 3 - (n+1)x & , \quad x \in (2(n+1)^{-1}, 3(n+1)^{-1}], \\ 0 & , \quad otherwise \end{cases}$$

$$v_n(x) := \begin{cases} (n+1)(x-1) & , \quad x \in [1, 1+(n+1)^{-1}], \\ 2 - (n+1)(x-1) & , \quad x \in (1+(n+1)^{-1}, 1+2(n+1)^{-1}], \\ 0 & , \quad otherwise \end{cases}$$

for all $n \in \mathbb{N}_0$. *Then,* $u_n^-(\lambda) = (\lambda+1)/(n+1)$, $u_n^+(\lambda) = (3-\lambda)/(n+1)$, $v_n^-(\lambda) = 1+\lambda/(n+1)$ *and* $v_n^+(\lambda) = 1+(2-\lambda)/(n+1)$. *By using the level sets for* $\lambda = 0$, *we have* $u_n^+(0) = 3/(n+1)$, $u_{n+1}^+(0) = 3/(n+2)$ *and* $v_n^+(0) = (n+3)/(n+1)$, $v_{n+1}^+(0) = (n+4)/(n+2)$. *One can easily observe that the conditions of the fuzzy ratio comparison test are satisfied. Because, the fuzzy sequence* $v_n \to \overline{1}$ *as* $n \to \infty$ *and the series* $\sum_{n=1}^{\infty} v_n$ *is convergent then* $\sum_{n=1}^{\infty} u_n$ *is convergent by the comparison test.*

Theorem 6.3.26 (Kummer's test). [219, Theorem 4.15] *Let (u_n) be a sequence of non-negative fuzzy numbers and let $\sum_n 1/v_n^+(0)$ be a divergent series. Define κ by $\kappa = \lim_{n\to\infty} \left[v_n^+(0) - \dfrac{u_{n+1}^+(0)}{u_n^+(0)} v_{n+1}^+(0) \right]$. If $\kappa > 0$, then $\sum_n u_n$ is convergent and if $\kappa < 0$, then $\sum_n u_n$ is divergent.*

Kummer's test is valuable because they are extremely versatile. Indeed, it is not difficult to show that a series is convergent or divergent for a suitable choice of $v_n^+(0)$. This test has a high practical value in proving some important basic tests. Now, we give an important convergence test with respect to Kummer's test.

Theorem 6.3.27 (Raabe's test). [219, Theorem 4.16] *Let $\sum_n u_n$ be a series of non-negative fuzzy numbers and $n\left[\dfrac{u_n^+(0)}{u_{n+1}^+(0)} - 1 \right] \to \beta$, as $n \to \infty$. Then, $\sum_n u_n$ is convergent if $\beta > 1$ and $\sum_n u_n$ is divergent if $\beta < 1$.*

Example 6.3.28. [219, Example 4.17] *Define* $(u_n) \in \omega(F)$ *by*

$$u_n(x) := \begin{cases} 1 - (n+1)^2 x & , \quad x \in [0, (n+1)^{-2}], \\ 0 & , \quad otherwise. \end{cases}$$

$u_n^-(0) = 0$ and $u_n^+(0) = (n+1)^{-2}$. *By using Theorem 6.3.27*

$$\lim_{n \to \infty} n \left[\frac{(n+2)^2}{(n+1)^2} - 1 \right] = \lim_{n \to \infty} \frac{n(2n+3)}{(n+1)^2} = 2.$$

Therefore, $\beta = 2 > 1$ and $\sum_n u_n$ is convergent by Raabe's test. One can easily see that ratio and root tests failed for this series.

We give three more results concerning the series of functions which are needed in the rest of the text.

Lemma 6.3.29 (Dirichlet's test). [221, p. 217] *If the sequence of partial sums of the series $\sum_n f_n$ of functions is uniformly bounded on the set A and the monotonic increasing sequence (g_n) of functions is uniformly convergent to zero on the set A, then the series $\sum_n f_n g_n$ of functions is uniformly convergent on the set A.*

Lemma 6.3.30 (Abel's test). [221, p. 219] *If the series $\sum_n f_n$ of functions is uniformly convergent on the set A and the monotonic sequence (g_n) of functions is uniformly bounded on the set A, then the series $\sum_n f_n g_n$ of functions is uniformly convergent on the set A.*

Lemma 6.3.31 (Dedekind's test). [219, Lemma 4.20] *If the series $\sum_n f_n$ and $\sum_n |g_n - g_{n+1}|$ of functions are uniformly convergent on the set A and the sequence (g_n) of functions is uniformly bounded on the set A, then the series $\sum_n f_n g_n$ of functions is uniformly convergent on the set A.*

Lemma 6.3.32. [219, Lemma 4.21] *Let $a, b, c, d \in \mathbb{R}$. Then, we have:*

$$\max\{a, b, c, d\} \tag{6.3.5}$$
$$= \frac{1}{4} \left\{ a + b + |a - b| + c + d + |c - d| + |a + b + |a - b| - c - d - |c - d|| \right\},$$
$$\min\{a, b, c, d\} \tag{6.3.6}$$
$$= \frac{1}{4} \left\{ a + b - |a - b| + c + d - |c - d| - |a + b - |a - b| - c - d + |c - d|| \right\}.$$

Proof. Since (6.3.6) can be established by the similar way, we consider only (6.3.5). It is immediate by the following relations that

$$\max\{a, b, c, d\} = \max\{\max(a, b), \max(c, d)\}$$
$$= \max \left\{ \frac{a + b + |a - b|}{2}, \frac{c + d + |c - d|}{2} \right\},$$

the equality (6.3.5) holds. □

Theorem 6.3.33. [219, Theorem 4.22] *Let $(u_n), (v_n) \in \omega(F)$. Then, $\sum_n u_n v_n$ is convergent if $\sum_n u_n$ is convergent, the bounded sequence (v_n) is monotonic and $\sum_n [v_n^+(0) - v_n^-(0)]$ is convergent.*

Proof. Suppose that the series $\sum_n u_n$ is convergent and (v_n) is a bounded monotonic sequence with $u_n, v_n \in L(\mathbb{R})$ for all $n \in \mathbb{N}_0$. Then, the series $\sum_n u_n^-(\lambda)$ and $\sum_n u_n^+(\lambda)$ are uniformly convergent on $[0, 1]$. Boundedness of the sequence (v_n) implies the uniform boundedness of the sequences $\{v_n^-(\lambda)\}$ and $\{v_n^+(\lambda)\}$ of functions on $[0, 1]$. Monotonicity of the sequence (v_n) implies the monotonicity of the sequences $\{v_n^-(\lambda)\}$ and $\{v_n^+(\lambda)\}$ of functions for each $\lambda \in [0, 1]$. For simplicity in the notation, we write

$$\begin{aligned} a_n(\lambda) &= u_n^-(\lambda)v_n^-(\lambda), & b_n(\lambda) &= u_n^-(\lambda)v_n^+(\lambda), \\ c_n(\lambda) &= u_n^+(\lambda)v_n^-(\lambda), & d_n(\lambda) &= u_n^+(\lambda)v_n^+(\lambda). \end{aligned}$$

Therefore, since

$$(u_n v_n)^+(\lambda) = \max\{u_n^-(\lambda)v_n^-(\lambda), u_n^-(\lambda)v_n^+(\lambda), u_n^+(\lambda)v_n^-(\lambda), u_n^+(\lambda)v_n^+(\lambda)\},$$

if the series

(1) $\sum_n a_n(\lambda), \sum_n b_n(\lambda), \sum_n c_n(\lambda), \sum_n d_n(\lambda)$

(2) $\sum_n |a_n(\lambda) - b_n(\lambda)|$

(3) $\sum_n |c_n(\lambda) - d_n(\lambda)|$

(4) $\sum_n \big| a_n(\lambda) + b_n(\lambda) + |a_n(\lambda) - b_n(\lambda)| - c_n(\lambda) - d_n(\lambda) - |c_n(\lambda) - d_n(\lambda)| \big|$

are uniformly convergent on $[0, 1]$, then the series $\sum_n (u_n v_n)^+(\lambda)$ is uniformly convergent on $[0, 1]$, by Lemma 6.3.32.

The series $\sum_n a_n(\lambda), \sum_n b_n(\lambda), \sum_n c_n(\lambda), \sum_n d_n(\lambda)$ are uniformly convergent on $[0, 1]$, by Abel's test. Since $(v_n) \in \ell_\infty(F)$, there is a $K > 0$ such that $D(v_n, \overline{0}) < K$ for all $n \in \mathbb{N}_0$. The convergence of the series $\sum_n u_n$ implies that $u = (u_n) \in \ell_\infty(F)$, i.e., there is a $M > 0$ such that $D(u_n, \overline{0}) < M$ for all $n \in \mathbb{N}_0$. Now, we show that the series in Parts (2)–(4) are uniformly convergent on $[0, 1]$, respectively. One can observe that the inequalities

$$\sum_n |a_n(\lambda) - b_n(\lambda)| \leq M \sum_n |v_n^-(\lambda) - v_n^+(\lambda)| \leq M \sum_n \left[v_n^+(0) - v_n^-(0) \right],$$

$$\sum_n |c_n(\lambda) - d_n(\lambda)| \leq M \sum_n |v_n^-(\lambda) - v_n^+(\lambda)| \leq M \sum_n \left[v_n^+(0) - v_n^-(0) \right]$$

hold for all $\lambda \in [0, 1]$. Since the series $\sum_n \left[v_n^+(0) - v_n^-(0) \right]$ is convergent, Weierstrass's M test yields the uniform convergence of the series in Parts (2) and (3), on the interval $[0, 1]$. Additionally,

$$\sum_n \big| a_n(\lambda) + b_n(\lambda) + |a_n(\lambda) - b_n(\lambda)| - c_n(\lambda) - d_n(\lambda) - |c_n(\lambda) - d_n(\lambda)| \big|$$

$$\leq \sum_n |a_n(\lambda) - b_n(\lambda)| + \sum_n |c_n(\lambda) - d_n(\lambda)| + \sum_n |a_n(\lambda) - c_n(\lambda)|$$

$$+ \sum_n |b_n(\lambda) - d_n(\lambda)|$$

holds for all $\lambda \in [0,1]$ which gives that

$$\sum_n |a_n(\lambda) - c_n(\lambda)| \leq K \sum_n |u_n^-(\lambda) - u_n^+(\lambda)| \leq K \sum_n \left[u_n^+(0) - u_n^-(0) \right],$$

$$\sum_n |b_n(\lambda) - d_n(\lambda)| \leq K \sum_n |u_n^-(\lambda) - u_n^+(\lambda)| \leq K \sum_n \left[u_n^+(0) - u_n^-(0) \right].$$

Since the series $\sum_n u_n$ is convergent, the series $\sum_n [u_n^+(0) - u_n^-(0)]$ is also convergent. So, the series in Part (4) is uniformly convergent on $[0,1]$, by Weierstrass's M test. Since those series in Parts (1), (2), (3) and (4) are uniformly convergent on $[0,1]$, the series $\sum_n (u_n v_n)^+(\lambda)$ is also uniformly convergent on $[0,1]$. By the similar way, it can be showed that the series $\sum_n (u_n v_n)^-(\lambda)$ is also uniformly convergent on $[0,1]$. Hence, $\sum_n u_n v_n$ is convergent. $\square$

Example 6.3.34. [219, Example 4.23] *Consider the sequences* $u = (u_n)$, $v = (v_n) \in \omega(F)$ *in Example 6.3.18. It is trivial that* $u_n^-(\lambda) = (\lambda + 1)/[n(n+1)]$ *and* $u_n^+(\lambda) = (3 - \lambda)/[n(n+1)]$ *for all* $\lambda \in [0,1]$. *Therefore, we see that* $\sum_{n \in \mathbb{N}_1} (u_n)_\lambda^- = \lambda + 1$ *and* $\sum_{n \in \mathbb{N}_1} (u_n)_\lambda^+ = 3 - \lambda$. *Then, it is conclude that* $\sum_n u_n$ *converges. Similarly,* $v_n^-(\lambda) = (\lambda + 1)/(n+1)^2$ *and* $v_n^+(\lambda) = (3 - \lambda)/(n+1)^2$ *for all* $\lambda \in [0,1]$. *It is trivial that, the sequence* (v_n) *is monotonic. Further, the series* $\sum_n [v_n^+(0) - v_n^-(0)] = \sum_n \dfrac{2}{(n+1)^2}$ *is convergent. Hence,* $\sum_n u_n v_n$ *is also convergent.*

Theorem 6.3.35. [219, Theorem 4.24] *Let* $(u_n), (v_n) \in \omega(F)$. *Then* $\sum_n u_n v_n$ *is convergent if the series* $\sum_n u_n$ *is bounded, the monotonic decreasing sequence* $(v_n) \in c_0(F)$ *and the series* $\sum_n [v_n^+(0) - v_n^-(0)]$ *is convergent.*

Proof. Suppose that the series $\sum_n u_n$ is bounded. Then, the sequence $(s_n) = (\sum_{k=0}^n u_k)$ of partial sums of the series $\sum_n u_n$ is bounded. Hence, the sequences $\{s_n^-(\lambda)\} = \{\sum_{k=0}^n u_k^-(\lambda)\}$ and $\{s_n^+(\lambda)\} = \{\sum_{k=0}^n u_k^+(\lambda)\}$ of functions are uniformly bounded on $[0,1]$. Monotonicity of the sequence (v_n) implies the monotonicity of the sequences $\{v_n^-(\lambda)\}$ and $\{v_n^+(\lambda)\}$ for each $\lambda \in [0,1]$. Since $(v_n) \in c_0(F)$, the sequences $\{v_n^-(\lambda)\}$ and $\{v_n^+(\lambda)\}$ are uniformly convergent to zero. Since $v_n \succ \overline{0}$ for all $n \in \mathbb{N}_0$, we have

$$(u_n v_n)^-(\lambda) = \min\{a_n(\lambda), b_n(\lambda)\} = \frac{1}{2}[a_n(\lambda) + b_n(\lambda) - |a_n(\lambda) - b_n(\lambda)|],$$

$$(u_n v_n)^+(\lambda) = \max\{c_n(\lambda), d_n(\lambda)\} = \frac{1}{2}[c_n(\lambda) + d_n(\lambda) + |c_n(\lambda) - d_n(\lambda)|]$$

for all $\lambda \in [0,1]$. It is clear by Drichlet's test that the series $\sum_n a_n(\lambda)$, $\sum_n b_n(\lambda)$, $\sum_n c_n(\lambda)$ and $\sum_n d_n(\lambda)$ of functions are uniformly convergent on $[0,1]$. Additionally, since $(s_n) \in \ell_\infty(F)$ there is a $M > 0$ such that $D(s_n, \overline{0}) < M$ for all $n \in \mathbb{N}_0$. So,

$$D(u_n, \overline{0}) = D(s_n, s_{n-1}) \leq D(s_n, \overline{0}) + D(s_{n-1}, \overline{0}) \leq 2M$$

for all $n \in \mathbb{N}_0$. Therefore, the series $\sum_n |a_n(\lambda) - b_n(\lambda)|$ and $\sum_n |c_n(\lambda) - d_n(\lambda)|$ are uniformly convergent on $[0,1]$, by Theorem 6.3.33. Hence, $\sum_n u_n v_n$ is convergent. $\qquad \square$

Example 6.3.36. [219, Example 4.25] *Consider* $u = (u_n)$ *and* $v = (v_n) \in \omega(F)$ *defined by*

$$
u_n(t) := \begin{cases} t2^n - 1 & , \quad \dfrac{1}{2^n} \leq t \leq \dfrac{2}{2^n}, \\[2mm] 1 & , \quad \dfrac{2}{2^n} < t \leq \dfrac{4}{2^n}, \\[2mm] 2 - t2^{n-2} & , \quad \dfrac{4}{2^n} < t \leq \dfrac{8}{2^n}, \\[2mm] 0 & , \quad otherwise, \end{cases}
$$

$$
v_n(t) := \begin{cases} 6n(n+1)t - 3 & , \quad \dfrac{1}{2n(n+1)} \leq t \leq \dfrac{2}{3n(n+1)}, \\[2mm] 1 & , \quad \dfrac{2}{3n(n+1)} < t \leq \dfrac{3}{4n(n+1)}, \\[2mm] 4 - 4n(n+1)t & , \quad \dfrac{3}{4n(n+1)} < t \leq \dfrac{1}{n(n+1)}, \\[2mm] 0 & , \quad otherwise \end{cases}
$$

for all $n \in \mathbb{N}_1$. *It is obvious that* $u_n^-(\lambda) = (\lambda + 1)/2^n$ *and* $u_n^+(\lambda) = 4(2 - \lambda)/2^n$ *for all* $\lambda \in [0,1]$. *Then,* $\sum_n (u_n)_\lambda^- = 2(\lambda + 1)$ *and* $\sum_n (u_n)_\lambda^+ = 8(2 - \lambda)$. *Therefore, it is conclude that* $\sum_n u_n$ *is bounded. Similarly,* $v_n^-(\lambda) = (\lambda + 3)/[6n(n+1)]$ *and* $v_n^+(\lambda) = (4 - \lambda)/[4n(n+1)]$ *for all* $\lambda \in [0,1]$. *It is trivial that the sequence* (v_n) *is a monotonic decreasing sequence such that* $v_n \to \bar{0}$, *as* $n \to \infty$. *Further, the series* $\displaystyle\sum_{n=1}^{\infty} [v_n^+(0) - v_n^-(0)] = \sum_{n=1}^{\infty} \dfrac{1}{2n(n+1)}$ *is convergent. Hence,* $\displaystyle\sum_{n=1}^{\infty} u_n v_n$ *is convergent.*

Theorem 6.3.37. [219, Theorem 4.26] *Let* $(u_n), (v_n) \in \omega(F)$. *If the series* $\sum_n u_n$ *and* $\sum_n D(v_n, v_{n+1})$ *are convergent,* $(v_n) \in \ell_\infty(F)$ *and* $\sum_n [v_n^+(0) - v_n^-(0)]$ *is convergent, then* $\sum_n u_n v_n$ *is convergent.*

Proof. Suppose that $\sum_n u_n$ is convergent. Then, the series $\sum_n u_n^-(\lambda)$ and $\sum_n u_n^+(\lambda)$ of functions are uniformly convergent on $[0,1]$. Suppose that the series $\sum_n D(v_n, v_{n+1})$ is convergent. Then, since

$$
D(v_n, v_{n+1}) = \sup_{\lambda \in [0,1]} \max\{|v_n^-(\lambda) - v_{n+1}^-(\lambda)|, |v_n^+(\lambda) - v_{n+1}^+(\lambda)|\},
$$

the series $\sum_n |v_n^-(\lambda) - v_{n+1}^-(\lambda)|$ and $\sum_n |v_n^+(\lambda) - v_{n+1}^+(\lambda)|$ are uniformly convergent on $[0,1]$. Since the sequence $(v_n) \in \ell_\infty(F)$, the sequences $\{v_n^-(\lambda)\}$ and $\{v_n^+(\lambda)\}$ of functions are uniformly bounded on $[0,1]$. Therefore, by Lemma 6.3.31, the series $\sum_n a_n(\lambda)$, $\sum_n b_n(\lambda)$, $\sum_n c_n(\lambda)$ and $\sum_n d_n(\lambda)$ are uniformly convergent on $[0,1]$. Now, it is not hard to show by the similar way used in the proof of Theorem 6.3.33 that $\sum_n u_n v_n$ is also convergent. $\qquad \square$

By the following example, we give a sequence of fuzzy numbers satisfying the conditions of previous theorems:

Example 6.3.38. [219, Example 4.27] *Consider* $v = (v_n) \in \omega(F)$ *defined by* $v_n := \chi_{[n^{-1}, 2^{-n}+n^{-1}]}$ *for all* $n \in \mathbb{N}_1$. *Then, we have* $v_n^+(\lambda) = n^{-1} + 2^{-n}$ *and* $v_n^-(\lambda) = n^{-1}$. *It is trivial that* (v_n) *is a monotonic decreasing sequence such that* $v_n \to \bar{0}$, *as* $n \to \infty$. *Further, the series* $\sum_{n=1}^{\infty} [v_n^+(0) - v_n^-(0)] = \sum_{n=1}^{\infty} \frac{1}{2^n}$ *is convergent. Since* $D(v_n, v_{n+1}) = 2^{-(n+1)} + 1/[n(n+1)]$, *the series* $\sum_{n=1}^{\infty} D(v_n, v_{n+1})$ *is also convergent.*

6.4 Power Series of Fuzzy Numbers

In this section, following Kadak and Başar [105, 104], we essentially deal with the power series of fuzzy numbers. We give two examples on the geometric series of fuzzy numbers and by using the four different cases of u^n of nth power of a fuzzy number u, we interest in the convergence of a power series of fuzzy numbers. Finally, we introduce the concept of a power series of fuzzy numbers with fuzzy coefficients.

The main purpose of this section is to present the beginning concepts on the power series of fuzzy numbers with real and fuzzy coefficients. Prior to giving a corollary and two examples on the geometric series of fuzzy numbers, four different cases of u^n of nth power of a fuzzy number u is examined in a basic lemma. By using these situations, the convergence of a power series of fuzzy numbers is investigated. Finally, the concept of a power series of fuzzy numbers with fuzzy coefficients is introduced. The different cases of power series of fuzzy numbers to be convergent is considered and a theorem on the term by term differentiation of a power series of fuzzy numbers is stated and proved. The final result of the present section is devoted to the Taylor expansion of a fuzzy valued function.

Definition 6.4.1. [213, Definition 2.14] *Let* $\{f_k(\lambda)\}$ *be a sequence of functions defined on* $[a, b]$ *and* $\lambda_0 \in (a, b]$. *Then,* $\{f_k(\lambda)\}$ *is said to be eventually equi-left-continuous at* λ_0 *if for any* $\varepsilon > 0$ *there exist* $n_0 \in \mathbb{N}_0$ *and* $\delta > 0$ *such that* $|f_k(\lambda) - f_k(\lambda_0)| < \varepsilon$ *whenever* $\lambda \in (\lambda_0 - \delta, \lambda_0]$ *and* $k \geq n_0$. *Similarly, eventually equi-right-continuity at* $\lambda_0 \in [a, b)$ *of* $\{f_k(\lambda)\}$ *can be defined.*

Theorem 6.4.2. [213, Theorem 2.15] *Let* (u_k) *be a fuzzy valued sequence, such that* $u_k^-(\lambda) \to u^-(\lambda)$ *and* $u_k^+(\lambda) \to u^+(\lambda)$, *as* $k \to \infty$, *for each* $\lambda \in [0, 1]$. *Then, the pair of functions* u^- *and* u^+ *determine a fuzzy number if and only if the sequences of functions* $\{u_k^-(\lambda)\}$ *and* $\{u_k^+(\lambda)\}$ *are eventually equi-left-continuous at each* $\lambda \in [0, 1]$ *and eventually equi-right-continuous at* $\lambda = 0$.

Thus, it is deduced that the series $\sum_k u_k^-(\lambda) = u^-(\lambda)$ and $\sum_k u_k^+(\lambda) = u^+(\lambda)$ define a fuzzy number if the sequences

$$\{s_n^-(\lambda)\} = \left\{\sum_{k=0}^n u_k^-(\lambda)\right\} \quad \text{and} \quad \{s_n^+(\lambda)\} = \left\{\sum_{k=0}^n u_k^+(\lambda)\right\}$$

satisfy the conditions of Theorem 6.4.2. Of course, this is a weaker condition than the uniform convergence.

Example 6.4.3. [214, Example 2.16] *As an example for convergent series, consider the series $\sum_k u_k$ with*

$$u_k(t) := \begin{cases} 1 - (k+1)^2 t & , \quad 0 \le t \le \dfrac{1}{(k+1)^2}, \\ 0 & , \quad otherwise, \end{cases}$$

for all $k \in \mathbb{N}_0$. It is trivial that $u_k^-(\lambda) = 0$ and $u_k^+(\lambda) = (1-\lambda)/(k+1)^2$ for all $\lambda \in [0,1]$. Therefore, we see that $\sum_k u_k^-(\lambda) = 0$ and $\sum_k u_k^+(\lambda) = \sum_k (1-\lambda)/(k+1)^2 = (1-\lambda)\pi^2/6$. Then, it is concluded that $\sum_k u_k = u$, where

$$u(t) := \begin{cases} 1 - \dfrac{6}{\pi^2} t & , \quad 0 \le t \le \dfrac{\pi^2}{6}, \\ 0 & , \quad otherwise. \end{cases}$$

Theorem 6.4.4. [110, Lemma 2.1] *Let $u, v \in L(\mathbb{R})$ with $\alpha \in \mathbb{R}$ for all $\lambda \in [0,1]$ and $[u]_\lambda = [u_\lambda^-, u_\lambda^+]$ and $[v]_\lambda = [v_\lambda^-, v_\lambda^+]$. Then, the following statements hold:*

(i) $[u]_\lambda^{-1} = [1/u_\lambda^+, 1/u_\lambda^-]$.

(ii) $[|u|]_\lambda = [\max\{0, u_\lambda^-, -u_\lambda^+\}, \max\{|u_\lambda^-|, |u_\lambda^+|\}]$.

(iii) $|\alpha u|_\lambda = \alpha [u]_\lambda$.

(iv) $[u/v]_\lambda = [u_\lambda^-/v_\lambda^+, u_\lambda^+/v_\lambda^-]$.

6.4.1 Power Series of Fuzzy Numbers with Real or Fuzzy Coefficients

Kadak and Başar [104] have recently studied some power series of fuzzy numbers and their properties. They have also given some relations concerning these series and showed the convergence of those series in the different cases and gave the theorem on the term-by-term differentiation of power series of fuzzy numbers with level sets. Finally, they have a result on the Taylor expansion of a fuzzy valued function. Throughout this section, we suppose that the H-difference $u - u_0$ exists. In this section, we summarize the results concerning with the power series of fuzzy numbers with real or fuzzy coefficients.

Definition 6.4.5. [104, Definition 4.2] *Let u be any element and u_0 be a fixed element in the space $L(\mathbb{R})$ of fuzzy numbers. Then, the power series of fuzzy numbers with real coefficients a_n is in the form with respect to Hukuhara difference*

$$\sum_{n=1}^{\infty} a_n(u - u_0)^n = a_1(u - u_0) + a_2(u - u_0)^2 + \cdots + a_n(u - u_0)^n + \cdots \quad (6.4.1)$$

For simplicity in notation, we write w instead of the fuzzy number $u \ominus u_0 = u - u_0$. Therefore, we have $[u - u_0]_\lambda = [w]_\lambda = [w_\lambda^-, w_\lambda^+]$ with $w_\lambda^- = u^-(\lambda) - u_0^-(\lambda)$ and $w_\lambda^+ = u^+(\lambda) - u_0^+(\lambda)$ from Hukuhara difference given by Proposition 6.2.16. Then, the power series in (6.4.1) is reduced to

$$
\begin{aligned}
\sum_{n=1}^{\infty} a_n[w]_\lambda^n &= a_1[w]_\lambda + a_2[w]_\lambda^2 + \cdots + a_n[w]_\lambda^n + \cdots \\
&= a_1[w_\lambda^-, w_\lambda^+] + a_2[(w_\lambda^-)^2, (w_\lambda^+)^2] + \cdots + a_n[(w_\lambda^-)^n, (w_\lambda^+)^n] + \cdots
\end{aligned}
$$

and the radius of convergence R is defined by

$$R = \frac{1}{\limsup_{n\to\infty} \sqrt[n]{|a_n|}} \quad \left(= \lim_{n\to\infty} \frac{|a_n|}{|a_{n+1}|} \right),$$

which is also given by the right-hand side provided the limit exists, where $0 \leq R \leq \infty$.

Remark 6.4.6. [104, Remark 5.2]

$$\liminf_{n\to\infty} \frac{|a_{n+1}|}{|a_n|} \leq \liminf_{n\to\infty} \sqrt[n]{|a_n|} \leq \limsup_{n\to\infty} \sqrt[n]{|a_n|} \leq \limsup_{n\to\infty} \frac{|a_{n+1}|}{|a_n|}$$

and if

$$\lim_{n\to\infty} \frac{|a_{n+1}|}{|a_n|} \quad \text{exists, then} \quad \liminf_{n\to\infty} \frac{|a_{n+1}|}{|a_n|} = \limsup_{n\to\infty} \frac{|a_{n+1}|}{|a_n|}$$

and so $\lim_{n\to\infty} \sqrt[n]{|a_n|}$ also exists and

$$\lim_{n\to\infty} \frac{|a_{n+1}|}{|a_n|} = \lim_{n\to\infty} \sqrt[n]{|a_n|}.$$

Now, we can give the result concerning with the four choices of $w \in L(\mathbb{R})$ in the power series, as follows:

Proposition 6.4.7. [105, Proposition 4.6] *Consider the power series of fuzzy numbers $\sum_{n=1}^{\infty} a_n[w]_\lambda^n$. Therefore,*

(i) If $w_\lambda^- < 0$ and $w_\lambda^+ > 0$, then

$$\sum_{n=1}^{\infty} a_n[w]_\lambda^n = \left[\sum_{n=1}^{\infty} a_n[(w_\lambda^+)^{n-1}](w_\lambda^-), \sum_{n=1}^{\infty} a_n(w_\lambda^+)^n \right].$$

(ii) Let $w_\lambda^- < 0$ and $w_\lambda^+ < 0$. Then,

    (a) If n is even, we have

$$\sum_{n=1}^{\infty} a_{2n}[w]_\lambda^{2n} = \left[\sum_{n=1}^{\infty} a_{2n}(w_\lambda^+)^{2n}, \sum_{n=1}^{\infty} a_{2n}(w_\lambda^-)^{2n} \right].$$

    (b) If n is odd, we have

$$\sum_{n=1}^{\infty} a_{2n-1}[w]_\lambda^{2n-1} = \left[\sum_{n=1}^{\infty} a_{2n-1}(w_\lambda^-)^{2n-1}, \sum_{n=1}^{\infty} a_{2n-1}(w_\lambda^+)^{2n-1} \right].$$

(iii) Let $w_\lambda^- < 0$ and $w_\lambda^+ = 0$. Then,

    (a) If n is even, we have

$$\sum_{n=1}^{\infty} a_{2n}[w]_\lambda^{2n} = \left[0, \sum_{n=1}^{\infty} a_{2n}(w_\lambda^-)^{2n} \right].$$

    (b) If n is odd, we have

$$\sum_{n=1}^{\infty} a_{2n-1}[w]_\lambda^{2n-1} = \left[\sum_{n=1}^{\infty} a_{2n-1}(w_\lambda^-)^{2n-1}, 0 \right].$$

(iv) If $w_\lambda^- > 0$ and $w_\lambda^+ > 0$, then we have

$$\sum_{n=1}^{\infty} a_n[w]_\lambda^n = \left[\sum_{n=1}^{\infty} a_n(w_\lambda^-)^n, \sum_{n=1}^{\infty} a_n(w_\lambda^+)^n \right].$$

Proof. (i) Let us consider the power series of fuzzy numbers with $w_\lambda^- < 0$ and $w_\lambda^+ > 0$. Then, the straightforward calculation leads us to the consequence that

$$
\begin{aligned}
\sum_{n=1}^{\infty} a_n[w]_\lambda^n &= a_1[w]_\lambda + a_2[w]_\lambda^2 + \cdots a_n[w]_\lambda^n + \cdots \\
&= a_1[w_\lambda^-, w_\lambda^+] + a_2[(w_\lambda^-)^2, (w_\lambda^+)^2] + \cdots + a_n[(w_\lambda^+)^{n-1}w_\lambda^-, (w_\lambda^+)^n] + \cdots \\
&= \left[\sum_{n=1}^{\infty} a_n(w_\lambda^+)^{n-1}(w_\lambda^-), \sum_{n=1}^{\infty} a_n(w_\lambda^+)^n \right].
\end{aligned}
$$

(ii) Given the power series of fuzzy numbers with $w_\lambda^- < 0$ and $w_\lambda^+ < 0$. Then, one can immediately see that

$$
\sum_{n=1}^{\infty} a_n[w]_\lambda^n = a_1[w]_\lambda + a_2[w]_\lambda^2 + \cdots + a_n[w]_\lambda^n + \cdots
$$

$$
= \begin{cases} \sum_{n=1}^{\infty} a_{2n}[(w_\lambda^+)^{2n}, (w_\lambda^-)^{2n}] & , \quad n \text{ even}, \\[2ex] \sum_{n=1}^{\infty} a_{2n-1}[(w_\lambda^-)^{2n-1}, (w_\lambda^+)^{2n-1}] & , \quad n \text{ odd}. \end{cases}
$$

a) If n is even, then we have

$$\sum_{n=1}^{\infty} a_{2n}[w]_{\lambda}^{2n} = \left[\sum_{n=1}^{\infty} a_{2n}(w_{\lambda}^+)^{2n}, \sum_{n=1}^{\infty} a_{2n}(w_{\lambda}^-)^{2n}\right].$$

b) If n is odd, then we have

$$\sum_{n=1}^{\infty} a_{2n-1}[w]_{\lambda}^{2n-1} = \left[\sum_{n=1}^{\infty} a_{2n-1}(w_{\lambda}^-)^{2n-1}, \sum_{n=1}^{\infty} a_{2n-1}(w_{\lambda}^+)^{2n-1}\right].$$

(iii) Given the power series of fuzzy numbers with $w_{\lambda}^- < 0$ and $w_{\lambda}^+ = 0$. Then, we conclude by the routine verification that

$$\sum_{n=1}^{\infty} a_n[w]_{\lambda}^n = a_1[w]_{\lambda} + a_2[w]_{\lambda}^2 + \cdots + a_n[w]_{\lambda}^n + \cdots$$

$$= \begin{cases} \sum_{n=1}^{\infty} a_{2n}[0,(w_{\lambda}^-)^{2n}] & , \quad n \text{ even}, \\ \sum_{n=1}^{\infty} a_{2n-1}[(w_{\lambda}^-)^{2n-1},0] & , \quad n \text{ odd}. \end{cases}$$

a) If n is even, then we have

$$\sum_{n=1}^{\infty} a_{2n}[w]_{\lambda}^{2n} = \left[0, \sum_{n=1}^{\infty} a_{2n}(w_{\lambda}^-)^{2n}\right].$$

b) If n is odd, then we have

$$\sum_{n=1}^{\infty} a_{2n-1}[w]_{\lambda}^{2n-1} = \left[\sum_{n=1}^{\infty} a_{2n-1}(w_{\lambda}^-)^{2n-1}, 0\right].$$

(iv) Let us consider the power series $\sum_{n=1}^{\infty} a_n[w]_{\lambda}^n$ of fuzzy numbers with $w_{\lambda}^- > 0$ and $w_{\lambda}^+ > 0$. Then, one can easily derive that

$$\sum_{n=1}^{\infty} a_n[w]_{\lambda}^n = a_1[w]_{\lambda} + a_2[w]_{\lambda}^2 + \cdots + a_n[w]_{\lambda}^n + \cdots$$

$$= a_1[w_{\lambda}^-, w_{\lambda}^+] + a_2[(w_{\lambda}^-)^2, (w_{\lambda}^+)^2] + \cdots + a_n[(w_{\lambda}^-)^n, (w_{\lambda}^+)^n] + \cdots$$

$$= \left[\sum_{n=1}^{\infty} a_n(w_{\lambda}^-)^n, \sum_{n=1}^{\infty} a_n(w_{\lambda}^+)^n\right].$$

$\square$

Now, we can give the definition of the power series of non-negative fuzzy numbers with the non-negative fuzzy coefficients.

Definition 6.4.8. [105, Definition 4.7] *u be any element and u_0 also be a fixed element in the space $L(\mathbb{R})$ of fuzzy numbers such that $u - u_0$ is non-negative fuzzy number, and (v_n) be a sequence of non-negative fuzzy numbers.*

Therefore, we have $[u - u_0]_\lambda = [w]_\lambda = [w^-(\lambda), w^+(\lambda)]$ *with* $w^-(\lambda) = u^-(\lambda) - u_0^-(\lambda)$ *and* $w^+(\lambda) = u^+(\lambda) - u_0^+(\lambda)$, *and* $[v]_\lambda = [v^-(\lambda), v^+(\lambda)]$. *Then, the power series of fuzzy numbers with the coefficients* v_n *is given by* $\sum_{n=1}^{\infty} v_n(u - u_0)^n$ *which can be expressed in terms of* λ-*level sets, as follows:*

$$
\begin{aligned}
\sum_{n=1}^{\infty} v_n[w]_\lambda^n &= v_1[w]_\lambda + v_2[w]_\lambda^2 + \cdots + v_n[w]_\lambda^n + \cdots \\
&= [v_1^-(\lambda), v_1^+(\lambda)][w^-(\lambda), w^+(\lambda)] + [v_2^-(\lambda), v_2^+(\lambda)][(w^-(\lambda))^2, (w^+(\lambda))^2] + \cdots \\
&= \left[\sum_{n=1}^{\infty} v_n^-(\lambda)[w^-(\lambda)]^n, \sum_{n=1}^{\infty} v_n^+(\lambda)[w^+(\lambda)]^n \right].
\end{aligned}
$$

Proposition 6.4.9. [104, Proposition 4.3] *Consider the power series of fuzzy numbers* $\sum_{n=1}^{\infty} a_n[w]_\lambda^n$. *If* $w_\lambda^- > 0$ *and* $w_\lambda^+ > 0$, *then the power series of fuzzy numbers with real coefficients* a_n *is in the form*

$$
\sum_{n=1}^{\infty} a_n[w]_\lambda^n = \left[\sum_{n=1}^{\infty} a_n(w_\lambda^-)^n, \sum_{n=1}^{\infty} a_n(w_\lambda^+)^n \right].
$$

Proof. Let us consider the power series of fuzzy numbers with $w_\lambda^- > 0$ and $w_\lambda^+ > 0$. Then, one can easily derive that

$$
\begin{aligned}
\sum_{n=1}^{\infty} a_n[w]_\lambda^n &= a_1[w]_\lambda + a_2[w]_\lambda^2 + \cdots + a_n[w]_\lambda^n + \cdots \\
&= a_1[w_\lambda^-, w_\lambda^+] + a_2[(w_\lambda^-)^2, (w_\lambda^+)^2] + \cdots + a_n[(w_\lambda^-)^n, (w_\lambda^+)^n] + \cdots \\
&= \left[\sum_{n=1}^{\infty} a_n(w_\lambda^-)^n, \sum_{n=1}^{\infty} a_n(w_\lambda^+)^n \right].
\end{aligned}
$$

$\square$

Definition 6.4.10. [159] *A sequence* $\{u_n(x)\}$ *of fuzzy-valued functions is said to be continuous at the point* x_0, *if for every* $\varepsilon > 0$, *there exists* $\delta > 0$ *such that* $|x - x_0| < \delta$ *we have* $D(u_n(x), u_n(x_0)) < \varepsilon$.

Definition 6.4.11. [159] *A sequence* $\{u_n(x)\}$ *of fuzzy-valued functions converges uniformly to* $u(x)$ *on a set* I *if for each* $\varepsilon > 0$ *there exists a number* n_0, *such that* $D(u_n(x), u(x)) < \varepsilon$ *for all* $x \in I$ *and* $n > n_0$.

It is clear that if $\{u_n(x)\}$ is uniformly convergent to u, then the sequence is pointwise convergent to u on I. But pointwise convergence of $\{u_n(x)\}$ to u on I does not imply uniform convergence of the sequence $\{u_n(x)\}$ on I.

Theorem 6.4.12. [159, Theorem 2.1] *Let* $\{u_n(x)\}$ *be a sequence of continuous functions on interval* I. *If* $\{u_n(x)\}$ *converges uniformly to a function* $u(x)$ *on* I, *then* u *is continuous on* I.

Proof. Fix any point $x_0 \in I$. We are going to show that u is continuous at the point x_0. For every $\varepsilon > 0$, since $\{u_n(x)\}$ converges uniformly to a function u on I, there exists an n_0 such that $D(u_n(x), u(x)) < \varepsilon/3$ for all $x \in I$ and $n > n_0$. Since $u_n(x)$ is continuous at the point x_0, there exists $\delta > 0$, which depends on x_0 and ε such that $|x - x_0| < \delta$ we have $D(u_n(x), u_n(x_0)) < \varepsilon/3$ and

$$
\begin{aligned}
D(u(x), u(x_0)) &\leq D(u(x), u_n(x)) + D(u_n(x), u(x_0)) \\
&\leq D(u(x), u_n(x)) + D(u_n(x), u_n(x_0)) + D(u_n(x_0), u(x_0)) \\
&< \frac{\varepsilon}{3} + \frac{\varepsilon}{3} + \frac{\varepsilon}{3} = \varepsilon.
\end{aligned}
$$

Thus, u is continuous at the point x_0. Since x_0 is arbitrary it follows that u is continuous on I. □

Definition 6.4.13. [159] *Let $\sum_{k=1}^{\infty} u_k(x)$ be a series of fuzzy mappings. If the sequence $\{s_n(x)\} = \{\sum_{k=1}^{n} u_k(x)\}_{n \in \mathbb{N}_1}$ converges to a fuzzy number $u(x)$, then we say that the series $\sum_{k=1}^{\infty} u_k(x)$ converges pointwise to $u \in L(\mathbb{R})$ on I. The series $\sum_{k=1}^{\infty} u_k(x)$ converges uniformly to $u \in L(\mathbb{R})$ on I if for each $\varepsilon > 0$, there exists a number n_0 such that $D(s_n(x), u(x)) < \varepsilon$ for all $x \in I$ and for all $n > n_0$. That is, the series $\sum_{k=1}^{\infty} u_k(x)$ converges uniformly to $u \in L(\mathbb{R})$ on I if the sequence $\{s_n(x)\}_{n \in \mathbb{N}_1}$ converges uniformly to $u \in L(\mathbb{R})$ on I.*

Theorem 6.4.14. [159, Theorem 3.2] *If the series $\sum_{k=1}^{\infty} u_k(x)$ converges uniformly on the set I and each of the terms $u_k(x)$ is continuous on I, then the sum of the series is continuous on I.*

Proof. Since uniform convergence of the series $\sum_{k=1}^{\infty} u_k(x)$ to $u(x) \in L(\mathbb{R})$ is equivalent to uniform convergence of the sequence $\{s_n(x)\}$ to $u(x) \in L(\mathbb{R})$ and each term $u_k(x)$ of the series is continuous on I, each term of the sequence $\{s_n(x)\}$ is also continuous on I. Thus, the continuity of $u(x)$ on I follows from Theorem 6.4.12. □

Theorem 6.4.15. (Cauchy Criterion) [159] *A fuzzy series of functions $\sum_{k=1}^{\infty} u_k(x)$ converges uniformly on a set I if and only if for every $\varepsilon > 0$ there exists an $n_0 = n_0(\varepsilon) \in \mathbb{N}_1$ such that*

$$
D\left(\sum_{k=n+1}^{m} u_k(x), \overline{0} \right) < \varepsilon \quad \text{for all } x \in I \text{ and for all } m > n > n_0.
$$

Definition 6.4.16. [104, Definition 5.1] *Let u be any element and u_0 be the fixed element of the space $L(\mathbb{R})$ of fuzzy numbers. Then, the power series of fuzzy numbers with real coefficients a_n is in the form with respect to Hukuhara difference for all $\lambda \in [0,1]$ for non-zero real numbers n by*

$$
\sum_{n=1}^{\infty} a_n [u - u_0]_\lambda^n = a_1 [u - u_0]_\lambda + a_2 [u - u_0]_\lambda^2 + \cdots + a_n [u - u_0]_\lambda^n + \cdots
$$

Now, we can define the radius of convergence R of a power series of fuzzy numbers with fuzzy coefficients v_n for $\lambda = 0$ by

$$R := \lim_{n \to \infty} \frac{v_n^+(0)}{v_{n+1}^+(0)}. \tag{6.4.2}$$

Definition 6.4.17. [104, Definition 5.3] *Consider a power series $\sum_{n=1}^{\infty} a_n(u - u_0)^n$ with the radius of convergence R given by (6.4.2). If the H-difference exists, then the set of points in an interval at which the series converges, is called the interval of convergence such that*

$$D(u, u_0) < R \Leftrightarrow u \in (u_0 - R, u_0 + R)$$

which must be either $(u_0 - R, u_0 + R)$, $(u_0 - R, u_0 + R]$, $[u_0 - R, u_0 + R)$ or $[u_0 - R, u_0 + R]$.

Theorem 6.4.18. [219, Theorem 5.6] *If $\sum_{n=1}^{\infty} a_n(u - u_0)^n$ is any given fuzzy power series with radius of convergence R, then the series converges for all $u \in L(\mathbb{R})$ satisfying $D(u, u_0) < R$ and the series diverges for all $u \in L(\mathbb{R})$ satisfying $D(u, u_0) > R$, where $0 \leq R \leq \infty$.*

Proof. By the definition, the radius of convergence is $|a_n|/|a_{n+1}| \to R$, as $n \to \infty$. The first part of the theorem follows from the ratio test, since

$$D(u, u_0) \times \lim_{n \to \infty} \left| \frac{a_{n+1}}{a_n} \right| \leq 1$$

$$D(u, u_0) \times \frac{1}{R} \leq R \times \frac{1}{R} = 1,$$

the series $\sum_{n=1}^{\infty} a_n(u - u_0)^n$ converges, if $D(u, u_0) < R$.

Now, we can prove the second part of the theorem. Suppose that $\sum_{n=1}^{\infty} a_n(u - u_0)^n$ converges with $D(u, u_0) > R$. Then, we must have $a_n(u - u_0)^n \to \bar{0}$, as $n \to \infty$. Let $\varepsilon = 1$. Then, there exists n_0 such that

$$|a_n(u - u_0)^n - 0| < 1, \ n > n_0 \Rightarrow |a_n(u - u_0)^n|^{1/n} < 1$$

$$\Rightarrow D(u, u_0) \times |a_n|^{1/n} < 1, \ n > n_0 \Rightarrow D(u, u_0) \leq \frac{1}{\sup\limits_{m \geq n} |a_n|^{1/n}}, \ m > n_0$$

$$\Rightarrow D(u, u_0) \leq \frac{1}{\varlimsup\limits_{n \to \infty} |a_n|^{1/n}} \Rightarrow R < D(u, u_0) \leq R,$$

which is a contradiction. Hence, the series $\sum_{n=1}^{\infty} a_n(u - u_0)^n$ must diverge for all $u \in L(\mathbb{R})$ satisfying $D(u, u_0) > R$.

This step completes the proof. $\square$

Lemma 6.4.19. [229] *Let (a_n) and (b_n) be the sequences such that $a_n \geq 0$, $\lim_{n \to \infty} b_n$ exists with $\lim_{n \to \infty} b_n \neq 0$. Then, $\limsup_{n \to \infty} a_n b_n = (\limsup_{n \to \infty} a_n)(\lim_{n \to \infty} b_n)$.*

Proof. Let $b_n \to B$, as $n \to \infty$. Given any $\varepsilon > 0$, there exists n_0 such that $|b_n - B| < \varepsilon$ for all $n > n_0$. Thus, since $a_n \geq 0$, $a_n(B - \varepsilon) < a_n b_n < a_n(B + \varepsilon)$ for $n > n_0$ and so

$$
\begin{aligned}
(B - \varepsilon) \limsup_{n \to \infty} a_n &= \limsup_{n \to \infty} a_n(B - \varepsilon) \leq \limsup_{n \to \infty} a_n b_n \\
&\leq \limsup_{n \to \infty} a_n(B + \varepsilon) = (B + \varepsilon) \limsup_{n \to \infty} a_n.
\end{aligned}
$$

Now, by letting $\varepsilon \to 0$, we have

$$
B \limsup_{n \to \infty} a_n \leq \limsup_{n \to \infty} a_n b_n \leq B \limsup_{n \to \infty} a_n
$$

$$
\limsup_{n \to \infty} a_n b_n = B \limsup_{n \to \infty} a_n = \left(\limsup_{n \to \infty} a_n \right) \left(\lim_{n \to \infty} b_n \right).
$$

This completes the proof. □

Theorem 6.4.20. [104, Theorem 6.5] *Suppose that $\sum_{n=1}^{\infty} a_n[u - u_0]_\lambda^n$ is a power series of fuzzy numbers with real coefficients for $\lambda \in [0,1]$ and the radius of convergence R given by (6.4.2), let the function $f : [a,b] \to L(\mathbb{R})$ be fuzzy differentiable on $[a,b]$. Then, we have*

$$
[f(u)]_\lambda = [f(u^-(\lambda)), f(u^+(\lambda))] = \sum_{n=1}^{\infty} a_n[u - u_0]_\lambda^n, \quad D(u, u_0) < R \quad (6.4.3)
$$

and the following statements hold:

(a) *The fuzzy power series $\sum_{n=1}^{\infty} n a_n[u - u_0]_\lambda^{n-1}$ also has the radius of convergence R.*

(b) *$[f'(u)]_\lambda = \sum_{n=1}^{\infty} n a_n[u - u_0]_\lambda^{n-1}$ is obtained by term-by-term differentiation of $\sum_{n=1}^{\infty} a_n[u - u_0]_\lambda^n$, which converges uniformly in u satisfying $D(u, u_0) < R$ for $\lambda \in [0,1]$.*

Proof. Suppose that $\sum_{n=1}^{\infty} a_n[u - u_0]_\lambda^n$ is a power series of fuzzy numbers with real coefficients for $\lambda \in [0,1]$ and the radius of convergence R given by (6.4.2). Let the function $f : [a,b] \to L(\mathbb{R})$ be fuzzy differentiable on $[a,b]$.

(a) Since,

$$
L = \limsup_{n \to \infty} \sqrt[n]{|n a_n|} = \limsup_{n \to \infty} \sqrt[n]{|a_n|} \sqrt[n]{n} = \limsup_{n \to \infty} \sqrt[n]{|a_n|} \lim_{n \to \infty} \sqrt[n]{n}
$$

$$
= \limsup_{n \to \infty} \sqrt[n]{|a_n|} = \frac{1}{R}
$$

for the fuzzy power series $\sum_{n=1}^{\infty} n a_n(u - u_0)^{n-1}$ by Lemma 6.4.19, it has a radius of convergence R.

(b) For any $\rho \in (0, R]$ and $M \in L(\mathbb{R})$, the series of functions $\sum_{n=1}^{\infty} n a_n(u - u_0)^{n-1}$ converges uniformly on $D(u, u_0) < \rho$ by the uniform convergence of fuzzy series theorem the closed interval $[u_0 - \rho, u_0 + \rho] \subseteq [u_0 - R, u_0 + R]$ and using ratio test and Weierstrass's M-test for fuzzy series, then $\sum_{n=1}^{\infty} a_n(u - u_0)^n$ converges, $a_n\rho^n \to 0$, as $n \to \infty$, and there exists $M \geq 0$ such that $|a_n\rho^n| \leq M$ for each $n \in \mathbb{N}_0$, and by using Theorem 6.4.4

$$\sum_{n=1}^{\infty} \left| na_n(u - u_0)^{n-1} \right| \leq \frac{M}{\rho} \sum_{n=1}^{\infty} n \left| \frac{u - u_0}{\rho} \right|^{n-1}.$$

Since $|(u - u_0)/\rho| < 1$ by Part (iv) of Theorem 6.4.4, the series $\sum_{n=1}^{\infty} n \left| \frac{u-u_0}{\rho} \right|^{n-1}$ converges by the ratio test. It follows that $\sum_{n=1}^{\infty} na_n(u - u_0)^{n-1}$ converges uniformly when $D(u, u_0) < R$.

This completes the proof of the theorem. $\square$

Theorem 6.4.21. [20, Theorem 3.1] *Let f be fuzzy differentiable for any order n and $[\alpha, \beta] \subseteq [a, b] \subseteq \mathbb{R}$. By using fuzzy Taylor formulae with integral remainder and the λ-level sets, we have*

$$[f(\beta)]_\lambda = f(\alpha) + f'^{(n-1)}(\alpha)\frac{(\alpha - \beta)^{n-1}}{(n - 1)!} + \frac{1}{(n - 1)!}\left[(FH) \int_\alpha^\beta (\beta - t)^{n-1} f^{(n)}(t)dt \right]_\lambda.$$

The estimate remainder $R_n(\beta)$ is given by

$$R_n(\beta) = \frac{1}{(n - 1)!}\left[(FH) \int_\alpha^\beta (\beta - t)^{n-1} f^{(n)}(t)dt \right]_\lambda.$$

The integral remainder is fuzzy continuous in β.

Corollary 6.4.22. [104, Corollary 6.13] *Given a power series $\sum_{k=0}^{\infty} a_k[u - u_0]_\lambda^k$ of fuzzy numbers with fuzzy coefficients a_k, $[f]_\lambda : [a, b] \to L(\mathbb{R})$ be a fuzzy differentiable function on $[a, b]$, radius of convergence R, $[f(u)]_\lambda = \sum_{k=0}^{\infty} a_k[u - u_0]_\lambda^k$ for all $u \in [a, b] \subset L(\mathbb{R})$ satisfying $D(u, u_0) < R$. Then, one can derive applying term by term differentiation successively that*

$$[f^{(n)}(u)]_\lambda = \sum_{k=n}^{\infty} \left(\prod_{j=k}^{k-n+1} j \right) a_k[u - u_0]_\lambda^{k-n}$$

for all $n \in \mathbb{N}_0$, $\lambda \in [0, 1]$ and any fixed point $[u_0]_\lambda$ such that $[u]_\lambda = [u_0]_\lambda$ which gives that $a_k = f^{(k)}[u_0]_\lambda/k! \in L(\mathbb{R})$ for all $k \in \mathbb{N}_0$.

Therefore, by using (6.4.3), we obtain the Taylor expansion of $[f]_\lambda$ at $[u_0]_\lambda$ as

$$f(u) = \sum_{k=0}^{n-1} \frac{f^{(k)}[u_0]_\lambda}{k!}[u - u_0]_\lambda^k + R_n(u), \tag{6.4.4}$$

where $R_n(u)$ denotes the estimate remainder for all $u \in L(\mathbb{R})$ satisfying $D(u, u_0) < R$. It is natural that the Taylor expansion of the fuzzy-valued function f in (6.4.4) is reduced to the Maclaurin expansion in the special case $u_0 = \overline{0}$ and by using (6.5.8) that

$$[f(u)]_\lambda = \sum_{k=0}^{n-1} \frac{f^{(k)}(\overline{0})}{k!} [u]_\lambda^k + R_n(u)$$

for all $\lambda \in [0, 1]$, where $R_n(u)$ estimate remainder and $u \in L(\mathbb{R})$ such that $D(u, u_0) < R$.

6.5 Alternating and Binomial Series of Fuzzy Numbers with the Level Sets

In this section, in the light of Kadak and Başar [108], the power series of fuzzy numbers with real or fuzzy coefficients and the results concerning with the power series of fuzzy numbers are summarized. Additionally, the results on the fuzzy Dirichlet test, fuzzy alternating series test and estimation theorem together with the binomial identity for the fuzzy numbers are stated and proved. In the final subsection, the significance of alternating and binomial series of fuzzy numbers are noted, and some further suggestions are recorded.

6.5.1 Fuzzy Alternating Series

Following Kadak and Başar [108], we focus almost exclusively on series of fuzzy numbers with positive terms up to this point. In this subsection, we begin to introduce into series of fuzzy numbers with both positive and negative terms, presenting a test which works for the series of fuzzy numbers whose terms alternate in sign.

Now, we state the results on the fuzzy Dirichlet test, fuzzy alternating series test and estimation theorem.

Definition 6.5.1. [108, Definition 4.1] *Let $(u_n)_{n=1}^{\infty}$ be a sequence of non-negative fuzzy numbers. Then, the series $\sum_{n=1}^{\infty}(-1)^n u_n$ whose terms are alternately positive and negative is called alternating series. If we write $[u_n]_\lambda = [u_n^-(\lambda), u_n^+(\lambda)]$ for all $\lambda \in [0, 1]$, by using λ-level set, then the following statement holds:*

$$\sum_{n=1}^{\infty}(-1)^n [u_n]_\lambda = -[u_1]_\lambda + [u_2]_\lambda - [u_3]_\lambda + \cdots . \tag{6.5.1}$$

Then, the given series is called an alternating series of fuzzy numbers with real coefficients for all $\lambda \in [0, 1]$.

Following Kadak and Başar [105], we define the set $cs(F)$ consisting of all convergent series, that is,

$$cs(F) := \left\{ u = (u_k) \in \omega(F) : \exists l \in L(\mathbb{R}) \ni \lim_{n \to \infty} D\left(\sum_{k=0}^{n} u_k, l \right) = 0 \right\}.$$

Theorem 6.5.2. [108, Theorem 4.2] *If the sequence of partial sums of the series $\sum_n f_n$ of fuzzy valued functions with $f_n : [a, b] \to L(\mathbb{R})$ is bounded ($f_n^{\pm}$ is uniformly bounded) on $[a, b]$ and the monotonic increasing sequence (g_n) of fuzzy valued functions with $g_n : [a, b] \to L(\mathbb{R})$, is convergent to $\overline{0}$ ($g_n^{\pm}$ is uniformly convergent to 0) on the set $[a, b]$, then the series $\sum_n f_n g_n$ is convergent on $[a, b]$.*

Proof. To prove the theorem, we derive the formula of the summation by part. Given the sequences of fuzzy valued functions (f_n) and (g_n), consider $[F_n(t)]_\lambda = \sum_{m=1}^{n} [f_m(t)]_\lambda$ for all $\lambda \in [0, 1]$, $t \in [a, b]$, we write

$$[s_n(t)]_\lambda = \sum_{k=1}^{n} [f_k(t) g_k(t)]_\lambda = [F_1(t) g_1(t)]_\lambda + \sum_{k=2}^{n} [(F_k(t) - F_{k-1}(t)) \, g_k(t)]_\lambda$$

$$= [F_1(t) g_1(t)]_\lambda + \sum_{k=2}^{n} [F_k(t) g_k(t)]_\lambda - \sum_{k=2}^{n} [F_{k-1}(t) g_k(t)]_\lambda$$

$$= \sum_{k=1}^{n} [F_k(t) g_k(t)]_\lambda - \sum_{k=1}^{n-1} [F_k(t) g_{k+1}(t)]_\lambda$$

$$= \sum_{k=1}^{n} [F_k(t)(g_k(t) - g_{k+1}(t))]_\lambda + [F_n(t) g_{n+1}(t)]_\lambda.$$

So if $m > n$, the difference between m^{th} and n^{th} partial sums is

$$[s_m(t) - s_n(t)]_\lambda$$

$$= \sum_{k=n+1}^{m} [F_k(t)(g_k(t) - g_{k+1}(t))]_\lambda + [F_m(t) g_{m+1}(t)]_\lambda - [F_n(t) g_{n+1}(t)]_\lambda.$$

If $[M]_\lambda = [M^-(\lambda), M^+(\lambda)] = \sup\{[F_n(t)]_\lambda \mid n \in \mathbb{N}, \lambda \in [0, 1], t \in [a, b]\}$, we have

$$[|s_m(t) - s_n(t)|]_\lambda$$

$$\leq [M]_\lambda \sum_{k=n+1}^{m} [g_k(t) - g_{k+1}(t)]_\lambda + [M]_\lambda [g_{m+1}(t)]_\lambda + [M]_\lambda [g_{n+1}(t)]_\lambda$$

$$= [M]_\lambda [g_{n+1}(t) - g_{m+1}(t)]_\lambda + [M]_\lambda [g_{m+1}(t)]_\lambda + [M]_\lambda [g_{n+1}(t)]_\lambda$$

$$= 2[M]_\lambda [g_{n+1}(t)]_\lambda. \tag{6.5.2}$$

Since $[g_{m+1}(t)]_\lambda \geq [0]_\lambda$, $[g_{n+1}(t)]_\lambda \geq [0]_\lambda$ and $[g_k(t) - g_{k+1}(t)]_\lambda \geq [0]_\lambda$ for all $k \in \mathbb{N}_1$. For each $t \in [a, b]$, $g_{n+1}(t) \to \overline{0}$, as $n \to \infty$. So, $\{s_n(t)\}$ is a Cauchy sequence and hence converges, say to $s(t)$, $t \in [a, b]$. By taking limit on (6.5.2), as $m \to \infty$, one can see that $[s(t) - s_n(t)]_\lambda \leq 2[M]_\lambda[g_{n+1}(t)]_\lambda$. Since $\{g_{n+1}^\pm(t)\}$ converges uniformly to zero as $n \to \infty$, we conclude that $\{s_n^\pm(t)\}$ converges uniformly to $s^\pm(t)$, as $n \to \infty$. Therefore, the series $\sum_n f_n g_n$ is convergent on $[a, b]$. □

Theorem 6.5.3. [108, Theorem 4.3] *Given the alternating series $\sum_{n=1}^\infty (-1)^{n+1} u_n$ of fuzzy numbers, suppose that the following two conditions are satisfied:*

(i) $u_n \succeq u_{n+1}$ with $u_n \succeq \bar{0}$,

(ii) $\lim_{n \to \infty} u_n^\pm(t) = 0$, uniformly in $t \in [a, b]$

for all $n \in \mathbb{N}_1$ and $\lambda \in [0, 1]$. Then, the alternating series $\sum_{n=1}^\infty (-1)^{n+1} u_n$ is convergent, i.e., $\sum_{n=1}^\infty (-1)^{n+1} u_n^\pm(\lambda)$ is uniformly convergent.

Proof. By using λ-level set, we take $[f_k(t)]_\lambda = [(-1)^{k+1}]_\lambda$ and $[F_n(t)]_\lambda = \sum_{k=1}^n [f_k(t)]_\lambda$ for each t, for all $\lambda \in [0, 1]$ and $n \in \mathbb{N}_1$. Then,

$$[F_n(t)]_\lambda = [1]_\lambda + [-1]_\lambda + \cdots + [(-1)^n]_\lambda + [(-1)^{n+1}]_\lambda = \begin{cases} [0]_\lambda & , \quad n \text{ even,} \\ [1]_\lambda & , \quad n \text{ odd.} \end{cases}$$

Thus, $[F_n(t)]_\lambda \leq [1]_\lambda$ is bounded for each t for all $\lambda \in [0, 1]$ and $n \in \mathbb{N}_1$. The fuzzy Dirichlet test leads to the desired result.

This completes the proof. □

Example 6.5.4. [108, Example 1] *Define the sequence $\{u_n(x)\} \in \omega(F)$ by*

$$u_n(x) := \begin{cases} n^2 x & , \quad 0 \leq x \leq \dfrac{1}{n^2}, \\ 2 - n^2 x & , \quad \dfrac{1}{n^2} < x \leq \dfrac{2}{n^2} \end{cases}$$

for all $n \in \mathbb{N}_1$. By using the fact $[u_n]_\lambda = [u_n^-(\lambda), u_n^+(\lambda)] = [\lambda/n^2, (2 - \lambda)/n^2]$, we can write for the alternating series of fuzzy numbers $\sum_{n=1}^\infty (-1)^{n+1} u_n$ with real coefficients for all $\lambda \in [0, 1]$ that

(i) $u_1^-(\lambda) \geq u_2^-(\lambda) \geq \cdots \geq u_n^-(\lambda)$ and $u_1^+(\lambda) \geq u_2^+(\lambda) \geq \cdots \geq u_n^+(\lambda)$. Then, the condition $u_1 \succeq u_2 \succeq \cdots \succeq u_n \succeq \bar{0}$ is satisfied.

(ii) By using Part (i) of Theorem 6.3.4, we say that $u_n \to \bar{0}$ uniformly, as $n \to \infty$, if the series $\sum_n u_n$ of fuzzy numbers converges to $\bar{0}$ and write $\sum_n u_n = \bar{0}$ which implies that $\sum_n u_n^-(\lambda) = \sum_n \lambda n^{-2} \to 0$ and $\sum_n u_n^+(\lambda) = \sum_n (2 - \lambda) n^{-2} \to 0$ uniformly in $\lambda \in [0, 1]$, as $n \to \infty$.

Therefore, the fuzzy alternating series test says that the alternating series $\sum_{n=1}^\infty (-1)^{n+1} u_n$ of fuzzy numbers with real coefficients converges.

Example 6.5.5. [108, Example 2] *Define the sequence $\{u_n(x)\}$ of fuzzy numbers by*

$$u_n(x) := \begin{cases} \dfrac{n-1}{2n} & , \quad x \in (1/2, 1], \\ 1 & , \quad x \in [0, 1/2], \\ 0 & , \quad \text{otherwise} \end{cases}$$

for all $n \in \mathbb{N}_1$. In this situation, $u_n^-(\lambda) \to 0$, as $n \to \infty$, and since

$$u_n^+(\lambda) := \begin{cases} 1 & , \quad \lambda \in [0, (n-1)/2n], \\ 1/2 & , \quad \lambda \in ((n-1)/2n, 1], \end{cases}$$

$u_n^+(\lambda) \not\to 0$, as $n \to \infty$. That is, $u_n \not\to \bar{0}$, as $n \to \infty$. Hence, the series $\sum_{n=1}^{\infty}(-1)^{n+1}u_n$ diverges by the alternating series test.

Theorem 6.5.6. [108, Theorem 4.4] *Let S be the sum of an alternating series of fuzzy numbers with real coefficients. The difference between S and the nth partial sum S_n of the alternating series of fuzzy numbers with the level sets, $(-1)^n[S - S_n]_\lambda$ is an estimate error and less than the magnitude of the next term $[u_{n+1}]_\lambda$ in the series $\sum_{k=1}^{\infty}(-1)^{k+1}[u_k]_\lambda$, i.e., $[0]_\lambda < (-1)^n[S - S_n]_\lambda < [u_{n+1}]_\lambda$.*

Proof. Let S be the sum of an alternating series $\sum_{k=1}^{\infty}(-1)^k u_k$ of fuzzy numbers with real coefficients and $[S_n]_\lambda$ be its nth partial sum. Therefore, one can see that

$$\begin{aligned} [S - S_n]_\lambda &= \left\{ (-1)^{n+1}\left[u_{n+1}^-(\lambda), u_{n+1}^+(\lambda)\right] + (-1)^{n+2}\left[u_{n+2}^-(\lambda), u_{n+2}^+(\lambda)\right] + \cdots \right\} \\ &= (-1)^{n+2}\left[u_{n+1}^-(\lambda) - u_{n+2}^-(\lambda) + \cdots, u_{n+1}^+(\lambda) - u_{n+2}^+(\lambda) + \cdots\right], \end{aligned}$$

which leads by multiplying $(-1)^n$ to

$$\begin{aligned} (-1)^n[S &- S_n]_\lambda \\ &= (-1)^{2n+2}\left[u_{n+1}^-(\lambda) - u_{n+2}^+(\lambda) + \cdots, u_{n+1}^+(\lambda) - u_{n+2}^-(\lambda) + \cdots\right] \\ &= \left[u_{n+1}^-(\lambda) - u_{n+2}^+(\lambda) + u_{n+3}^-(\lambda)\cdots, u_{n+1}^+(\lambda) - u_{n+2}^-(\lambda) + u_{n+3}^+(\lambda)\cdots\right] \\ &= \left\{u_{n+1}^-(\lambda) - \left[u_{n+2}^+(\lambda) - u_{n+3}^-(\lambda)\right] - \cdots, u_{n+1}^+(\lambda) - \left[u_{n+2}^-(\lambda) - u_{n+3}^+(\lambda)\right] - \cdots\right\} \\ &< [u_{n+1}]_\lambda. \end{aligned}$$

This shows that the estimate error is $[0]_\lambda < (-1)^n[S - S_n]_\lambda < [u_{n+1}]_\lambda$. □

In Example 6.5.4, one can apply the fuzzy alternating series test. Now, we calculate the estimation error for $n = 100$, the difference between S and the 100th partial sum of the alternating series, $(-1)^2[S - S_{100}]_\lambda$ is an estimate error and less than the magnitude of the next term $[u_{101}]_\lambda$ in the series so, $[0]_\lambda < [S - S_{100}]_\lambda < [u_{101}]_\lambda$.

$$\begin{aligned} [S - S_{100}]_\lambda &= \left[u_{101}^-(\lambda) - u_{102}^+(\lambda) + \cdots, u_{101}^+(\lambda) - u_{102}^-(\lambda) + \cdots\right] \\ &= \left[\lambda/(101^2) - (2-\lambda)/(102^2) + \cdots, (2-\lambda)/(101^2) - \lambda/(102^2)\cdots\right] \\ &< \left[\lambda/(101^2), (2-\lambda)/(102^2)\right] \end{aligned}$$

By choosing the value $\lambda = 0.01$, we have the estimation error as

$$[S - S_{100}]_\lambda < [0.000009, 0.000195].$$

6.5.2 Fuzzy Binomial Identity

The binomial theorem which gives the expansion of $(a+b)^k$, was known by Chinese mathematicians many centuries before the time of Newton for the case, where the exponent k is a positive integer. In 1665, Newton was the first to discover the infinite series expansion of $(a+b)^k$ with a positive or negative fractional exponent.

In this subsection, we give the fuzzy analogous of the binomial identity and binomial series with level sets.

Theorem 6.5.7. [108, Theorem 5.1] *Let u,v be two non-negative fuzzy numbers and k also be a positive integer. Then, the binomial identity with level sets*

$$[u+v]_\lambda^k = \sum_{j=0}^{k} \binom{k}{j} [u]_\lambda^{k-j}[v]_\lambda^j \ \text{holds for all } \lambda \in [0,1]. \tag{6.5.3}$$

Proof. Let u,v be two non-negative fuzzy numbers and k be a positive integer. Combining the concept of λ-level set with Part (i) of Basic Lemma, we obtain that $[u+v]_\lambda^k = \left[\left(u_\lambda^- + v_\lambda^- \right)^k, \left(u_\lambda^+ + v_\lambda^+ \right)^k \right]$, where

$$\left(u_\lambda^\pm + v_\lambda^\pm \right)^k = \sum_{j=0}^{k} \binom{k}{j} \left(u_\lambda^\pm \right)^{k-j} \left(v_\lambda^\pm \right)^j.$$

Therefore, we have

$$[u+v]_\lambda^k = \left[\sum_{j=0}^{k} \binom{k}{j} \left(u_\lambda^- \right)^{k-j} \left(v_\lambda^- \right)^j, \sum_{j=0}^{k} \binom{k}{j} \left(u_\lambda^+ \right)^{k-j} \left(v_\lambda^+ \right)^j \right],$$

which leads by using level sets for all $\lambda \in [0,1]$ to (6.5.3) and is proved by mathematical induction, as follows:

(i) (6.5.3) holds for $k=1$, since $[u+v]_\lambda^1 = \sum_{j=0}^{1} \binom{1}{j} [u]_\lambda^{1-j}[v]_\lambda^j = [u+v]_\lambda$.

(ii) Suppose $k>1$ and (6.5.3) holds for $k-1$, that is

$$[u+v]_\lambda^{k-1} = \sum_{j=0}^{k-1} \binom{k-1}{j} [u]_\lambda^{k-1-j}[v]_\lambda^j \ \text{holds for all } \lambda \in [0,1]. \tag{6.5.4}$$

(iii) By considering (6.5.4) we prove that (6.5.3) holds for $k \in \mathbb{N}_1$. A direct calculation yields that

$$[u+v]_\lambda^{k-1} = \left[\sum_{j=0}^{k-1} \binom{k-1}{j} \left(u_\lambda^- \right)^{k-1-j} \left(v_\lambda^- \right)^j, \sum_{j=0}^{k-1} \binom{k-1}{j} \left(u_\lambda^+ \right)^{k-1-j} \left(v_\lambda^+ \right)^j \right],$$

since $[u+v]_\lambda^{k-1} = \left[\left(u_\lambda^- + v_\lambda^-\right)^{k-1}, \left(u_\lambda^+ + v_\lambda^+\right)^{k-1} \right]$, where

$$\left(u_\lambda^\pm + v_\lambda^\pm\right)^{k-1} = \sum_{n=0}^{k-1} \binom{k-1}{n} \left(u_\lambda^\pm\right)^{k-1-n} \left(v_\lambda^\pm\right)^n.$$

Taking into account this fact one can see that

$$
\begin{aligned}
[u+v]_\lambda^k &= [u+v]_\lambda [u+v]_\lambda^{k-1} \\
&= \left[u_\lambda^- + v_\lambda^-, u_\lambda^+ + v_\lambda^+ \right] \left[\left(u_\lambda^- + v_\lambda^-\right)^{k-1}, \left(u_\lambda^+ + v_\lambda^+\right)^{k-1} \right] \\
&= \left[\left(u_\lambda^- + v_\lambda^-\right)\left(u_\lambda^- + v_\lambda^-\right)^{k-1}, \left(u_\lambda^+ + v_\lambda^+\right)\left(u_\lambda^+ + v_\lambda^+\right)^{k-1} \right] \\
&= \left[u_\lambda^- \left(u_\lambda^- + v_\lambda^-\right)^{k-1} + v_\lambda^- \left(u_\lambda^- + v_\lambda^-\right)^{k-1}, u_\lambda^+ \left(u_\lambda^+ + v_\lambda^+\right)^{k-1} \right. \\
&\quad \left. + v_\lambda^+ \left(u_\lambda^+ + v_\lambda^+\right)^{k-1} \right] \\
&= [u]_\lambda [u+v]_\lambda^{k-1} + [v]_\lambda [u+v]_\lambda^{k-1} = \sum_{j=0}^{k} \binom{k}{j} [u]_\lambda^{k-j} [v]_\lambda^j,
\end{aligned}
$$

as desired.

$\square$

The fuzzy binomial series extends to the fuzzy binomial theorem when k is a positive integer. In particular, if we put $u = \overline{1}$, by using H-derivative when the difference exists and Corollary 6.4.22, $f : [a,b] \subseteq L(\mathbb{R}) \to L(\mathbb{R})$ such that the fuzzy valued function $[f(u)]_\lambda = [\overline{1} + u]_\lambda^k$ for each $[u]_\lambda = [u_\lambda^-, u_\lambda^+] \in [a,b] \subseteq L(\mathbb{R})$ and for all $\lambda \in [0,1]$, we get

$$[f(u)]_\lambda = [\overline{1} + u]_\lambda^k = \sum_{j=0}^{k} \binom{k}{j} [u]_\lambda^j.$$

According to Corollary 6.4.22, to find this series by using H-derivative, we compute the Maclaurin expansion of fuzzy valued function with the level sets $[f(u)]_\lambda = [\overline{1} + u]_\lambda^p$

$$[f(u)]_\lambda = \sum_{p=0}^{\infty} \frac{f^{(p)}(0)}{p!} [u]_\lambda^p$$

and we have

$$[f(u)]_\lambda = [\overline{1} + u]_\lambda^p = \sum_{n=0}^{p-1} \frac{p(p-1)(p-2)\cdots(p-n+1)}{n!} [u]_\lambda^n + [R_n(u)]_\lambda,$$

where $R_n(u)$ is the estimation remainder in Theorem 6.4.21. We say that

$$[f(u)]_\lambda = \sum_{n=0}^{p-1} \binom{p}{n} [u]_\lambda^n + [R_n(u)]_\lambda.$$

This series is called a fuzzy binomial series. Now, we search convergence of binomial series of fuzzy numbers with real coefficients. Firstly, taking into account $R_n(u) \to \bar{0}$, as $n \to \infty$. By using the concept of the radius of convergence given by Definition 6.4.16 and binomial coefficients $a_n = \binom{p}{n}$, we therefore have

$$\lim_{n \to \infty} \left| \frac{a_{n+1}}{a_n} \right| = \lim_{n \to \infty} \frac{|p - n|}{|n + 1|} = \lim_{n \to \infty} \frac{\left| 1 - \frac{p}{n} \right|}{1 + \frac{1}{n}} = 1.$$

Since $R = 1$, the fuzzy binomial series converges if $D(u, \bar{0}) < 1$ and diverges if $D(u, \bar{0}) > 1$.

6.5.3 Examples on the Radius of Convergence of Fuzzy Power Series

Example 6.5.8. [104, Example 5.5] *Find the radius of convergence and interval of convergence for the fuzzy power series*

$$\sum_{n=1}^{\infty} \frac{n(n+2)}{1 + (n+2)^3} (u - u_0)^n. \tag{6.5.5}$$

From the ratio test, when $u - u_0 \neq \bar{0}$ one can observe that

$$\begin{aligned}
\frac{1}{R} &= \lim_{n \to \infty} \frac{|a_{n+1}|}{|a_n|} \\
&= \lim_{n \to \infty} \frac{(n+1)(n+3)[1 + (n+2)^3]}{n(n+2)[1 + (n+3)^3]} \\
&= \lim_{n \to \infty} \frac{(n+1)[1 + (n+2)^3]}{n[1 + (n+3)^3]} = 1
\end{aligned}$$

which gives that the radius of convergence is $R = 1$. Therefore, it is immediate that the series is convergent if $D(u, u_0) < 1$ and is divergent if $D(u, u_0) > 1$. Hence, the interval of convergence is $(u_0 - \bar{1}, u_0 + \bar{1})$.

Let us consider the power series given by (6.5.5) and investigate its convergence at $u = v \in L(\mathbb{R})$. In this situation, since $\sum_{n=1}^{\infty} |a_n(u - u_0)^n|$ is the series of fuzzy numbers with positive terms, the ratio test for the series of fuzzy numbers can be applied to this series. Therefore, we observe that the series is convergent if $D(v, u_0) \times \lim_{n \to \infty} \left| \frac{a_{n+1}}{a_n} \right| < 1$, and the series is divergent if $D(v, u_0) \times \lim_{n \to \infty} \left| \frac{a_{n+1}}{a_n} \right| > 1$.

Example 6.5.9. [104, Example 5.7] *Consider the power series $\sum_{n=1}^{\infty} n(u - u_0)^{2n}$ of fuzzy numbers with real coefficients. Then, one can see by the ratio test that*

$$D(u, u_0) \times \lim_{n \to \infty} \left| \frac{a_{n+1}}{a_n} \right| = D(u, u_0) \times \lim_{n \to \infty} \left| 1 + \frac{1}{n} \right| = D(u, u_0).$$

Therefore, the series $\sum_{n=1}^{\infty} n(u - u_0)^{2n}$ is convergent if $D(u, u_0) < 1$ and is divergent if $D(u, u_0) > 1$.

Theorem 6.5.10. [104, Theorem 5.8] *Let $\sum_{n=1}^{\infty} a_n(u - u_0)^n$ be a power series with real coefficients a_n and radius R of convergence, and the interval I of convergence, and u, c, d be any elements in the space $L(\mathbb{R})$ of fuzzy numbers, and u_0 be the fixed element. Then, the fuzzy power series of functions $\sum_{n=1}^{\infty} a_n(u - u_0)^n$ converges uniformly on any closed interval $[c, d] \subseteq I$.*

Proof. There are three cases:

(i) Let $c \preceq u_0 \preceq d$. Since $\sum_{n=1}^{\infty} a_n(d - u_0)^n$ and $\sum_{n=1}^{\infty} a_n(c - u_0)^n$, and the fuzzy power series $\sum_{n=1}^{\infty} a_n(u - u_0)^n$ converges uniformly on $c \preceq u \preceq u_0$ and $u_0 \preceq u \preceq d$ and therefore on the union $[c, d] = [c, u_0] \cup [u_0, d]$, where $d - u_0 \succeq \overline{0}$.

(ii) Let $u_0 \prec c \preceq d$. Since $\sum_{n=1}^{\infty} a_n(d - u_0)^n$ converges, the fuzzy power series $\sum_{n=1}^{\infty} a_n(u - u_0)^n$ converges uniformly on $[u_0, d]$ and hence on the sub-interval $[c, d] \subseteq [u_0, d]$.

(iii) Let $c \preceq d \prec u_0$. Since $\sum_{n=1}^{\infty} a_n(c - u_0)^n$ converges, the fuzzy power series $\sum_{n=1}^{\infty} a_n(u - u_0)^n$ converges uniformly on $[c, u_0]$ and hence on the sub-interval $[c, d] \subseteq [c, u_0]$.

This concludes the proof. $\qquad \square$

Proposition 6.5.11. *Given a fuzzy power series $\sum_{n=1}^{\infty} a_n(u - u_0)^n$ with the radius of convergence R and the interval of convergence I. Then, the following statements hold:*

(i) The series converges uniformly over the closed interval $[-R, R]$ whenever $I = [-R, R]$.

(ii) The series converges uniformly over any closed interval $[a, R]$ for all $a \in (-R, R)$ whenever $I = (-R, R]$.

(iii) The series converges uniformly over any closed interval $[-R, b]$ for all $b \in (-R, R)$ whenever $I = [-R, R)$.

(iv) The series converges uniformly over any closed interval $[a, b] \subset (-R, R)$ whenever $I = (-R, R)$.

Proof. Consider the power series $\sum_{n=1}^{\infty} a_n(u-u_0)^n$ with radius of convergence R and the interval of convergence I. We use Theorem 6.5.10, the power series of fuzzy numbers $\sum_{n=1}^{\infty} a_n(u-u_0)^n$ converges uniformly over any closed interval $[a,b] \subseteq I$ and $[a,R] \subset (-R,R]$ for all $a \in (-R,R)$ and $[-R,b] \subset [-R,R)$ for all $b \in (-R,R)$ then the series converges uniformly on the closed intervals $[a,R] \subseteq I$ and $[-R,b] \subseteq I$.

Since $[a,b] \subset (-R,R) \subset [-R,R]$, the series converges uniformly on the closed interval $[a,b]$.

This completes the proof. $\qquad\qquad\qquad\qquad\qquad\qquad\qquad\qquad\square$

6.5.4 Differentiation of Fuzzy Power Series

The concept of differentiability comes from a generalization of the Hukuhara difference for compact convex sets. Combining the concepts of strongly and weakly generalized differentiability, we obtain very simple formulations of the concepts and results with weakly generalized Hukuhara derivative (H-derivative) by means of H-difference. It is also mentioned that this concept has a very intuitive interpretation too. The presented derivative concept is slightly more general than the notion of strongly generalized (Hukuhara) differentiability for the case of interval-valued functions, it is actually equivalent to the concept of weakly generalized (Hukuhara) differentiability. We prove several properties of the derivative considered here.

Definition 6.5.12. [207, Definition 8] *We say that $f : [a,b] \to I$ is strongly generalized (Hukuhara) differentiable at $x \in [a,b]$ if:*

(1) $f'(x) \in I$ exists such that, for all $h > 0$ sufficiently near to 0, there are $f(x+h) \ominus f(x)$ and $f(x) \ominus f(x-h)$:

$$\lim_{h \to 0^+} \frac{f(x+h) \ominus f(x)}{h} = \lim_{h \to 0^+} \frac{f(x) \ominus f(x-h)}{h} = f'(x)$$

or

(2) $f'(x) \in I$ exists such that, for all $h < 0$ sufficiently near to 0, there are $f(x+h) \ominus f(x)$ and $f(x) \ominus f(x-h)$:

$$\lim_{h \to 0^-} \frac{f(x+h) \ominus f(x)}{h} = \lim_{h \to 0^-} \frac{f(x) \ominus f(x-h)}{h} = f'(x).$$

Here the limit is taken in the metric space $(L(\mathbb{R}), D)$.

Definition 6.5.13. [207, Definition 10] *The H-derivative $f'(x)$ of a function $f : [a,b] \to I$ at x is defined as*

$$f'(x) = \lim_{h \to 0} \frac{f(x+h) \ominus f(x)}{h},$$

where $x, x+h \in [a,b]$. If $f'(x) \in I$ exists, we say that f is generalized Hukuhara differentiable (H-differentiable) at x.

Theorem 6.5.14. [207, Theorem 17] *Let* $u : [a,b] \to I$ *with* $u(x) = [u^-(x), u^+(x)]$. *The function* u *is H-differentiable if and only if* u^- *and* u^+ *are differentiable real-valued functions. Additionally,*

$$f'(x) = [\min\{(u^-)'(x), (u^+)'(x)\}, \max\{(u^-)'(x), (u^+)'(x)\}]. \qquad (6.5.6)$$

Proof. To do this, we follow Chalco-Cano and Román-Flores [55]. If u is H-differentiable, then u^- and u^+ are differentiable and (6.5.6) holds. In the case of interval-valued functions, the H-difference always exists. Analyzing all of the possible cases of existence of the H-difference on the left and right sides, we observe that u is H-differentiable and (6.5.6) holds whenever u^- and u^+ are differentiable. $\qquad \square$

Theorem 6.5.15. [104, Theorem 6.5] *Suppose that* $\sum_{n=1}^{\infty} a_n(u - u_0)^n$ *is a power series of fuzzy numbers with real coefficients with the radius of convergence* R *such that* $f(u) = \sum_{n=1}^{\infty} a_n(u - u_0)^n$, $D(u, u_0) < R$. *Then, the following statements hold:*

(a) *The fuzzy power series* $\sum_{n=1}^{\infty} na_n(u - u_0)^{n-1}$ *also has the radius* R *of convergence.*

(b) $f'(u) = \sum_{n=1}^{\infty} na_n(u-u_0)^{n-1}$ *is obtained by term by term differentiation of* $\sum_{n=1}^{\infty} a_n(u - u_0)^n$, *which converges uniformly for all* u *satisfying* $D(u, u_0) < R$.

Proof. Let $\sum_{n=1}^{\infty} a_n(u - u_0)^n$ be a power series of fuzzy numbers with real coefficients with the radius of convergence R such that $f(u) = \sum_{n=1}^{\infty} a_n(u - u_0)^n$.

(a) Since,

$$L = \limsup_{n \to \infty} \sqrt[n]{|na_n|} \quad = \quad \limsup_{n \to \infty} \sqrt[n]{|a_n|} \sqrt[n]{n} = \limsup_{n \to \infty} \sqrt[n]{|a_n|} \lim_{n \to \infty} \sqrt[n]{n}$$

$$= \quad \limsup_{n \to \infty} \sqrt[n]{|a_n|} = \frac{1}{R}$$

for the fuzzy power series $\sum_{n=1}^{\infty} na_n(u - u_0)^{n-1}$ by Lemma 6.4.19, it has a radius R of convergence.

(b) For any $\rho \in (0, R)$ and $M \in L(\mathbb{R})$, the power series $\sum_{n=1}^{\infty} na_n(u - u_0)^{n-1}$ converges uniformly on $D(u, u_0) < \rho$ by the uniform convergence of fuzzy series theorem on the closed interval $[u_0 - \rho, u_0 + \rho] \subseteq [u_0 - R, u_0 + R]$ and using ratio test and Weierstrass's M-test for fuzzy series, then $\sum_{n=1}^{\infty} a_n(u - u_0)^n$ converges, $a_n \rho^n \to 0$, as $n \to \infty$, and there exists $M \geq 0$ such that $|a_n \rho^n| \leq M$ for each $n \in \mathbb{N}_1$, and by Theorem 6.4.4 we have

$$\sum_{n=1}^{\infty} |na_n(u - u_0)^{n-1}| \leq \frac{M}{\rho} \sum_{n=1}^{\infty} n \left| \frac{u - u_0}{\rho} \right|^{n-1}.$$

Since $|(u - u_0)/\rho| < 1$ by Part (iv) of Theorem 6.4.4, the series $\sum_{n=1}^{\infty} n \, |(u - u_0)/\rho|^{n-1}$ converges by the ratio test. It follows that $\sum_{n=1}^{\infty} n a_n (u - u_0)^{n-1}$ converges uniformly when $D(u, u_0) < R$.

This completes the proof of the theorem. $\qquad\qquad\qquad\qquad\square$

Definition 6.5.16. [20, Definition 1.6] *Let* $u, v \in L(\mathbb{R})$. *If there exists a* $w \in L(\mathbb{R})$ *such that* $u = v \oplus w$, *then* w *is called H-difference of* u *and* v, *and is denoted by* $w = u - v$.

Definition 6.5.17. [20, Definition 1.9] *Let* $f : [a, b] \to L(\mathbb{R})$. *We say that* f *is Fuzzy-Riemann integrable to* $H \in L(\mathbb{R})$ *if for any* $\epsilon > 0$, *there exists* $\delta > 0$ *such that for any partition* $P = \{[a, b]; \xi\}$ *of* $[a, b]$ *with the norms* $\Delta(P) < \delta$, *we have* $D\left(\sum(v - u)f(\xi), H\right) < \epsilon$. *We write in this situation that* $H := (FR) \int_a^b f(x) dx$ *and also say that* f *is* (FR)-*integrable.*

Corollary 6.5.18. [20, Corollary 1.12] *If* $f : [a, b] \to L(\mathbb{R})$ *has a continuous H-derivative* f' *on* $[a, b]$, *then* f' *is* (FR)-*integrable over* $[a, b]$ *and we have for any* $u \geq u_0$ *that*

$$f(u) = f(u_0) \oplus (FR) \int_{u_0}^u f'(x) dx.$$

Lemma 6.5.19. [20, Lemma 1.15] *Let* $f : [a, b] \to L(\mathbb{R})$ *be fuzzy continuous function. Then,* $(FR) \int_a^x f(t) dt$ *is a continuous function in* $x \in [a, b]$.

Theorem 6.5.20. [90] *Let* $f : [a, b] \to L(\mathbb{R})$ *be a continuous function. Then,* $(FR) \int_a^b f(x) dx$ *exists and belongs to* $L(\mathbb{R})$. *Furthermore,*

$$\left[(FR) \int_a^b f(x) dx \right]_\lambda = \left[\int_a^b f(x)_\lambda^- dx, \int_a^b f(x)_\lambda^+ dx \right]$$

holds for all $\lambda \in [0, 1]$.

Theorem 6.5.21. [90] *Let* $f : [a, b] \to L(\mathbb{R})$ *be H-fuzzy differentiable function and* $t \in [a, b]$, $0 \leq \lambda \leq 1$. *Let us write* $[f(t) dt]_\lambda = [f(t)_\lambda^- dt, f(t)_\lambda^+ dt] \subseteq \mathbb{R}$ *with* $[f]_\lambda = [f_\lambda^-, f_\lambda^+]$. *Then,* f_λ^- *and* f_λ^+ *are differentiable and* $[f]_\lambda' = [(f_\lambda^-)', (f_\lambda^+)']$.

Theorem 6.5.22. [20, Theorem 2.1] *We say that* $f^{(n)} : [a, b] \to I$ *are H-differentiable at* $x \in [a, b]$ *if:*

(1) If $f^{(n)}(x) \in I$ *exists such that, for all* $h > 0$ *sufficiently near to* 0, *then* $f^{(n)}(x + h) - f^{(n)}(x)$ *and* $f^{(n)}(x) - f^{(n)}(x - h)$ *have the limits in the metric* D *for which*

$$\lim_{h \to 0^+} \frac{f^{(n)}(x + h) - f^{(n)}(x)}{h} = \lim_{h \to 0^+} \frac{f^{(n)}(x) - f^{(n)}(x - h)}{h} = f^{(n)}(x) \quad (6.5.7)$$

or

(2) If $f^{(n)}(x) \in I$ exists such that, for all $h < 0$ sufficiently near to 0, then $f^{(n)}(x+h) - f^{(n)}(x)$ and $f^{(n)}(x) - f^{(n)}(x-h)$ have the limits in the metric D for which

$$\lim_{h \to 0^-} \frac{f^{(n)}(x+h) - f^{(n)}(x)}{h} = \lim_{h \to 0^-} \frac{f^{(n)}(x) - f^{(n)}(x-h)}{h} = f^{(n)}(x).$$

Also, we assume that $f^{(n)}$'s are fuzzy continuous in D. Then, for $u, u_0 \in [a, b]$ with $u \geq u_0$ and by using (6.5.7), we obtain that

$$f(u) = f(u_0) \oplus f'(u_0)(u - u_0) \oplus f''(u_0)\frac{(u-u_0)^2}{2!} \oplus \cdots \oplus f^{(n-1)}(u_0)\frac{(u-u_0)^{n-1}}{(n-1)!}$$

$$\oplus R_n(u_0, u), \tag{6.5.8}$$

$$R_n(u_0, u) = (FR)\int_{u_0}^u \left(\int_{u_0}^{u_1} \left(\cdots \left(\int_{u_0}^{u_{n-1}} f^{(n)}(u_n)du_n \right) du_{n-1} \cdots \right) du_1 \right).$$

Here, $R_n(u_0, u)$ is fuzzy continuous on $[a, b]$ as a function of u.

We can apply the same argument to the differentiated series for differentiating once more. By taking into account Theorem 6.5.15, we derive by differentiating from $f(u) = \sum_n a_n(u - u_0)^n$ that

$$f'(u) = \sum_{n=1}^{\infty} n a_n (u - u_0)^{n-1}. \tag{6.5.9}$$

Similarly, by differentiating (6.5.9) we obtain the formula for $f''(u)$, as follows;

$$f''(u) = \sum_{n=2}^{\infty} n(n-1) a_n (u - u_0)^{n-2}.$$

Let us express explicitly the formulas of $f(u)$, $f'(u)$ and $f''(u)$, as follows:

$$\begin{aligned}
f(u) &= a_0 + a_1(u - u_0) + a_2(u - u_0)^2 + a_3(u - u_0)^3 + \cdots \\
f'(u) &= a_1 + 2a_2(u - u_0) + 3a_3(u - u_0)^2 + \cdots \\
f''(u) &= 2a_2 + 3!a_3(u - u_0) + \cdots
\end{aligned}$$

These expressions are valid in the interval $(u_0 - R, u_0 + R)$. Therefore, we obtain for $u = u_0$ that

$$\begin{aligned}
f(u_0) &= a_0, \\
f'(u_0) &= a_1, \\
f''(u_0) &= 2a_2.
\end{aligned}$$

If we continue in this way, we can obtain the fuzzy power series expansions of the fuzzy valued function f. Now, applying term-by-term differentiation repeatedly we have the following:

Corollary 6.5.23. [104, Corollary 6.13] *Given a power series $\sum_k a_k(u - u_0)^k$ of fuzzy numbers with real coefficients with the radius of convergence R such that $f(u) = \sum_k a_k(u - u_0)^k$ for all $u \in L(\mathbb{R})$ satisfying $D(u, u_0) < R$. Then, one can derive successively applying term by term differentiation that*

$$f^{(n)}(u) = \sum_{k=n}^{\infty} \left(\prod_{j=k}^{k-n+1} j \right) a_k(u - u_0)^{k-n}$$

for all $n \in \mathbb{N}_0$ which gives that $a_k = f^{(k)}(u_0)/k!$ for all $k \in \mathbb{N}_0$. Therefore, we obtain the Taylor expansion of f at u_0 as

$$f(u) = \sum_k \frac{f^{(k)}(u_0)}{k!}(u - u_0)^k \tag{6.5.10}$$

for all $u \in L(\mathbb{R})$ satisfying $D(u, u_0) < R$. It is natural that the Taylor expansion of the fuzzy valued function f in (6.5.10) is reduced to the Maclaurin expansion in the special case $u_0 = \overline{0}$ that

$$f(u) = \sum_k \frac{f^{(k)}(\overline{0})}{k!} u^k$$

for all $u \in L(\mathbb{R})$ satisfying $D(u, u_0) < R$.

6.6 On Fourier Series of Fuzzy-Valued Functions

Fourier series were introduced by Joseph Fourier (1768-1830) for the purpose of solving the heat equation in a metal plate, and it has long provided one of the principal methods of analysis for mathematical physics, engineering, and signal processing. While the original theory of Fourier series applies to the periodic functions occurring in wave motion, such as with light and sound, its generalizations often relate to wider settings, such as the time-frequency analysis underlying the recent theories of wavelet analysis and local trigonometric analysis. Additionally, the idea of Fourier was to model a complicated heat source as a superposition (or linear combination) of simple sine and cosine waves, and to write the solution as a superposition of the corresponding eigen solutions. This superposition or linear combination is called the Fourier series.

Due to the rapid development of the fuzzy theory, however, some of these basic concepts have been modified and improved. One of them set mapping operations to the case of interval-valued fuzzy sets. To accomplish this we need to introduce the idea of the level sets of interval fuzzy sets and the related formulation of a representation of an interval-valued fuzzy set in terms of its level sets. Once having these structures we then can provide the desired extension to interval-valued fuzzy sets. The effectiveness of level sets comes from not only their required memory capacity for fuzzy sets, but also their two valued nature. This nature contributes to an effective derivation of the fuzzy-inference algorithm based on the families of the level sets. Besides this, the definition of fuzzy sets by level sets offers advantages over membership functions, especially when the fuzzy sets are in universes of discourse with many elements.

In this section, following Kadak and Başar [107], we study the Fourier series of periodic fuzzy-valued functions. Using a different approach, it can be shown that the Fourier series with fuzzy coefficients converges. Applying this idea we establish some connections between the Fourier series and Fourier series of fuzzy-valued functions with the level sets.

The rest of this section is organized, as follows: In subsection 6.6.2, we give some required definitions and consequences related with the fuzzy numbers, sequences and series of fuzzy numbers. We also report the most relevant and recent literature in this section. In subsection 6.6.3, first the definition of periodic fuzzy-valued function is given, which will be used in the proof of the main results. In this subsection, Hukuhara differentiation and Henstock integration are presented according to fuzzy-valued functions depend on $x, t \in [a, b]$. In the final subsection, we assert that the Fourier series of a fuzzy-valued function with 2π period converges and especially prove that the convergence about a discontinuity point by using Dirichlet kernel and one-sided limits.

6.6.1   Fuzzy-Valued Functions with the Level Sets

In this subsection, we consider sequences and series of fuzzy-valued functions and develop uniform convergence, Hukuhara differentiation and Henstock integration. Additionally, we present characterizations of uniform convergence signs in sequences of fuzzy-valued functions.

Definition 6.6.1. (cf. [104]) *Consider the function f^t from $[a, b]$ into $L(\mathbb{R})$ with respect to a membership function μ_{f^t} which is called trapezoidal fuzzy number and is interpreted as follows:*

$$\mu_{f^t}(x) := \begin{cases} \dfrac{x - f_1(t)}{f_2(t) - f_1(t)} & , \quad f_1(t) \le x \le f_2(t), \\ 1 & , \quad f_2(t) \le x \le f_3(t), \\ \dfrac{f_4(t) - x}{f_4(t) - f_3(t)} & , \quad f_3(t) \le x \le f_4(t), \\ 0 & , \quad f_4(t) < x < f_1(t). \end{cases}$$

Then, the membership function turns out $f^t(x) = [f_\lambda^-(t), f_\lambda^+(t)] = [(f_2(t) - f_1(t))\lambda + f_1(t), f_4(t) - (f_4(t) - f_3(t))\lambda] \in L(\mathbb{R})$ consisting of each function f_λ^-, f_λ^+ depend on $t \in [a, b]$ for all $\lambda \in [0, 1]$. Then, the function f^t is said to be a fuzzy-valued function on $[a, b]$ for all $x, t \in [a, b]$.

Remark 6.6.2. [107, Remark 14] *The functions f_i with $i \in \{1, 2, 3, 4\}$ given in Definition 6.6.1 are also defined for all $t \in [a, b]$ as $f_i(t) = k$, where k is any constant.*

Now, following Kadak [103], we give the sets $C_F[a, b]$ and $B_F[a, b]$ consisting of the continuous and bounded fuzzy-valued functions, that is,

$$C_F[a, b] := \left\{ f^t \big| f^t : [a, b] \longrightarrow L(\mathbb{R}); f^t \text{ is fuzzy-valued continuous function } \forall x, t \in [a, b] \right\},$$

$$B_F[a, b] := \left\{ f^t \big| f^t : [a, b] \longrightarrow L(\mathbb{R}); f^t \text{ is fuzzy-valued bounded function } \forall x, t \in [a, b] \right\}.$$

Obviously, from Theorem 6.2.1, each function f_λ^-, f_λ^+ depend on $t \in [a, b]$ is left continuous on $\lambda \in (0, 1]$ and right continuous at $\lambda = 0$. It was shown that $C_F[a, b]$ and $B_F[a, b]$ are complete with the metric D_{F_∞} on $L(\mathbb{R})$ defined by means of the Hausdorff metric d as

$$
\begin{aligned}
D_{F_\infty}(f^t, g^t) &:= \sup_{x \in [a,b]} \left\{ D(f^t(x), g^t(x)) \right\} = \sup_{x \in [a,b]} \left\{ \sup_{\lambda \in [0,1]} d([f^t(x)]_\lambda, [g^t(x)]_\lambda) \right\} \\
&:= \max \left\{ \sup_{\lambda \in [0,1]} \sup_{t \in [a,b]} \left| f_\lambda^-(t) - g_\lambda^-(t) \right|, \sup_{\lambda \in [0,1]} \sup_{t \in [a,b]} \left| f_\lambda^+(t) - g_\lambda^+(t) \right| \right\},
\end{aligned}
$$

where $f^t = f^t(x)$ and $g^t = g^t(x)$ are the elements of the sets $C_F[a, b]$ or $B_F[a, b]$ with $x, t \in [a, b]$.

6.6.2 Generalized Hukuhara Differentiation

The concept of fuzzy differentiability comes from a generalization of the Hukuhara difference for compact convex sets. We prove several properties of the derivative of fuzzy-valued functions considered here. As a continuation of Hukuhara derivatives for real fuzzy-valued functions [99], we can define H-differentiation of a fuzzy-valued function f^t with respect to level sets. For short, throughout the chapter, we write H instead of "Hukuhara sense."

Definition 6.6.3. [107, Definition 15] *A fuzzy-valued function $f^t : [a, b] \to L(\mathbb{R})$ is said to be generalized H-differentiable with respect to the level sets at $x, t \in [a, b]$ if:*

(1) *$(f^t)'(x) \in L(\mathbb{R})$ exists such that, for all $h > 0$ sufficiently near to 0, the H-difference $f^t(x + h) \ominus f^t(x)$ exists then the H-derivative $(f^t)'(x)$ is given as follows:*

$$
\begin{aligned}
(f^t)'(x) &= \lim_{h \to 0^+} \left[\frac{f^t(x + h) \ominus f^t(x)}{h} \right]_\lambda \\
&= \left[\lim_{h \to 0^+} \frac{f_\lambda^-(t + h) - f_\lambda^-(t)}{h}, \lim_{h \to 0^+} \frac{f_\lambda^+(t + h) - f_\lambda^+(t)}{h} \right] \\
&= \left[(f_\lambda^-(t))', (f_\lambda^+(t))' \right]
\end{aligned}
$$

or

(2) $(f^t)'(x) \in L(\mathbb{R})$ exists such that, for all $h < 0$ sufficiently near to 0, the H-difference $f^t(x+h) \ominus f^t(x)$ exists then the H-derivative $(f^t)'(x)$ is given as follows:

$$(f^t)'(x) = \lim_{h \to 0^-} \left[\frac{f^t(x+h) \ominus f^t(x)}{h} \right]_\lambda$$

$$= \left[\lim_{h \to 0^-} \frac{f_\lambda^-(t+h) - f_\lambda^-(t)}{h}, \lim_{h \to 0^-} \frac{f_\lambda^+(t+h) - f_\lambda^+(t)}{h} \right]$$

$$= \left[(f_\lambda^-)'(t), (f_\lambda^+)'(t) \right]$$

for all $x, t \in [a, b]$ and $\lambda \in [0, 1]$.

From here, we remind that the H-derivative of f^t at $x, t \in [a, b]$ depends on the value t and the choice of a constant $\lambda \in [0, 1]$.

Corollary 6.6.4. [107, Corollary 16] *A fuzzy-valued function f^t is H-differentiable if and only if f_λ^- and f_λ^+ are differentiable functions, in the usual sense.*

Definition 6.6.5. (Periodicity) [107, Definition 17] *A fuzzy-valued function f^t is called periodic if there exists a positive constant T for which $f^t(x+T) = f^t(x)$ for any $x, t \in [a, b]$. Thus, it can easily be seen that the conditions $f_\lambda^-(t+T) = f_\lambda^-(t)$ and $f_\lambda^+(t+T) = f_\lambda^+(t)$ hold for all $t \in [a, b]$ and $\lambda \in [0, 1]$. Such a smallest positive constant T is called a period of the function f^t.*

6.6.3 Generalized Fuzzy-Henstock Integration

Definition 6.6.6. [20, Definition 8.7] *A fuzzy-valued function f^t is said to be fuzzy-Henstock integrable, in short FH-integrable, if for any $\epsilon > 0$, there exists $\delta > 0$ such that*

$$D \left[\sum_P (v - u) f^t(\xi), I \right] \tag{6.6.1}$$

$$= \sup_{\lambda \in [0,1]} \max \left\{ \left| \sum_P (v - u) f_\lambda^-(t) - I_\lambda^- \right|, \left| \sum_P (v - u) f_\lambda^+(t) - I_\lambda^+ \right| \right\} < \epsilon$$

for any partition $P = \{[u, v]; \xi\}$ of $[a, b]$ with the norms $\triangle(P) < \delta$, where $I := (FH) \int_a^b f^t(x) dx$ and $t \in [a, b]$. One can conclude that $\sum_P$ in (6.6.1) denotes the usual Riemann sum for any partition P of $[a, b]$.

Theorem 6.6.7. [20, Theorem 8.8] *Let $f^t \in C_F[a, b]$ and FH-integrable on $[a, b]$. If there exists $x_0 \in [a, b]$ such that $f_\lambda^-(x_0) = f_\lambda^+(x_0) = 1$, then*

$$\left[(FH) \int_a^b f^t(x) dx \right]_\lambda = \left[\int_a^{x_0} f_\lambda^-(t) dt, \int_{x_0}^b f_\lambda^+(t) dt \right]. \tag{6.6.2}$$

Remark 6.6.8. [107, Remark 20] *We remark that the integrals $\int_a^b f_\lambda^\pm(t)dt$ in (6.6.2) exist in the usual sense for all $\lambda \in [0,1]$ and $t \in [a,b]$. It is easy to see that the pair of functions $f_\lambda^\pm : [a,b] \to \mathbb{R}$ are continuous.*

Remark 6.6.9. [107, Remark 21] *Note that if f^t is a periodic fuzzy-valued function and FH-integrable on any interval of length P, then it is FH-integrable on any other interval of the same length, and the value of the integral is the same, i.e.,*

$$\left[(FH) \int_a^{a+P} f^t(x)dx \right]_\lambda = \left[(FH) \int_b^{b+P} f^t(x)dx \right]_\lambda \qquad (6.6.3)$$

for all $x, t \in [a,b]$ and $\lambda \in [0,1]$.

This property is an immediate consequence of the interpretation of an integral as an area. In fact, each integral (6.6.3) equals the area bounded by the curves $f^\pm(x)$, the straight lines $x = a$ and $x = b$ and the closed interval $[a,b]$ of x-axis. In the present case, the areas represented by two integrals are the same, because of the periodicity of f^t. Hereafter, when we say that a fuzzy-valued function f^t with period P is FH-integrable, we mean that it is FH-integrable on an interval of length P. It follows from the property just proved that f^t is also FH-integrable on any interval of finite length.

Definition 6.6.10. [106] (Uniform convergence) *Let $\{f_n^t(x)\}$ be a sequence of fuzzy-valued functions defined on a set $A \subseteq \mathbb{R}$. We say that $\{f_n^t(x)\}$ converges pointwise on A if for each $x \in A$ the sequence $\{f_n^t(x)\}$ converges for all $x, t \in A$ and $\lambda \in [0,1]$. If a sequence $\{f_n^t(x)\}$ converges pointwise on a set A, then we can define $f^t : A \to L(\mathbb{R})$ by $f_n^t(x) \to f^t(x)$, as $n \to \infty$, for all $x, t \in A$. In other words, $\{f_n^t(x)\}$ converges to f^t on A if and only if for each $x \in A$ and for an arbitrary $\epsilon > 0$, there exists an integer $N = N(\epsilon, x)$ such that $D(f_n^t(x), f^t(x)) < \epsilon$ whenever $n > N$. The integer N in the definition of pointwise convergence may, in general, depend on both $\epsilon > 0$ and $x \in A$. If, however, one integer can be found that works for all points in A, then the convergence is said to be uniform. That is, a sequence of fuzzy-valued functions $\{f_n^t(x)\}$ converges uniformly to f^t on a set A if for each $\epsilon > 0$, there exists an integer $N(\epsilon)$, such that*

$$D(f_n^t(x), f^t(x)) < \epsilon \quad \text{whenever} \quad n > N(\epsilon) \quad \text{and for all} \quad x, t \in A.$$

Obviously, the sequence (f_n^t) of fuzzy-valued functions converges to a fuzzy-valued function f^t if and only if $\{(f_\lambda^-)_n(t)\}$ and $\{(f_\lambda^+)_n(t)\}$ converge uniformly to f_λ^- and f_λ^+ in $\lambda \in [0,1]$, respectively. Often, we say that f^t is the uniform limit of the sequence $\{f_n^t(x)\}$ on A and write $f_n^t \to f^t$, $n \to \infty$, uniformly on A.

Now, as a consequence of Definition 6.6.10, the following theorem characterizes the uniform convergence of fuzzy-valued sequences.

Theorem 6.6.11. [106] *Let $x, t \in A$ and $\lambda \in [0, 1]$. Then, the following statements hold:*

(i) *A sequence of fuzzy-valued functions $\{f_n^t(x)\}$ defined on a set $A \subseteq \mathbb{R}$ converges uniformly to a fuzzy-valued function f^t on A if and only if*

$$\delta_n = \sup_{x \in [a,b]} D(f_n^t(x), f^t(x)) = \sup_{x \in [a,b]} \left\{ \sup_{\lambda \in [0,1]} d\left([f_n^t(x)]_\lambda, [f^t(x)]_\lambda\right) \right\}$$

with $\delta_n \to \overline{0}$, as $n \to \infty$.

(ii) *The limit of a uniformly convergent sequence of continuous fuzzy-valued functions $\{f_n^t(x)\}$ on a set A is continuous. That is, for each $a \in A$,*

$$\lim_{x \to a} \left[\lim_{n \to \infty} f_n^t(x) \right] = \lim_{n \to \infty} \left[\lim_{x \to a} f_n^t(x) \right]. \tag{6.6.4}$$

Theorem 6.6.12. (Interchange of limit and integration) [107, Theorem 24]. *Suppose that $f_n^t \in C_F[a, b]$ for all $n \in \mathbb{N}_0$ such that $\{f_n^t(x)\}$ converges uniformly to $f^t(x)$ on $[a, b]$. Combining this fact by (6.6.4), it is obtained that the equalities*

$$\lim_{n \to \infty} \left[(FH) \int_a^b f_n^t(x) dx \right]_\lambda = \left[(FH) \int_a^b \lim_{n \to \infty} f_n^t(x) dx \right]_\lambda = \left[(FH) \int_a^b f^t(x) dx \right]_\lambda$$

hold, where the integral $(FH) \int_a^b f^t(x) dx$ exists for all $x, t \in [a, b]$ and $\lambda \in [0, 1]$. Also, for each $p \in [a, b]$, it is trivial that

$$\lim_{n \to \infty} \left[(FH) \int_a^p f_n^t(x) dx \right]_\lambda = \left[(FH) \int_a^p f^t(x) dx \right]_\lambda = \left[\int_a^p f_\lambda^-(t) dt, \int_a^p f_\lambda^+(t) dt \right]$$

and the convergence is uniform on $[a, b]$.

Proof. Note that by Part (ii) of Theorem 6.6.11, f^t is continuous on $[a, b]$, so that $(FH) \int_a^b f^t(x) dx$ exists. Let $\varepsilon > 0$ be given. Then, since $f_n^t \to f^t$ uniformly on $[a, b]$, there is an integer $N = N(\varepsilon)$ such that

$$D[f_n^t(x), f^t(x)] = \max \left\{ \sup_{\lambda \in [0,1]} \sup_{t \in [a,b]} \left| f_n(t)_\lambda^- - f_\lambda^-(t) \right|, \sup_{\lambda \in [0,1]} \sup_{t \in [a,b]} \left| f_n(t)_\lambda^+ - f_\lambda^+(t) \right| \right\}$$

$$< \frac{\varepsilon}{b - a}$$

for $n > N(\varepsilon)$. Again, since the distance function $D(f_n^t, f^t)$ is continuous on $[a, b]$, it follows

$$D\left[(FH) \int_a^b f_n^t(x) dx, (FH) \int_a^b f^t(x) dx \right] = \sup_{\lambda \in [0,1]} d\left(\left[(FH) \int_a^b f_n^t(x) dx \right]_\lambda, \right. \tag{6.6.5}$$

$$\left. \left[(FH) \int_a^b f^t(x) dx \right]_\lambda \right)$$

and the statement on the right-hand side of (6.6.5) is evaluated as

$$\sup_{\lambda \in [0,1]} d \left(\left[(FH) \int_a^b f_n^t(x) dx \right]_\lambda, \left[(FH) \int_a^b f^t(x) dx \right]_\lambda \right)$$

$$= \sup_{\lambda \in [0,1]} \max \left\{ \left| \int_a^b [f_n(t)_\lambda^- - f(t)_\lambda^-] dt \right|, \left| \int_a^b [f_n(t)_\lambda^+ - f(t)_\lambda^+] dt \right| \right\}$$

$$\leq \int_a^b \max \left\{ \sup_{\lambda \in [0,1]} \sup_{t \in [a,b]} \left| f_n(t)_\lambda^- - f(t)_\lambda^- \right|, \sup_{\lambda \in [0,1]} \sup_{t \in [a,b]} \left| f_n(t)_\lambda^+ - f(t)_\lambda^+ \right| \right\} dt$$

$$< \frac{\varepsilon}{b-a} (b-a) = \varepsilon$$

for $n > N(\varepsilon)$. Since ε is arbitrary, this step completes the proof. $\qquad \square$

The hypothesis of Theorem 6.6.12 is sufficient for our purposes and may be used to show the nonuniform convergence of the sequence $\{f_n^t(x)\}$ on $[a,b]$. Also, it is important to point out that a direct analogue of Theorem 6.6.12 does not hold for H-derivatives.

Remark 6.6.13. [107, Remark 25] *In Theorem 6.6.12, uniform convergence of $\{f_n^t(x)\}$ is sufficient but is not necessary. In other words the conclusion of Theorem 6.6.12 holds without $\{f_n^t(x)\}$ being convergent uniformly on $[a,b]$.*

Definition 6.6.14. [107, Definition 26] *The series $\sum_{k=1}^\infty f_k^t(x)$ is said to be uniformly convergent to a fuzzy-valued function $f^t(x)$ on A if the partial level sum $\{S_n^t(x)\}$ converges uniformly to $f^t(x)$ on A. That is, the series converges uniformly to $f^t(x)$ if given any $\varepsilon > 0$, there exists an integer $n_0(\varepsilon)$ such that*

$$D \left[\sum_{k=1}^\infty f_k^t(x), f^t(x) \right]$$

$$= \max \left\{ \sup_{\lambda \in [0,1]} \sup_{t \in [a,b]} \left| \sum_{k=1}^\infty f_k(t)_\lambda^- - f(t)_\lambda^- \right|, \sup_{\lambda \in [0,1]} \sup_{t \in [a,b]} \left| \sum_{k=1}^\infty f_k(t)_\lambda^+ - f(t)_\lambda^+ \right| \right\} < \varepsilon$$

for all $x, t \in A$ and $\lambda \in [0,1]$ whenever $n \geq n_0(\varepsilon)$.

Corollary 6.6.15. [107, Corollary 27] *If $\{f_k^t(x)\}$ is a sequence of fuzzy-valued continuous functions on $A \subseteq \mathbb{R}$ and $\sum_{k=1}^\infty f_k^t(x)$ is uniformly convergent to $f^t(x)$ on A, then f^t is continuous on A for all $x, t \in A$.*

Corollary 6.6.16. (Interchange of summation and integration) [107, Corollary 28] *Suppose that $\{f_k^t(x)\}$ is a sequence in $C_F[a,b]$ and $\sum_k f_k^t(x)$ converges uniformly to $f^t(x)$ on $[a,b]$. Then,*

$$\left[(FH) \sum_k \int_a^b f_k^t(x) dx \right]_\lambda = \left[(FH) \int_a^b \sum_k f^t(x) dx \right]_\lambda = \left[(FH) \int_a^b f_k^t(x) dx \right]_\lambda,$$

where $(FH) \int_a^b f^t(x) dx$ exists for all $x, t \in [a,b]$ and $\lambda \in [0,1]$.

Now, we give an important trigonometric system whose special case of one of the system of functions is applying to the well-known inequalities. By a trigonometric system, we mean the system of the periodic functions *cosine* and *sine* with period 2π which is given by

$$1, \cos x, \sin x, \cos 2x, \sin 2x, \ldots, \cos nx, \sin nx, \ldots \qquad (6.6.6)$$

for all $n \in \mathbb{N}_1$. We now prove some auxiliary formulas for any positive integer n such that $\int_{-\pi}^{\pi} \cos nx dx = \int_{-\pi}^{\pi} \sin nx dx = 0$. Therefore, one can see by using trigonometric identities that

$$\int_{-\pi}^{\pi} \cos mx \cos nx dx := \begin{cases} 0 & , \quad m \neq n, \\ 2\pi & , \quad m = n = 0, \\ \pi & , \quad m = n \neq 0, \end{cases}$$

$$\int_{-\pi}^{\pi} \sin mx \sin nx dx := \begin{cases} 0 & , \quad m \neq n, \\ \pi & , \quad m = n \neq 0. \end{cases}$$

It is known that the integral of a periodic function is the same over any interval whose length equals to its period. Therefore, the formulas are valid not only for the interval $[-\pi, \pi]$ but also for any interval $[a, a + 2\pi]$, i.e., the system (6.6.6) is orthogonal on every such interval, where $a \in \mathbb{R}$.

6.6.4 Fourier Series for Fuzzy-Valued Functions of Period 2π

Definition 6.6.17. [107, Definition 29] *Let f^t be a fuzzy-valued periodic function with period 2π on a set A. The Fourier series of fuzzy-valued periodic function f^t with period 2π is defined as follows*

$$f^t(x) \cong a_0 + \sum_{n=1}^{\infty} (a_n \cos nx + b_n \sin nx) \qquad (6.6.7)$$

with respect to the fuzzy coefficients a_n and b_n for all $n \in \mathbb{N}_1$ which converges uniformly in $\lambda \in [0, 1]$ and for all $x, t \in A$.

Now, we can calculate the Fourier coefficients a_0, a_n and b_n with respect to the level sets, i.e., $a_n = [(a_n)_\lambda^-, (a_n)_\lambda^+]$. We derive from (6.6.7) by FH-integrating over $[-\pi, \pi]$ that

$$\left[(FH) \int_{-\pi}^{\pi} f^t(x) dx \right]_\lambda = \left[(FH) \int_{-\pi}^{\pi} a_0 dx \right]_\lambda$$

$$+ \sum_{n=1}^{\infty} \left[(FH) \int_{-\pi}^{\pi} (a_n \cos nx \oplus b_n \sin nx) \, dx \right]_\lambda. \qquad (6.6.8)$$

As an extension of the relation (6.6.8) for writing with level sets we have

$$\left[\int_{-\pi}^{\pi} f_\lambda^-(t) dt, \int_{-\pi}^{\pi} f_\lambda^+(t) dt \right] = \left[\int_{-\pi}^{\pi} (a_0)_\lambda^-(t) dt, \int_{-\pi}^{\pi} (a_0)_\lambda^+(t) dt \right]$$

$$+ \sum_{n=1}^{\infty} \left[\int_{-\pi}^{\pi} (a_n)_\lambda^-(t) \cos nt dt, \int_{-\pi}^{\pi} (a_n)_\lambda^+(t) \cos nt dt \right]$$

$$+ \sum_{n=1}^{\infty} \left[\int_{-\pi}^{\pi} (b_n)_\lambda^-(t) \sin nt dt, \int_{-\pi}^{\pi} (b_n)_\lambda^+(t) \sin nt dt \right]$$

for each $\lambda \in [0,1]$ and $x, t \in [a,b]$. By taking into account the formulas of orthogonal system in (6.6.6) for each $m, n \in \mathbb{N}$ with $m \neq n$, to get a_n, multiplying (6.6.8) by $\cos mx$, we obtain by FH-integrating it over $[-\pi, \pi]$ that

$$\left[(FH) \int_{-\pi}^{\pi} f^t(x) dx \right]_{\lambda}$$

$$= \left[(FH) \int_{-\pi}^{\pi} a_0 dx \right]_{\lambda} + \sum_{n=1}^{\infty} \left[(FH) \int_{-\pi}^{\pi} (a_n \cos mx \cos nx + b_n \sin nx \cos mx) dx \right]_{\lambda}.$$

Similarly, to get b_n, multiplying (6.6.8) by $\sin mx$ and by FH-integrating it over $[-\pi, \pi]$ that the coefficients a_0, a_n and b_n with respect to the level sets are derived such that

$$
\begin{aligned}
a_n &= \frac{1}{\pi} \left[(FH) \int_{-\pi}^{\pi} f^t(x) \cos nx \, dx \right]_{\lambda} && (6.6.9) \\
&= \frac{1}{\pi} \left[\int_{-\pi}^{\pi} f_{\lambda}^-(t) \cos nt \, dt, \int_{-\pi}^{\pi} f_{\lambda}^+(t) \cos nt \, dt \right], (n \geq 0), \\
b_n &= \frac{1}{\pi} \left[(FH) \int_{-\pi}^{\pi} f^t(x) \sin nx \, dx \right]_{\lambda} && (6.6.10) \\
&= \frac{1}{\pi} \left[\int_{-\pi}^{\pi} f_{\lambda}^-(t) \sin nt \, dt, \int_{-\pi}^{\pi} f_{\lambda}^+(t) \sin nt \, dt \right], (n \geq 1).
\end{aligned}
$$

Combining the trigonometric identity $\cos(a - b) = \cos a \cos b + \sin a \sin b$ with $a = ns$ and $b = nx$, and substituting the formulas (6.6.9), and (6.6.10) in (6.6.7) one can observe that

$$f^t(x) \cong \frac{1}{2\pi} \left[(FH) \int_{-\pi}^{\pi} f^t(x) dx \right]_{\lambda} + \sum_{n=1}^{\infty} \frac{1}{\pi} \left[(FH) \int_{-\pi}^{\pi} f^t(x) \cos(ns - nx) dx \right]_{\lambda} \quad (6.6.11)$$

which is the desired alternate form of the Fourier series of fuzzy-valued function f^t on the interval $[-\pi, \pi]$ for each $\lambda \in [0,1]$.

Therefore, in looking for a trigonometric series of fuzzy-valued functions whose level sum is a given fuzzy-valued function f^t, it is natural to examine the series whose coefficients are given by (6.6.9) and (6.6.10). The trigonometric series with these coefficients is called the Fourier series of fuzzy-valued function f^t. Incidentally, we note that fuzzy coefficients involve FH-integrating of a fuzzy-valued function of period 2π. Therefore, the interval of integration can be replaced by any other interval of length 2π.

Remark 6.6.18. [107, Remark 30] *Let f^t be any fuzzy-valued function defined only on $[-\pi, \pi]$ in trigonometric series. In this case, nothing at all is said about the periodicity of f^t. In fact, if the Fourier series of fuzzy-valued functions turns out to converge to f^t, then, since it is a periodic function, the level sum of this automatically gives us the required periodic extension of f^t.*

Now, we give the definitions of the well-known two types of fuzzy numbers with the λ-level sets.

Definition 6.6.19 (Triangular fuzzy number). [137, Definition, p. 137] *Consider the triangular fuzzy number $u = (u_1, u_2, u_3)$ whose membership function $\mu_{(u)}$ is interpreted, as follows;*

$$\mu_{(u)}(x) := \begin{cases} \dfrac{x - u_1}{u_2 - u_1} & , \quad u_1 \leq x \leq u_2, \\ \dfrac{u_3 - x}{u_3 - u_2} & , \quad u_2 \leq x \leq u_3, \\ 0 & , \quad x < u_1, \quad x > u_3. \end{cases}$$

Then, we have $[u]_\lambda := [u^-(\lambda), u^+(\lambda)] = [(u_2 - u_1)\lambda + u_1, -(u_3 - u_2)\lambda + u_3]$ for each $\lambda \in [0, 1]$.

Definition 6.6.20 (Trapezoidal fuzzy number). [137, Definition, p. 145] *Consider the trapezoidal fuzzy number $u = (u_1, u_2, u_3, u_4)$ whose membership function $\mu_{(u)}$ is interpreted, as follows;*

$$\mu_{(u)}(x) := \begin{cases} \dfrac{x - u_1}{u_2 - u_1} & , \quad u_1 \leq x \leq u_2, \\ 1 & , \quad u_2 \leq x \leq u_3, \\ \dfrac{u_4 - x}{u_4 - u_3} & , \quad u_3 \leq x \leq u_4, \\ \\ 0 & , \quad x < u_1, \quad x > u_4. \end{cases}$$

Then, we have $[u]_\lambda := [u^-(\lambda), u^+(\lambda)] = [(u_2 - u_1)\lambda + u_1, -(u_4 - u_3)\lambda + u_4]$ for each $\lambda \in [0, 1]$.

Example 6.6.21. [107, Example 31] *Let f^t be a fuzzy valued periodic function with period 2π in the trapezoidal form defined by*

$$f^t(x) := \begin{cases} \dfrac{x + \pi}{t + \pi} & , \quad -\pi \leq x \leq t, \\ 1 & , \quad t \leq x \leq \pi - t, \\ \dfrac{\pi - x}{t} & , \quad \pi - t \leq x \leq \pi, \\ 0 & , \quad x < -\pi, \quad x > \pi \end{cases}$$

which is FH-integrable on $[-\pi, \pi]$ for each $x, t \in [a, b]$ and $\lambda \in [0, 1]$. By using Definition 6.6.20, the level set $[f^t]_\lambda$ of the membership function f^t can be written, as follows;

$$[f^t]_\lambda := [f^-_\lambda(t), f^+_\lambda(t)] = [t\lambda + \pi(\lambda - 1), \pi - t\lambda].$$

Therefore, we calculate the Fourier coefficients a_0, a_n and b_n, as follows:

$$a_0 = \frac{1}{2\pi} \left[\int_{-\pi}^{\pi} [t\lambda + \pi(\lambda - 1)]dt, \int_{-\pi}^{\pi} [\pi - t\lambda]dt \right] = [\pi(\lambda - 1), \pi],$$

$$a_n = \frac{1}{\pi} \left[\int_{-\pi}^{\pi} [t\lambda + \pi(\lambda - 1)] \cos ntdt, \int_{-\pi}^{\pi} [\pi - t\lambda] \cos ntdt \right] = [0, 0] = [0]_\lambda,$$

$$b_n = \frac{1}{\pi} \left[\int_{-\pi}^{\pi} [t\lambda + \pi(\lambda - 1)] \sin ntdt, \int_{-\pi}^{\pi} [\pi - t\lambda] \sin ntdt \right] = (-1)^n \left[-\frac{2\lambda}{n}, \frac{2\lambda}{n} \right].$$

By considering above coefficients in (6.6.7) and the condition $k[u_\lambda^-, u_\lambda^+] = [ku_\lambda^+, ku_\lambda^-]$ if $k < 0$, we have

$$f^t(x) \cong [\pi(\lambda - 1), \pi] \oplus \bigoplus_{n=1}^{\infty} (-1)^n \left[\frac{-2\lambda}{n}, \frac{2\lambda}{n}\right] \sin nx$$

$$= \left[\pi(\lambda - 1) + 2\lambda \sin x - \lambda \sin 2x + \frac{2\lambda}{3} \sin 3x - \cdots,\right.$$

$$\left. \pi - 2\lambda \sin x + \lambda \sin 2x - \frac{2\lambda}{3} \sin 3x + \cdots\right].$$

Definition 6.6.22. (Complex form) [107, Definition 32] *Let f^t be a fuzzy-valued function and FH-integrable on $[-\pi, \pi]$, and its Fourier series is in the form (6.6.7). By substituting the Euler's well-known formulas related to the trigonometric and exponential functions: $e^{ix} = \cos x + i\sin x$ and $\cos nx = (e^{inx} + e^{-inx})/2$, $\sin nx = (e^{inx} - e^{-inx})/2i$ in (6.6.7), the complex form of Fourier series of fuzzy-valued function f^t is given by*

$$f^t(x) \cong \frac{1}{2}a_0 + \sum_{n=1}^{\infty} \left[\frac{1}{2}(a_n + ib_n)e^{inx} + \frac{1}{2}(a_n \ominus ib_n)e^{-inx}\right], \quad (6.6.12)$$

where the H-difference $a_n \ominus ib_n$ exists for all $n \in \mathbb{N}_1$ and $x, t \in A$.

If we set

$$c_0 = \frac{1}{2}a_0, \quad c_n = \frac{1}{2}(a_n + ib_n), \quad c_{-n} = \frac{1}{2}(a_n \ominus ib_n), \quad (6.6.13)$$

then the Mth partial sum of the series (6.6.12), and hence of the series (6.6.7), can be written in the form

$$s^t{}_M(x) = c_0 + \sum_{n=1}^{M} \left(c_n e^{inx} + c_{-n}e^{-inx}\right).$$

Therefore, it is natural to write $f^t(x) \cong \sum_{n=-\infty}^{\infty} c_n e^{inx}$. The coefficients c_n's are given by (6.6.13) called the complex Fourier fuzzy coefficients and satisfy the following relation

$$c_n = \frac{1}{2\pi} \left[(FH) \int_{-\pi}^{\pi} f^t(x)e^{-inx}dx\right]_\lambda.$$

Definition 6.6.23. [107, Definition 33] *Let f^t be any fuzzy-valued function on $[a, b]$, defined either on the whole x-axis or on some interval. Then, f^t is said to be an even function if $f^t(-x) = f(x)$ for every x. Thus, the conditions $f_\lambda^-(-t) = f_\lambda^-(t)$ and $f_\lambda^+(-t) = f_\lambda^+(t)$ hold for all $t \in [a, b]$ and $\lambda \in [0, 1]$.*

Definition 6.6.24. [107, Definition 34] *Let f^t be an even function on $[-\pi, \pi]$, or else an even periodic function. Then, the Fourier coefficients of f^t are*

$$a_n = \frac{1}{\pi} \left[(FH) \int_{-\pi}^{\pi} f^t(x) \cos nx \, dx\right]_\lambda = \frac{2}{\pi} \left[(FH) \int_0^{\pi} f^t(x) \cos nx \, dx\right]_\lambda$$

and $b_n = [0]_\lambda$. Therefore, Fourier series of f^t consists of cosines, i.e.,

$$f^t(x) \cong a_0 + \sum_{n=1}^{\infty} a_n \cos nx.$$

Remark 6.6.25. [107, Remark 35] *By taking into account Definition 6.6.1, one can conclude that a fuzzy-valued function cannot be odd. Because the functions f^- and f^+ are given in Theorem 6.2.1 cannot be odd functions. Therefore, the Fourier series of fuzzy-valued function do not consist of the sines. However, we can define the Fourier series of f^t consists of sines without using the oddness property, as follows:*

Definition 6.6.26. [107, Definition 36] *Let f^t be a fuzzy-valued periodic function on a closed interval. If the Fourier coefficient $a_n = \overline{0}$, then its Fourier series consists of sines, i.e., $f^t(x) \cong \sum_{n=1}^{\infty} b_n \sin nx$.*

Definition 6.6.27. (One-sided H-derivatives) [107, Definition 37] *Let f^t be any fuzzy-valued function on A and continuous except possibly for a finite number of finite jumps. This means that f^t is permitted to be discontinuous at a finite number of points in each period, but at these points we assume that both the one-sided limits exist and finite. For convenience, we introduce this notation for these limits:*

$$f^t(x_0-) = \left[\lim_{t \to t_0 - 0} f_\lambda^-(t), \lim_{t \to t_0 - 0} f_\lambda^+(t) \right] = \left[f_\lambda^-(t_0-), f_\lambda^+(t_0-) \right] = \lim_{x \to x_0 - 0} f^t(x),$$

$$f^t(x_0+) = \left[\lim_{t \to t_0 + 0} f_\lambda^-(t), \lim_{t \to t_0 + 0} f_\lambda^+(t) \right] = \left[f_\lambda^-(t_0+), f_\lambda^+(t_0+) \right] = \lim_{x \to x_0 + 0} f^t(x)$$

for all $x, t \in A$. In addition, we suppose that the generalized left-hand H-derivative $(f_L^t)'(x_0)$ exists and is defined by

$$(f_L^t)'(x_0) = \lim_{h \to 0} \left[\frac{f^t(x_0 + h) \ominus f^t(x_0-)}{h} \right]_\lambda = \lim_{u \to 0} \left[\frac{f^t(x_0 - u) \ominus f^t(x_0-)}{-u} \right]_\lambda.$$

Thus, we can write

$$\begin{aligned}
(f_L^t)'(x_0) &= \left[\lim_{h \to 0^-} \frac{f_\lambda^-(t_0 + h) - f_\lambda^-(t_0-)}{h}, \lim_{h \to 0^-} \frac{f_\lambda^+(t_0 + h) - f_\lambda^+(t_0-)}{h} \right] \\
&= \left[(f_\lambda^-)'(t_0), (f_\lambda^+)'(t_0) \right]. \tag{6.6.14}
\end{aligned}$$

If f^t is continuous at x_0, then (6.6.14) coincides with the usual left-hand derivative. If f^t has a discontinuity at x_0, then we take care to use the left-hand instead of just writing $f^t(x_0)$.

  Symmetrically, we also assume that the generalized right-hand H-derivative $(f_R^t)'(x_0)$ exists and is defined by

$$\begin{aligned}
(f_R^t)'(x_0) &= \lim_{h \to 0} \left[\frac{f^t(x_0 + h) \ominus f^t(x_0+)}{h} \right]_\lambda \\
&= \left[\lim_{h \to 0^+} \frac{f_\lambda^-(t_0 + h) - f_\lambda^-(t_0+)}{h}, \lim_{h \to 0^+} \frac{f_\lambda^+(t_0 + h) - f_\lambda^+(t_0+)}{h} \right].
\end{aligned}$$

Now, we quote the following lemmas which are needed in proving the convergence of a Fourier series of fuzzy-valued functions at each point of discontinuity.

Lemma 6.6.28. (Dirichlet kernel) [49, Lemma 2.11.3] The Dirichlet kernel D_N is defined by $D_N(u) = \frac{1}{2} + \sum_{n=1}^{N} \cos nu$, where $N \in \mathbb{N}_1$. The Dirichlet kernel D_N has the following two properties: The first involves the definite integral of D_N on the interval $[0, \pi]$. That is,

$$\int_0^\pi D_N(u)du = \int_0^\pi \left[\frac{1}{2} + \sum_{n=1}^{N} \cos nu \right] du = \frac{\pi}{2},$$

and the second property is

$$D_N(u) = \frac{\sin \dfrac{(2N+1)u}{2}}{2 \sin \dfrac{u}{2}}. \tag{6.6.15}$$

Lemma 6.6.29. [107, Lemma 39] *Let $g^t \in C_F[0, \pi]$ and FH-integrable on $[0, \pi)$, then*

$$\lim_{n \to \infty} \left[(FH) \int_0^\pi g^t(u) \sin \left(nu + \frac{u}{2} \right) du \right]_\lambda = [0]_\lambda, \tag{6.6.16}$$

where n is a positive integer.

Proof. By taking into account FH-integration and the Dirichlet kernel defined in Lemma 6.6.28, the integral in (6.6.16) can be evaluated as

$$\left[\int_0^\pi g_\lambda^-(t) \left[\sin(t/2) \cos nt + \cos(t/2) \sin nt \right] dt, \int_0^\pi g_\lambda^+(t) \left[\sin(t/2) \cos nt + \cos(t/2) \sin nt \right] dt \right]$$

$$= \frac{\pi}{2} \left[(a_n)_\lambda^- + (b_n)_\lambda^-, (a_n)_\lambda^+ + (b_n)_\lambda^+ \right]$$

$$= \frac{\pi}{2} a_n + \frac{\pi}{2} b_n,$$

where $(a_n)_\lambda^-$ and $(a_n)_\lambda^+$ are the Fourier cosine coefficients of $g_\lambda^-(t) \sin(t/2)$ and $g_\lambda^+(t) \sin(t/2)$ on the interval $(0, \pi)$ in Definition 6.6.24, similarly $(b_n)_\lambda^-$ and $(b_n)_\lambda^+$ are the Fourier sine coefficients of $g_\lambda^-(t) \cos(t/2)$ and $g_\lambda^+(t) \cos(t/2)$ on the interval $(0, \pi)$ in Definition 6.6.26, respectively. Taking the limit on both sides and using orthogonal formulas, we have $a_n \to \overline{0}$ and $b_n \to \overline{0}$, as $n \to \infty$. Therefore, we have

$$\lim_{n \to \infty} \left[(FH) \int_0^\pi g^t(u) \sin \left(nu + \frac{u}{2} \right) du \right]_\lambda = \lim_{n \to \infty} \left(\frac{\pi}{2} a_n + \frac{\pi}{2} b_n \right) = [0]_\lambda$$

for all $u, t \in [0, \pi]$. $\qquad \square$

Lemma 6.6.30. [107, Lemma 40] *Suppose that $g^t \in C_F[0, \pi]$ and $(g_R^t)'(0)$ exists. Then,*

$$\lim_{N \to \infty} \left[(FH) \int_0^\pi g^t(u) D_N(u) du \right]_\lambda = \frac{\pi}{2} g^t(0+). \qquad (6.6.17)$$

Proof. Let $g^t \in C_F[0, \pi]$ and $(g_R^t)'(0)$ be exist. Then, we have from (6.6.17) that

$$\left[(FH) \int_0^\pi g^t(u) D_N(u) du \right]_\lambda$$

$$= \left[\int_0^\pi [g_\lambda^-(t) - g_\lambda^-(0+) + g_\lambda^-(0+)] D_N(t) dt, \int_0^\pi [g_\lambda^+(t) - g_\lambda^+(0+) + g_\lambda^+(0+)] D_N(t) dt \right]$$

and this equality turns out

$$\left[\int_0^\pi [g_\lambda^-(t) - g_\lambda^-(0+)] D_N(t) dt + \int_0^\pi g_\lambda^-(0+) D_N(t) dt \right.,$$
$$\left. \int_0^\pi [g_\lambda^+(t) - g_\lambda^+(0+)] D_N(t) dt + \int_0^\pi g_\lambda^+(0+) D_N(t) dt \right] \qquad (6.6.18)$$

for all $t \in [0, \pi]$ and $\lambda \in [0, 1]$. Each of the integrals on the right-hand side will be considered individually. First, using the second property of the Dirichlet kernel in (6.6.15) we get

$$\int_0^\pi [g_\lambda^\pm(t) - g_\lambda^\pm(0+)] D_N(t) dt = \int_0^\pi [g_\lambda^\pm(t) - g_\lambda^\pm(0+)] \frac{\sin\left(nt + \frac{t}{2}\right)}{2 \sin \frac{t}{2}} dt$$

$$= \int_0^\pi \left[\frac{g_\lambda^\pm(t) - g_\lambda^\pm(0+)}{2 \frac{t}{2}} \right] \frac{\frac{t}{2}}{\sin \frac{t}{2}} \sin\left(nt + \frac{t}{2}\right) dt$$

$$= \int_0^\pi \left[\frac{g_\lambda^\pm(t) - g_\lambda^\pm(0+)}{t - 0} \right] \frac{\frac{t}{2}}{\sin \frac{t}{2}} \sin\left(nt + \frac{t}{2}\right) dt.$$

Let h^t be a fuzzy-valued function defined by $h^t(u) = [g_\lambda^\pm(t) - g_\lambda^\pm(0+)]t/[2t \sin(t/2)]$ and continuous on $(0, \pi]$. For the sake of argument, it must be established that $h_\lambda^\pm$ are piecewise continuous on $(0, \pi)$. The piecewise continuity of $h_\lambda^\pm$ hinges on the right side limit at $t = 0$. Therefore,

$$\lim_{t \to 0+} h^t(u) = \lim_{t \to 0+} \left[\frac{g_\lambda^\pm(t) - g_\lambda^\pm(0+)}{t - 0} \right] \frac{\frac{t}{2}}{\sin \frac{t}{2}}$$

provided the individual limits at (6.6.18) exist. The continuity of h^t allows the application of Lemma 6.6.29, so that

$$\lim_{N \to \infty} \int_0^\pi [g_\lambda^\pm(t) - g_\lambda^\pm(0+)] D_N(t) dt = \lim_{N \to \infty} \left[h^t(u) \sin\left(nu + \frac{u}{2}\right) du \right]_\lambda = [0]_\lambda.$$

It follows for the second integral on (6.6.18) that

$$\lim_{N \to \infty} \int_0^\pi g_\lambda^\pm(0+) D_N(t) dt = \frac{\pi}{2} g_\lambda^\pm(0+).$$

Combining the results, it follows that

$$\lim_{N \to \infty} \left[(FH) \int_0^\pi g^t(u) D_N(u) du \right]_\lambda$$

$$= \left[\lim_{N \to \infty} \int_0^\pi [g_\lambda^-(t) - g_\lambda^-(0+)] D_N(t) dt + \lim_{N \to \infty} \int_0^\pi g_\lambda^-(0+) D_N(t) dt, \right.$$

$$\left. \lim_{N \to \infty} \int_0^\pi [g_\lambda^+(t) - g_\lambda^+(0+)] D_N(t) dt + \lim_{N \to \infty} \int_0^\pi g_\lambda^+(0+) D_N(t) dt \right]$$

$$= \left[\frac{\pi}{2} g_\lambda^-(0+), \frac{\pi}{2} g_\lambda^+(0+) \right]$$

$$= \frac{\pi}{2} g^t(0+),$$

which completes the proof. □

Theorem 6.6.31. [107, Theorem 41] *Let f^t be any fuzzy-valued continuous, periodic function with period 2π and H-differentiable on $[-\pi, \pi]$. The Fourier series of fuzzy-valued function converges to*

(i) *$f^t(x)$ for every value x, where $f^t \in C_F[-\pi, \pi]$ for each $\lambda \in [0, 1]$.*

(ii) *the arithmetic mean of the right-hand and left-hand limits $f^t(x-)$ and $f^t(x+)$ which are given in Definition 6.6.27, where the one-sided limits exist at each point of discontinuity.*

Proof. (i) Firstly, continuity and the existence of one-sided H-derivatives are sufficient for convergence. Secondly, if $f^t \in C_F[-\pi, \pi]$ at x , it follows that $f^t(x+) = f^t(x) = f^t(x-)$, so the Fourier series of fuzzy-valued function converges to $f^t(x)$ for all $x, t \in [-\pi, \pi]$ and each $\lambda \in [0, 1]$.

(ii) The continuity means Fourier fuzzy coefficients a_n and b_n exist for all appropriate values of n, the corresponding Fourier series for f^t is given by (6.6.11). The Nth partial level sum S_N of the series in (6.6.11) is

$$f^t(x) \cong \frac{1}{2\pi} \left[(FH) \int_{-\pi}^\pi f^t(x) dx \right]_\lambda \oplus \sum_{n=1}^N \frac{1}{\pi} \left[(FH) \int_{-\pi}^\pi f^t(x) \cos(ns - nx) dx \right]_\lambda. \quad (6.6.19)$$

Since the first property of Dirichlet kernel $D_N(s-x) = \frac{1}{2} + \sum_{n=1}^N \cos(ns-nx)$, using the partial level sum in (6.6.19), we get

$$S_N^t(x) = \frac{1}{\pi} \left[(FH) \int_{-\pi}^\pi f^t(x) D_N(s-x) dx \right]_\lambda$$

$$= \frac{1}{\pi} \left[\int_{-\pi}^\pi f_\lambda^-(t) D_N(s-t) dt, \int_{-\pi}^\pi f_\lambda^+(t) D_N(s-t) dt \right]$$

for $x, t \in [-\pi, \pi]$ and $s \in \mathbb{R}$. By using 2π-periodicity of f^t and the Dirichlet kernel in Lemma 6.6.28, we have

$$S_N^t(x) = \frac{1}{\pi} \left[\int_{t-\pi}^{t+\pi} f_\lambda^-(t) D_N(s-t) dt, \int_{t-\pi}^{t+\pi} f_\lambda^+(t) D_N(s-t) dt \right]. \quad (6.6.20)$$

The integral in (6.6.20) split into two following integrals

$$S_N^t(x) = \frac{1}{\pi} \int_{t-\pi}^{t} [f_\lambda^-(t) \, D_N(s-t) dt, f_\lambda^+(t) \, D_N(s-t) dt] \quad (6.6.21)$$

$$+ \frac{1}{\pi} \int_{t}^{t+\pi} [f_\lambda^-(t) \, D_N(s-t) dt, f_\lambda^+(t) \, D_N(s-t) dt].$$

Each integral on the right-hand side of (6.6.21) can be simplified by using Lemma 6.6.30, after making an appropriate change of variable. For the first integral, the change of variable will be $u = -t + s$ so that

$$\int_{t-\pi}^{t} f_\lambda^\pm(t) \, D_N(s-t) dt = - \int_{\pi}^{0} f_\lambda^\pm(s-u) \, D_N(u) du = \int_{0}^{\pi} f_\lambda^\pm(s-u) \, D_N(u) du. \quad (6.6.22)$$

Suppose that $[f_\lambda^-(s-u), f_\lambda^+(s-u)] = [f^{s-u}(t_0)]_\lambda = [g^u(t_0)]_\lambda = [g_\lambda^-(u), g_\lambda^+(u)]$ for all $t_0 \in [0, \pi]$ in (6.6.22). Since the functions $g_\lambda^\pm$ are piecewise continuous on $(0, \pi)$ and $g_R^u(0)$ exists, to establish the existence of the right-hand H-derivative of $g^u(t_0)$ at $t_0 = 0$, we have

$$(g^u)'_R(0) = \lim_{t \to 0+} \left[\frac{g^u(t) \ominus g^u(0+)}{t \ominus 0+} \right]_\lambda,$$

where

$$g^u(0+) = \lim_{t \to 0+} [g^u(t)]_\lambda = \lim_{t \to 0+} [f^{s-u}(t)]_\lambda = \lim_{s \to u} f^t(s-u) = f^t(x-).$$

Consequently, we get by deriving that

$$\lim_{N \to \infty} \int_{t-\pi}^{t} f_\lambda^\pm(t) \, D_N(s-t) dt = \lim_{N \to \infty} \int_{0}^{\pi} f_\lambda^\pm(s-u) \, D_N(u) du$$

$$= \lim_{N \to \infty} \int_{0}^{\pi} g_\lambda^\pm(u) \, D_N(u) du$$

$$= \frac{\pi}{2} g^u(0+) = \frac{\pi}{2} f^t(x-). \quad (6.6.23)$$

The second integral on the right-hand side of (6.6.21) is analyzed in a similar way. In this case, we use the change of variable $u = t - s$. If we take $[f_\lambda^-(s+u), f_\lambda^+(s+u)] = [g_\lambda^-(u), g_\lambda^+(u)]$, then

$$\lim_{N \to \infty} \int_{t}^{t+\pi} f_\lambda^\pm(t) \, D_N(t-s) du = \lim_{N \to \infty} \int_{0}^{\pi} g_\lambda^\pm(u) \, D_N(u) du \quad (6.6.24)$$

$$= \frac{\pi}{2} g^u(0+) = \frac{\pi}{2} f^t(x+).$$

By taking into account (6.6.23) and (6.6.24) if we let $N \to \infty$ in (6.6.21), then we have

$$\lim_{N\to\infty} \frac{1}{\pi} \Big[\int_{t-\pi}^{t} [f_\lambda^-(t) \, D_N(s-t)dt, f_\lambda^+(t) \, D_N(s-t)dt]$$

$$+ \int_{t}^{t+\pi} [f_\lambda^-(t) \, D_N(s-t)dt, f_\lambda^+(t) \, D_N(s-t)dt] \Big] = \frac{1}{2} \left[f^t(x-) + f^t(x+) \right].$$

This completes the proof. $\qquad\qquad\qquad\qquad\qquad\qquad\qquad\qquad\qquad\qquad$ □

We assume that the above results hold with respect to 2π-periodic fuzzy-valued functions. The similar results can be obtained for a continuous H-differentiable periodic fuzzy-valued function of an arbitrary period $P > 0$.

6.7 On the Slowly Decreasing Sequences of Fuzzy Numbers

Following Talo and Başar [217], in this section, we introduce the slowly decreasing condition for fuzzy valued sequences and prove that this is a Tauberian condition for the statistical convergence and the Cesáro convergence of a fuzzy valued sequence.

The concept of statistical convergence was introduced by Fast [83]. A sequence $(x_k)_{k\in\mathbb{N}_0}$ of real numbers is said to be *statistically convergent* to some number l if for every $\varepsilon > 0$ we have

$$\lim_{n\to\infty} \frac{1}{n+1} |\{k \le n : |x_k - l| \ge \varepsilon\}| = 0,$$

where by $|S|$, we denote the cardinal number of the set S. In this case, we write $\text{st}-\lim_{k\to\infty} x_k = l$.

A sequence (x_k) of real numbers is said C_1-convergent to l if its Cesàro transform $\{(C_1 x)_n\}$ of order one converges to l, as $n \to \infty$, where

$$(C_1 x)_n = \frac{1}{n+1} \sum_{k=0}^{n} x_k \quad \text{for all } n \in \mathbb{N}_0.$$

In this case, we write $x_k \to l(C_1)$, as $k \to \infty$.

We recall that a sequence (x_k) of real numbers is said to be *slowly decreasing* according to Schmidt [201] if

$$\lim_{\lambda\to 1^+} \liminf_{n\to\infty} \min_{n<k\le\lambda_n} (x_k - x_n) \ge 0, \tag{6.7.1}$$

where by λ_n, we denote the integral part of the product λn, that is, $\lambda_n := [\lambda n]$.

It is easy to see that (6.7.1) is satisfied if and only if for every $\varepsilon > 0$ there exist $n_0 = n_0(\varepsilon)$ and $\lambda = \lambda(\varepsilon) > 1$, as close to 1 as we wish, such that

$$x_k - x_n \geq -\varepsilon \quad \text{whenever} \quad n_0 \leq n < k \leq \lambda_n.$$

Lemma 6.7.1. [164, Lemma 1] *Let* (x_k) *be a sequence of real numbers. Condition (6.7.1) is equivalent to the following relation:*

$$\lim_{\lambda \to 1^-} \liminf_{n \to \infty} \min_{\lambda_n < k \leq n} (x_n - x_k) \geq 0. \tag{6.7.2}$$

A sequence (x_k) of real numbers is said to be *slowly increasing* if

$$\lim_{\lambda \to 1^+} \limsup_{n \to \infty} \max_{n < k \leq \lambda_n} (x_k - x_n) \leq 0.$$

Clearly, it is trivial that (x_k) is slowly increasing if and only if the sequence $(-x_k)$ is slowly decreasing.

Furthermore, if a sequence (x_k) of real numbers satisfies Landau's one-sided Tauberian condition (see [97, p. 121])

$$k(x_k - x_{k-1}) \geq -M \quad \text{for some} \quad M > 0 \quad \text{and every} \quad k \in \mathbb{N}_1,$$

then (x_k) is slowly decreasing. Móricz [164, Lemma 6] proved that if a sequence (x_k) is slowly decreasing, then

$$st - \lim_{k \to \infty} x_k = l \Rightarrow \lim_{k \to \infty} x_k = l. \tag{6.7.3}$$

Also, Hardy [97, Theorem 68] proved that if a sequence (x_k) is slowly decreasing, then

$$C_1 - \lim_{k \to \infty} x_k = l \Rightarrow \lim_{k \to \infty} x_k = l. \tag{6.7.4}$$

Maddox [146] defined a slowly decreasing sequence in an ordered linear space and proved implication (6.7.4) for slowly decreasing sequences in an ordered linear space.

One can extend the natural order relation on the real line to intervals, as; $A \preceq B$ if and only if $\underline{A} \leq \underline{B}$ and $\overline{A} \leq \overline{B}$. Also, the partial ordering relation on $L(\mathbb{R})$ is defined as follows:

$$u \preceq v \iff [u]_\alpha \preceq [v]_\alpha \iff u^-(\alpha) \leq v^-(\alpha) \text{ and } u^+(\alpha) \leq v^+(\alpha) \text{ for all } \alpha \in [0, 1].$$

We say that $u \prec v$ if $u \preceq v$ and there exists $\alpha_0 \in [0, 1]$ such that $u^-(\alpha_0) < v^-(\alpha_0)$ or $u^+(\alpha_0) < v^+(\alpha_0)$.

Lemma 6.7.2. [26, Lemma 6] *Let* $u, v \in L(\mathbb{R})$ *and* $\varepsilon > 0$. *The following statements are equivalent:*

(i) $D(u, v) \leq \varepsilon$.

(ii) $u - \varepsilon \preceq v \preceq u + \varepsilon$.

Lemma 6.7.3. [25, Lemma 5] *Let* $\mu, \nu \in L(\mathbb{R})$. *If* $\mu \preceq \nu + \varepsilon$ *for every* $\varepsilon > 0$, *then* $\mu \preceq \nu$.

Lemma 6.7.4. [138, Lemma 3.4] *Let* $u, v, w \in L(\mathbb{R})$. *Then, the following statements hold:*

(i) *If* $u \preceq v$ *and* $v \preceq w$, *then* $u \preceq w$.

(ii) *If* $u \prec v$ *and* $v \prec w$, *then* $u \prec w$.

Theorem 6.7.5. [138, Teorem 4.9] *Let* $u, v, w, e \in L(\mathbb{R})$. *Then, the following statements hold:*

(i) *If* $u \preceq w$ *and* $v \preceq e$, *then* $u + v \preceq w + e$.

(ii) *If* $u \succeq \overline{0}$ *and* $v \succ w$, *then* $uv \succeq uw$.

Nanda [174] introduced the concept of Cauchy sequence of fuzzy numbers and showed that every convergent fuzzy valued sequence is Cauchy.

Statistical convergence of a fuzzy valued sequence was introduced by Nuray and Savaş [178]. Nuray and Savaş [178] proved that if a sequence (u_k) is convergent, then (u_k) is statistically convergent. However, the converse is false, in general.

Lemma 6.7.6. [17, Remark 3.7] *If* $(u_k) \in \omega(F)$ *is statistically convergent to some* μ, *then there exists a sequence* (v_k) *which is convergent in the ordinary sense to* μ *and*

$$\lim_{n \to \infty} \frac{1}{n+1} |\{k \leq n : u_k \neq v_k\}| = 0. \tag{6.7.5}$$

Basic results on statistical convergence of fuzzy valued sequences can be found in [25, 24, 136, 200].

The Cesàro convergence of a fuzzy valued sequence is defined in [210], as follows: The sequence (u_k) is said to be *Cesàro convergent* (in short, C_1-convergent) to a fuzzy number μ if

$$\lim_{n \to \infty} (C_1 u)_n = \mu. \tag{6.7.6}$$

Talo and Çakan [218, Theorem 2.1] have recently proved that if a sequence (u_k) of fuzzy numbers is convergent, then (u_k) is C_1-convergent. However, the converse is false, in general.

Definition 6.7.7. [17] *A sequence* (u_k) *of fuzzy numbers is said to be slowly oscillating if*

$$\inf_{l > 1} \limsup_{n \to \infty} \max_{n < k \leq l_n} D(u_k, u_n) = 0. \tag{6.7.7}$$

It is easy to see that (6.7.7) is satisfied if and only if for every $\varepsilon > 0$ *there exist* $n_0 = n_0(\varepsilon)$ *and* $\lambda = \lambda(\varepsilon) > 1$, *as close to 1 as we wish, such that* $D(u_k, u_n) \leq \varepsilon$ *whenever* $n_0 \leq n < k \leq \lambda_n$.

Talo and Çakan [218, Corollary 2.7] proved that if a sequence (u_k) of fuzzy numbers is slowly oscillating, then the implication (6.7.4) holds.

In this section, we define the slowly decreasing sequence over $L(\mathbb{R})$ which is partial ordered and is not a linear space. Also, we prove that if $(u_k) \in \omega(F)$ is slowly decreasing, then the implications (6.7.3) and (6.7.4) hold.

6.7.1 The Main Results

Definition 6.7.8. [217, Definition 10] *A sequence (u_k) of fuzzy numbers is said to be slowly decreasing if for every $\varepsilon > 0$ there exist $n_0 = n_0(\varepsilon)$ and $\lambda = \lambda(\varepsilon) > 1$, as close to 1 we wish, such that for every $n > n_0$ $u_k \succeq u_n - \varepsilon$ whenever $n < k \leq \lambda_n$. Similarly, (u_k) is slowly increasing if for every $\varepsilon > 0$ there exist $n_0 = n_0(\varepsilon)$ and $\lambda = \lambda(\varepsilon) > 1$, as close to 1 we wish, such that for every $n > n_0$ $u_k \preceq u_n + \varepsilon$ whenever $n < k \leq \lambda_n$.*

Remark 6.7.9. [217, Remark 11] *Each slowly oscillating fuzzy valued sequence is slowly decreasing. On the other hand, we define the sequence $(u_n) = \left(\sum_{k=0}^n v_k\right)$, where*

$$v_k(t) := \begin{cases} 1 - t\sqrt{k+1} & , \quad 0 \leq t \leq \dfrac{1}{\sqrt{k+1}}, \\ \overline{0} & , \quad \text{otherwise} \end{cases}$$

for all $k \in \mathbb{N}_0$. Then, for each $\alpha \in [0,1]$ since

$$u_n^-(\alpha) = 0 \quad \text{and} \quad u_n^+(\alpha) = (1 - \alpha) \sum_{k=0}^n \frac{1}{\sqrt{k+1}},$$

(u_n) is increasing. Therefore, (u_n) is slowly decreasing. However, it is not slowly oscillating because for each $n \in \mathbb{N}_0$ and $\lambda > 1$ we get $\alpha = 0$ and $k = \lambda_n$; then $k \leq \lambda n$ and

$$\begin{aligned} \lim_{n \to \infty} \left[u_k^+(0) - u_n^+(0) \right] &= \lim_{n \to \infty} \left(\sum_{j=n+1}^k \frac{1}{\sqrt{j+1}} \right) \\ &\geq \lim_{n \to \infty} \frac{k - n}{\sqrt{k+1}} \\ &\geq \lim_{n \to \infty} \frac{\lambda n - 1 - n}{\sqrt{\lambda n + 1}} \\ &\geq \lim_{n \to \infty} \left[\frac{n(\lambda - 1)}{\sqrt{\lambda n + 1}} - \frac{1}{\sqrt{\lambda n + 1}} \right] = \infty. \end{aligned}$$

Lemma 6.7.10. [217, Lemma 12] *Let (u_n) be a fuzzy valued sequence. If (u_n) is slowly decreasing, then for every $\varepsilon > 0$ there exist $n_0 = n_0(\varepsilon)$ and $\lambda = \lambda(\varepsilon) < 1$, as close to 1 we wish, such that for every $n > n_0$*

$$u_n \succeq u_k - \varepsilon \quad \text{whenever } \lambda_n < k \leq n.$$

Proof. We prove the lemma by an indirect way. Assume that the sequence (u_n) is slowly decreasing and there exists some $\varepsilon > 0$ such that for all $\lambda < 1$ and $m \geq 1$ there exist integers k and $n \geq m$ for which

$$u_n \npreceq u_k - \varepsilon \quad \text{whenever} \quad \lambda_n < k \leq n.$$

Therefore, there exists $\alpha_0 \in [0,1]$ such that

$$u_n^-(\alpha_0) < u_k^-(\alpha_0) - \varepsilon \quad \text{or} \quad u_n^+(\alpha_0) < u_k^+(\alpha_0) - \varepsilon.$$

For the sake of definiteness, we only consider the case $u_n^-(\alpha_0) < u_k^-(\alpha_0) - \varepsilon$. Clearly, (6.7.2) is not satisfied by $\{u_n^-(\alpha_0)\}$. That is, $\{u_n^-(\alpha_0)\}$ is not slowly decreasing. This contradicts the hypothesis that (u_n) is slowly decreasing. $\square$

Theorem 6.7.11. [217, Theorem 13] *Let (u_n) be a sequence of fuzzy numbers. If (u_n) is statistically convergent to some $\mu \in L(\mathbb{R})$ and slowly decreasing, then (u_n) is convergent to μ.*

Proof. Let us start by setting $n = l_m$ in (6.7.5), where $0 \leq l_0 < l_1 < l_2 < \cdots$ is a subsequence of those indices k for which $u_k = v_k$. Therefore, we have

$$\lim_{m \to \infty} \frac{1}{l_m + 1} |\{k \leq l_m : u_k \neq v_k\}| = \lim_{m \to \infty} \frac{m+1}{l_m + 1} = 1.$$

Consequently, it follows that

$$\lim_{m \to \infty} \frac{l_{m+1}}{l_m} = \lim_{m \to \infty} \frac{l_{m+1}}{m+1} \times \frac{m+1}{m} \times \frac{m}{l_m} = 1. \tag{6.7.8}$$

By the definition of the subsequence (l_m), we have

$$\lim_{m \to \infty} u_{l_m} = \lim_{m \to \infty} v_{l_m} = \mu. \tag{6.7.9}$$

Since (u_n) is slowly decreasing for every $\varepsilon > 0$ there exist $n_0 = n_0(\varepsilon)$ and $\lambda = \lambda(\varepsilon) > 1$, as close to 1 we wish, such that for every $n > n_0$

$$u_k \succeq u_n - \frac{\varepsilon}{2} \quad \text{whenever} \quad n < k \leq \lambda_n.$$

For every large enough m

$$u_k \succeq u_{l_m} - \frac{\varepsilon}{2} \quad \text{whenever} \quad l_m < k \leq \lambda l_m.$$

By (6.7.8), we have $l_{m+1} \leq \lambda l_m$ for every large enough m, whence it follows

$$u_k \succeq u_{l_m} - \frac{\varepsilon}{2} \quad \text{whenever} \quad l_m < k < l_{m+1}. \tag{6.7.10}$$

By (6.7.9) and Lemma 6.7.2, for every large enough m we have

$$\mu - \frac{\varepsilon}{2} \prec u_{l_m} \prec \mu + \frac{\varepsilon}{2}. \tag{6.7.11}$$

Combining (6.7.10) and (6.7.11) we can see that

$$u_k \succ \mu - \varepsilon \quad \text{whenever} \quad l_m < k < l_{m+1}. \tag{6.7.12}$$

On the other hand, by virtue of Lemma 6.7.10 for every $\varepsilon > 0$ there exist $n_0 = n_0(\varepsilon)$ and $\lambda = \lambda(\varepsilon) < 1$ such that for every $n > n_0$

$$u_n \succeq u_k - \frac{\varepsilon}{2} \quad \text{whenever} \quad \lambda_n < k \leq n.$$

For every large enough m

$$u_{l_{m+1}} \succeq u_k - \frac{\varepsilon}{2} \quad \text{whenever} \quad \lambda l_{m+1} < k \leq l_{m+1}.$$

By (6.7.8), we have $\lambda l_{m+1} \leq l_m$ for every large enough m, whence it follows

$$u_{l_{m+1}} \succeq u_k - \frac{\varepsilon}{2} \quad \text{whenever} \quad l_m < k < l_{m+1}. \tag{6.7.13}$$

By (6.7.9) and Lemma 6.7.2, for every large enough m we have

$$\mu - \frac{\varepsilon}{2} \prec u_{l_{m+1}} \prec \mu + \frac{\varepsilon}{2}. \tag{6.7.14}$$

Therefore, (6.7.13) and (6.7.14) lead us to the consequence that

$$u_k \prec \mu + \varepsilon \quad \text{whenever} \quad l_m < k < l_{m+1}$$

which yields with (6.7.12) for each $\varepsilon > 0$ and Lemma 6.7.2 that

$$D(u_k, \mu) < \varepsilon \quad \text{whenever} \quad l_m < k < l_{m+1}. \tag{6.7.15}$$

(6.7.15) gives together with (6.7.9) that the whole sequence (u_k) is convergent to μ. $\qquad\square$

Lemma 6.7.12. [217, Lemma 14] *Let $\mu, \nu, w \in L(\mathbb{R})$. If $\mu + w \preceq \nu + w$, then $\mu \preceq \nu$.*

Proof. Let $\mu, \nu, w \in L(\mathbb{R})$. If $\mu + w \preceq \nu + w$, then

$$\mu^-(\alpha) + w^-(\alpha) \leq \nu^-(\alpha) + w^-(\alpha) \quad \text{and} \quad \mu^+(\alpha) + w^+(\alpha) \leq \nu^+(\alpha) + w^+(\alpha)$$

for all $\alpha \in [0, 1]$. Therefore, we have $\mu^-(\alpha) \leq \nu^-(\alpha)$ and $\mu^+(\alpha) \leq \nu^+(\alpha)$ for all $\alpha \in [0, 1]$. This means that $\mu \preceq \nu$, as desired. $\qquad\square$

Theorem 6.7.13. [217, Theorem 15] *Let $(u_n) \in \omega(F)$. If (u_n) is C_1-convergent to some $\mu \in L(\mathbb{R})$ and slowly decreasing, then (u_n) is convergent to μ.*

Proof. Assume that $(u_n) \in \omega(F)$ satisfies (6.7.6) and is slowly decreasing. Then, for every $\varepsilon > 0$ there exist $n_0 = n_0(\varepsilon)$ and $\lambda = \lambda(\varepsilon) > 1$, as close to 1 we wish, such that for every $n > n_0$

$$u_k \succeq u_n - \frac{\varepsilon}{3} \quad \text{whenever} \quad n < k \leq \lambda_n.$$

If n is large enough in the sense that $\lambda_n > n$, then

$$\frac{\lambda_n + 1}{\lambda_n - n}(C_1 u)_{\lambda_n} + (C_1 u)_n = \frac{\lambda_n + 1}{\lambda_n - n}(C_1 u)_n + \frac{1}{\lambda_n - n}\sum_{k=n+1}^{\lambda_n} u_k. \quad (6.7.16)$$

For every large enough n, since $\dfrac{\lambda_n + 1}{\lambda_n - n} \leq \dfrac{2\lambda}{\lambda - 1}$ we have

$$\lim_{n \to \infty} D\left[\frac{\lambda_n + 1}{\lambda_n - n}(C_1 u)_{\lambda_n}, \frac{\lambda_n + 1}{\lambda_n - n}(C_1 u)_n\right] = \lim_{n \to \infty} \frac{\lambda_n + 1}{\lambda_n - n} D[(C_1 u)_{\lambda_n}, (C_1 u)_n]$$

$$\leq \lim_{n \to \infty} \frac{2\lambda}{\lambda - 1} D\left[(C_1 u)_{\lambda_n}, (C_1 u)_n\right] = 0.$$

By Lemma 6.7.2, we obtain for large enough n that

$$\frac{\lambda_n + 1}{\lambda_n - n}(C_1 u)_n - \frac{\varepsilon}{3} \preceq \frac{\lambda_n + 1}{\lambda_n - n}(C_1 u)_{\lambda_n} \preceq \frac{\lambda_n + 1}{\lambda_n - n}(C_1 u)_n + \frac{\varepsilon}{3}. \quad (6.7.17)$$

By (6.7.6), for large enough n we obtain

$$\mu - \frac{\varepsilon}{3} \preceq (C_1 u)_n \preceq \mu + \frac{\varepsilon}{3}. \quad (6.7.18)$$

Since (u_n) is slowly decreasing, we have

$$\frac{1}{\lambda_n - n}\sum_{k=n+1}^{\lambda_n} u_k \succeq u_n - \frac{\varepsilon}{3}. \quad (6.7.19)$$

Combining (6.7.17), (6.7.18) and (6.7.19) we obtain by (6.7.16) for each $\varepsilon > 0$ that

$$\frac{\lambda_n + 1}{\lambda_n - n}(C_1 u)_n + \frac{\varepsilon}{3} + \mu + \frac{\varepsilon}{3} \succeq \frac{\lambda_n + 1}{\lambda_n - n}(C_1 u)_n + u_n - \frac{\varepsilon}{3}.$$

By Lemma 6.7.12, we have

$$\mu + \varepsilon \succeq u_n. \quad (6.7.20)$$

On the other hand, by virtue of Lemma 6.7.10 for every $\varepsilon > 0$ there exist $n_0 = n_0(\varepsilon)$ and $\lambda = \lambda(\varepsilon) < 1$ such that for every $n > n_0$

$$u_n \succeq u_k - \frac{\varepsilon}{3} \quad \text{whenever} \quad \lambda_n < k \leq n.$$

If n is large enough in the sense that $\lambda_n < n$, then

$$\frac{\lambda_n + 1}{n - \lambda_n}(C_1 u)_{\lambda_n} + \frac{1}{n - \lambda_n} \sum_{k=\lambda_n+1}^{n} u_k = \left(\frac{\lambda_n + 1}{n - \lambda_n} + 1 \right) (C_1 u)_n.$$

For large enough n, since $\dfrac{\lambda_n + 1}{n - \lambda_n} \leq \dfrac{2\lambda}{1 - \lambda}$ we have

$$\lim_{n \to \infty} D \left[\frac{\lambda_n + 1}{n - \lambda_n}(C_1 u)_{\lambda_n}, \frac{\lambda_n + 1}{n - \lambda_n}(C_1 u)_n \right] = 0.$$

Using the similar argument above, we conclude that

$$u_n \succeq \mu - \varepsilon. \tag{6.7.21}$$

Therefore, combining (6.7.20) and (6.7.21) for each $\varepsilon > 0$ and large enough n it is obtained that $D(u_n, \mu) \leq \varepsilon$.

This completes the proof. $\qquad\qquad\qquad\qquad\qquad\qquad\qquad\qquad \square$

Now, we give the lemma on the Landau's one-sided Tauberian condition for fuzzy valued sequences.

Lemma 6.7.14. [217, Lemma 16] *If a sequence $(u_n) \in \omega(F)$ satisfies the one-sided Tauberian condition*

$$nu_n \succeq nu_{n-1} - \overline{H} \quad \text{for some} \ \ H > 0 \ \ \text{and every} \ \ n \in \mathbb{N}_1, \tag{6.7.22}$$

then (u_n) is slowly decreasing.

Proof. A fuzzy-valued sequence (u_k) satisfies $nu_n \succeq nu_{n-1} - \overline{H}$ for $n \in \mathbb{N}_1$, where $H > 0$ is suitably chosen. Therefore, for all $\alpha \in [0, 1]$ we have

$$u_n^-(\alpha) - u_{n-1}^-(\alpha) \geq \frac{-H}{n} \quad \text{and} \quad u_n^+(\alpha) - u_{n-1}^+(\alpha) \geq \frac{-H}{n}.$$

For all $n < k$ and $\alpha \in [0, 1]$, we obtain that

$$u_k^-(\alpha) - u_n^-(\alpha) \geq \sum_{j=n+1}^{k} [u_j^-(\alpha) - u_{j-1}^-(\alpha)] \geq \sum_{j=n+1}^{k} \frac{-H}{j} \geq -H \left(\frac{k - n}{n} \right).$$

Hence, for each $\varepsilon > 0$ and $1 < \lambda \leq 1 + \varepsilon/H$ we get for all $n < k \leq \lambda_n$ that

$$u_k^-(\alpha) - u_n^-(\alpha) \geq -H \left(\frac{k}{n} - 1 \right) \geq -H(\lambda - 1) \geq -\varepsilon. \tag{6.7.23}$$

Similarly, for all $n < k \leq \lambda_n$ and $\alpha \in [0, 1]$ we have

$$u_k^+(\alpha) - u_n^+(\alpha) \geq -\varepsilon. \tag{6.7.24}$$

Combining (6.7.23) and (6.7.24), one can see that $u_k \succeq u_n - \varepsilon$ which proves that (u_k) is slowly decreasing. $\qquad\qquad\qquad\qquad\qquad\qquad\quad \square$

By Theorem 6.7.11, Theorem 6.7.13 and Lemma 6.7.14, we derive the following two consequences:

Corollary 6.7.15. [217, Corollary 17] *If* $(u_k) \in \omega(F)$ *is statistically convergent to* $\mu_0 \in L(\mathbb{R})$ *and satisfies (6.7.22), then* $u_k \to \mu_0$*, as* $k \to \infty$.

Corollary 6.7.16. [217, Corollary 18] *If* $(u_k) \in \omega(F)$ *is* C_1*-convergent to* $\mu_0 \in L(\mathbb{R})$ *and satisfies (6.7.22), then* $u_k \to \mu_0$*, as* $k \to \infty$.

Lemma 6.7.17. [217, Lemma 19] *If* $(u_n) \in \omega(F)$ *satisfies (6.7.22), then*

$$n(C_1 u)_n \succeq n(C_1 u)_{n-1} - \overline{H} \quad \text{for some} \quad H > 0 \quad \text{and for every} \quad n \in \mathbb{N}_1.$$

Proof. Assume that the sequence $(u_n) \in \omega(F)$ satisfies (6.7.22). Then, for all $\alpha \in [0, 1]$ we have

$$n \left[u_n^-(\alpha) - u_{n-1}^-(\alpha) \right] \geq -H \quad \text{and} \quad n \left[u_n^+(\alpha) - u_{n-1}^+(\alpha) \right] \geq -H.$$

By the proof of Theorem 2.3 in [84], we obtain

$$n \left[(C_1 u)_n^-(\alpha) - (C_1 u)_{n-1}^-(\alpha) \right] \geq -H \quad \text{and} \quad n \left[(C_1 u)_n^+(\alpha) - (C_1 u)_{n-1}^+(\alpha) \right] \geq -H.$$

This means that $n(C_1 u)_n \succeq n(C_1 u)_{n-1} - \overline{H}$, as desired. $\square$

Corollary 6.7.18. [217, Corollary 20] *If the sequence* $(u_n) \in \omega(F)$ *satisfies (6.7.22), then*

$$\text{st} - \lim_{n \to \infty} (C_1 u)_n = \mu_0 \quad \text{implies} \quad \lim_{n \to \infty} u_n = \mu_0.$$

Proof. By Lemma 6.7.17, $n(C_1 u)_n \succeq n(C_1 u)_{n-1} - \overline{H}$ which is a Tauberian condition for statistical convergence by Corollary 6.7.15. Therefore, $\text{st} - \lim_{n \to \infty} (C_1 u)_n = \mu_0$ implies that $(C_1 u)_n \to \mu_0$, as $n \to \infty$. Then, Corollary 6.7.16 yields that $u_n \to \mu_0$, as $n \to \infty$. $\square$

6.8 Determination of the Duals of Classical Sets of Sequences of Fuzzy Numbers and Related Matrix Transformations

The convergence of a series of fuzzy sets was examined via Zadeh's Extension Principle by M. Stojaković and Z. Stojaković [208]. Since the utilization of this approach is quite difficult in practice, we prefer the idea of using the sum of the series of λ-level sets. Following Talo and Başar [214], the main purpose of the present section is to determine the alpha-, beta- and gamma-duals of the classical sets of fuzzy valued sequences, and is to give the necessary and sufficient conditions on an infinite matrix of fuzzy numbers transforming one of the classical sets to the another one.

6.8.1   Introduction

By $\mathbb{N}_k$, we denote the set of integers which are greater than or equal to the integer k. We begin with some required definitions and consequences on the sequences and series of fuzzy numbers.

We denote the classical sets of bounded, convergent, null and absolutely p-summable fuzzy valued sequences by $\ell_\infty(F)$, $c(F)$, $c_0(F)$ and $\ell_p(F)$, respectively, that is,

$$\ell_\infty(F) \;\; := \;\; \left\{ (u_k) \in \omega(F) : \sup_{k \in \mathbb{N}_0} D(u_k, \overline{0}) < \infty \right\},$$

$$c(F) \;\; := \;\; \left\{ (u_k) \in \omega(F) : \exists l \in L(\mathbb{R}) \ni \lim_{k \to \infty} D(u_k, l) = 0 \right\},$$

$$c_0(F) \;\; := \;\; \left\{ (u_k) \in \omega(F) : \lim_{k \to \infty} D(u_k, \overline{0}) = 0 \right\},$$

$$\ell_p(F) \;\; := \;\; \left\{ (u_k) \in \omega(F) : \sum_k D(u_k, \overline{0})^p < \infty \right\}.$$

For simplicity in notation, here and in what follows, the summation without limits runs from 0 to ∞. Throughout the text, we also suppose that $1 \leq p < \infty$ with $p^{-1} + q^{-1} = 1$. In [174], it was shown that $c(F)$ and $\ell_\infty(F)$ are complete metric spaces with the Hausdorff metric D_∞ defined by

$$D_\infty(u, v) = \sup_{k \in \mathbb{N}_0} D(u_k, v_k),$$

where $u = (u_k)$, $v = (v_k)$ are the elements of the sets $c(F)$ or $\ell_\infty(F)$. Of course, $c_0(F)$ is also a complete metric space with respect to the Hausdorff metric D_∞. Besides this, Nanda [174] introduced and proved that the set $\ell_p(F)$ is a complete metric space with the Hausdorff metric D_p defined by

$$D_p(u, v) = \left\{ \sum_k [D(u_k, v_k)]^p \right\}^{1/p},$$

where $u = (u_k)$, $v = (v_k)$ are the points of $\ell_p(F)$. Mursaleen and Başarır [170] have recently introduced some new sets of fuzzy valued sequences generated by a non-negative regular matrix A some of which reduced to the Maddox spaces $\ell_\infty(p, F)$, $c(p, F)$, $c_0(p, F)$ and $\ell(p, F)$ of fuzzy valued sequences for the special cases of that matrix A. Altın et al. [16] have recently defined the concepts of lacunary statistical convergence and lacunary strongly convergence of generalized difference sequences of fuzzy numbers. They have also given some relations related to these concepts and showed that lacunary $\triangle^m$-statistical convergence and lacunary strong $\triangle^m_{(p)}$-convergence are equivalent for $\triangle^m$-bounded sequences of fuzzy numbers. Quite recently; Talo and Başar [213] have extended the main results of Başar and Altay [35] to the fuzzy numbers. Also, Talo and Başar [216] have recently studied the normed quasi-linearity of the classical sets $\ell_\infty(F)$, $c(F)$, $c_0(F)$ and $\ell_p(F)$ of fuzzy valued sequences and derived some related results. Finally, Talo and Başar [215] have introduced the sets $\ell_\infty(F, f)$, $c(F, f)$, $c_0(F, f)$ and $\ell_p(F, f)$ of fuzzy valued sequences defined by a modulus function and given some topological properties of the sets together with some inclusion relations.

The main purpose of the present section is to study the corresponding sets $\ell_\infty(F)$, $c(F)$, $c_0(F)$ and $\ell_p(F)$ of fuzzy valued sequences to the classical spaces ℓ_∞, c, c_0 and ℓ_p of sequences with real or complex terms. We essentially proceed with some classes of matrix transformations between the classical sets of sequences of fuzzy numbers. To do this, since it is needed to the beta dual of the sets $\ell_\infty(F)$, $c(F)$, $c_0(F)$ and $\ell_p(F)$ we find them together with their alpha- and gamma-duals. Furthermore, we emphasize the solidness of the sets $\ell_\infty(F)$, $c_0(F)$ and $\ell_p(F)$.

The rest of this section is organized, as follows:

Subsection 6.8.2 is devoted to the calculation of the alpha-, beta- and gamma-duals of the classical sets $\ell_\infty(F)$, $c(F)$, $c_0(F)$ and $\ell_p(F)$. It is also established in subsection 6.8.2 that the classical sets $\ell_\infty(F)$, $c_0(F)$ and $\ell_p(F)$ are solid. In subsection 6.8.3, the classes $(\mu(F) : \ell_\infty(F))$, $(c_0(F) : c(F))$, $(c_0(F) : c_0(F))$, $(c(F) : c(F), p)$, $(\ell_p(F) : c(F))$, $(\ell_p(F) : c_0(F))$ and $(\ell_\infty(F) : c_0(F))$ of infinite matrices of fuzzy numbers are characterized, where μ denotes any one of the classical sequence spaces ℓ_∞, c, c_0 or ℓ_p. Furthermore, four examples concerning the matrix transformations of fuzzy valued sequences are constructed. In the final subsection, the results are summarized, open problems and further suggestions are noted.

Lemma 6.8.1. [214, Lemma 2.6] *The following statements hold:*

(i) $D(uv, \overline{0}) \le D(u, \overline{0})D(v, \overline{0})$ *for all* $u, v \in L(\mathbb{R})$.

(ii) *If* $u_k \to u$, *as* $k \to \infty$ *then* $D(u_k, \overline{0}) \to D(u, \overline{0})$, *as* $k \to \infty$; *where* $(u_k) \in \omega(F)$.

Proof. (i) It is trivial that the inequalities $|u^-(\lambda)| \le D(u, \overline{0})$ and $|u^+(\lambda)| \le D(u, \overline{0})$ hold for all $\lambda \in [0, 1]$. By considering these facts, one can see that

$$D(uv, \overline{0}) = \sup_{\lambda \in [0,1]} \max\{|(uv)^-(\lambda)|, |(uv)^+(\lambda)|\}$$

$$\le \sup_{\lambda \in [0,1]} \max\{|u^-(\lambda)||v^-(\lambda)|, |u^-(\lambda)||v^+(\lambda)|, |u^+(\lambda)||v^-(\lambda)|, |u^+(\lambda)||v^+(\lambda)|\}$$

$$\le \sup_{\lambda \in [0,1]} \max\{D(u, \overline{0})|v^-(\lambda)|, D(u, \overline{0})|v^+(\lambda)|, D(u, \overline{0})|v^-(\lambda)|, D(u, \overline{0})|v^+(\lambda)|\}$$

$$= D(u, \overline{0}) \sup_{\lambda \in [0,1]} \max\{|v^-(\lambda)|, |v^+(\lambda)|\}$$

$$= D(u, \overline{0})D(v, \overline{0})$$

which completes the proof of Part (i).

(ii) This is trivial by using the fact given by Part (v) of Proposition 6.2.10. $\qquad\square$

Definition 6.8.2. [214, Definition 2.8] *An infinite matrix $A = (a_{nk})$ of fuzzy numbers is a double sequence of fuzzy numbers defined by a function A from the set $\mathbb{N}_0 \times \mathbb{N}_0$ into the set $L(\mathbb{R})$. The fuzzy number a_{nk} denotes the value of the function at $(n, k) \in \mathbb{N}_0 \times \mathbb{N}_0$ and is called as the element of the matrix which stands on the nth row and k^{th} column.*

If $u_k \preceq u_{k+1}$ for every $k \in \mathbb{N}_0$, then (u_k) is said to be a monotonic increasing sequence.

Lemma 6.2.13 provides that uniform convergence does not necessary in order for defining a fuzzy number by the series $\sum_k u_k^-(\lambda) = u^-(\lambda)$ and $\sum_k u_k^+(\lambda) = u^+(\lambda)$. This can be seen from the criteria related with the level convergence of a sequence of fuzzy numbers given by Fang and Huang [81], as follows:

Theorem 6.8.3. [214, Theorem 2.18] *If $\sum_k u_k$ and $\sum_k v_k$ converge, then $D(\sum_k u_k, \sum_k v_k) \leq \sum_k D(u_k, v_k)$.*

Proof. Let $\sum_k u_k = u$ and $\sum_k v_k = v$, i.e., $D(\sum_{k=0}^n u_k, u) \to 0$ and $D(\sum_{k=0}^n v_k, v) \to 0$, as $n \to \infty$. It is obvious that there is no problem in the case $\sum_k D(u_k, v_k) = \infty$.

Suppose that $\sum_k D(u_k, v_k) < \infty$. Then, by using the properties of the metric D given by Proposition 6.2.10, we derive that

$$
\begin{aligned}
D(u, v) &= D\left(u + \sum_{k=0}^n u_k + \sum_{k=0}^n v_k, v + \sum_{k=0}^n u_k + \sum_{k=0}^n v_k\right) \\
&\leq D\left(\sum_{k=0}^n u_k, u\right) + D\left(\sum_{k=0}^n v_k, v\right) + D\left(\sum_{k=0}^n u_k, \sum_{k=0}^n v_k\right) \\
&\leq D\left(\sum_{k=0}^n u_k, u\right) + D\left(\sum_{k=0}^n v_k, v\right) + \sum_{k=0}^n D\left(u_k, v_k\right)
\end{aligned}
$$

which leads us by letting $n \to \infty$ that $D(u, v) = D(\sum_k u_k, \sum_k v_k) \leq \sum_k D(u_k, v_k)$. $\qquad\square$

We conclude this subsection by summarizing the concerning results given by M. Stojaković and Z. Stojaković in [208]. Let $(Z, \| \cdot \|)$ be a real Banach space and X, Y be two subsets of Z. Then, the Hausdorff distance h is defined by

$$
h(X : Y) := \max \left\{ \sup_{x \in X} \inf_{y \in Y} \|x - y\|, \sup_{y \in Y} \inf_{x \in X} \|x - y\| \right\}.
$$

Denote $h(X, \{0\}) = \sup_{x \in X} \|x\|$ by $|X|$. The mapping $u : Z \to [0, 1]$ is a fuzzy set on Z. By $F(Z)$, let us denote the set of fuzzy sets such that λ-level sets $[u]_\lambda = \{x \in Z : u(x) \geq \lambda\}$ are non-empty for every $\lambda \in (0, 1]$.

If $u_k \in F(Z)$ for $k \in \mathbb{N}_0$, then it is defined that

$$\left(\sum_k u_k\right)(x) = \sup\left\{\inf\{u_k(x_k)\}_{k \in \mathbb{N}_0}, \; x = \sum_k x_k\right\}, \; (x \in Z). \quad (6.8.1)$$

Theorem 6.8.4. [214, Theorem 2.15] *Let (u_k) be a sequence of fuzzy numbers such that $\lim_{k \to \infty} u_k^-(\lambda) = \alpha(\lambda)$ and $\lim_{k \to \infty} u_k^+(\lambda) = \beta(\lambda)$ for each $\lambda \in [0,1]$. Then, the pair of functions α and β determines a fuzzy number if and only if the sequences of functions $\{u_k^-(\lambda)\}$ and $\{u_k^+(\lambda)\}$ are eventually equi-left-continuous at each $\lambda \in (0,1]$ and eventually equi-right-continuous at $\lambda = 0$.*

Thus, it is deduced that the series $\sum_k u_k^-(\lambda) = u^-(\lambda)$ and $\sum_k u_k^+(\lambda) = u^+(\lambda)$ define a fuzzy number if the sequences

$$\{s_k^-(\lambda)\} = \left\{\sum_{k=0}^n u_k^-(\lambda)\right\} \quad \text{and} \quad \{s_k^+(\lambda)\} = \left\{\sum_{k=0}^n u_k^+(\lambda)\right\}$$

satisfy the conditions of Theorem 6.8.4. Of course, this is a weaker condition than the uniform convergence.

Theorem 6.8.5. [214, Theorem 2.19] *If λ-level sets of u_k are compact for all $k \in \mathbb{N}_0$ and for all $\lambda \in (0,1]$, and if $\sum_k |[u_k]_\lambda| < \infty$ for all $\lambda \in (0,1]$, then*

$$\left[\sum_k u_k\right]_\lambda = \sum_k [u_k]_\lambda \quad (6.8.2)$$

for every $\lambda \in (0,1]$.

Theorem 6.8.6. [214, Theorem 2.20] *If λ-level sets of u_k are upper semicontinuous fuzzy sets with compact supports for all $k \in \mathbb{N}_0$ and if $\sum_k |\text{supp } u_k| < \infty$, then (6.8.2) holds.*

Theorem 6.8.7. [214, Theorem 2.21] *If $u_k : \mathbb{R} \to [0,1]$ are fuzzy sets for $k \in \mathbb{N}_0$ such that λ-level sets $[u_k]_\lambda$ are non-empty, bounded and closed for every $\lambda \in (0,1]$, then the λ-level sets of sum are non-empty, bounded and closed, and*

(i) $[\sum_{k=0}^n u_k]_\lambda = \sum_{k=0}^n [u_k]_\lambda$,

(ii) if $\sum_k |[u_k]_\lambda| < \infty$ for every $\lambda \in (0,1]$, then (6.8.2) holds.

Theorem 6.8.8. [214, Theorem 2.22] *If λ-level sets of u_k are compact and convex for all $k \in \mathbb{N}_0$ and for all $\lambda \in (0,1]$, and if $\sum_k |[u_k]_\lambda| < \infty$, then the λ-level set of $\sum_k u_k$ is also compact and convex for every $\lambda \in (0,1]$.*

If $u \in L(\mathbb{R})$, then since $[u]_\lambda$ is compact and convex for $\lambda \in (0,1]$ one can see that Theorem 6.8.7 and Theorem 6.8.8 are satisfied. Additionally, the set $\text{supp } u_k = \overline{\{t \in R : u(t) > 0\}}$ is compact and u is an upper semicontinuous function. Thus, Theorem 6.8.7 is also satisfied. Therefore, it can conclude that $\sum_k |[u_k]_\lambda| \leq \sum_k |\text{supp } u_k|$. Thus, we have:

Corollary 6.8.9. [214, Corollary 2.23] *Let $u_k \in L(\mathbb{R})$ for all $k \in \mathbb{N}_0$. If $\sum_k |supp\ u_k| < \infty$, then (6.8.2) holds.*

Proof. Let $u_k \in L(\mathbb{R})$ for all $k \in \mathbb{N}_0$. Then, since we use the real field it is obtained from here that

$$|supp\ u_k| = \sup_{x \in [u_k]_0} |x| = \sup_{x \in [u_k^-(0), u_k^+(0)]} |x| = \max\{|u_k^-(0)|, |u_k^+(0)|\} = D(u_k, \overline{0}).$$

This leads us to the consequence that if $u = (u_k) \in \ell_1(F)$, then (6.8.2) holds. Additionally, since $[u]_\lambda = [u^-(\lambda), u^+(\lambda)]$ for all $\lambda \in [0, 1]$ when $u \in L(\mathbb{R})$, it follows that

$$\sum_k [u_k]_\lambda = \left[\sum_k u_k^-(\lambda), \sum_k u_k^+(\lambda)\right].$$

Since $|u_k^-(\lambda)| \leq D(u_k, \overline{0})$ and $|u_k^+(\lambda)| \leq D(u_k, \overline{0})$ for all $k \in \mathbb{N}_0$, Weierstrass's criterion yields the uniform convergence of the series $\sum_k u_k^-(\lambda)$ and $\sum_k u_k^+(\lambda)$.

Finally, by using the fact that convergence is uniform one can see for the case $\lambda = 0$ that

$$\left[\sum_k u_k\right]_0 = \lim_{\lambda \to 0^+} \left[\sum_k u_k\right]_\lambda = \lim_{\lambda \to 0^+} \sum_k [u_k]_\lambda = \sum_k \lim_{\lambda \to 0^+} [u_k]_\lambda = \sum_k [u_k]_0.$$

This step completes the proof. □

M. Stojaković and Z. Stojaković [208] used the concept of convergence of series of fuzzy numbers defined by Zadeh's Extension Principle. In the present section, we emphasized the fact that $u = \sum u_k$ if and only if $\sum_k u_k^-(\lambda) = u^-(\lambda)$ and $\sum_k u_k^+(\lambda) = u^+(\lambda)$, uniformly in λ. Therefore, in the case $\sum_k D(u_k, \overline{0}) < \infty$ our definition coincides with the convergence of a series given by (6.9.1). However, it is possible to define a fuzzy number by using (6.9.1) even if the condition $\sum D(u_k, \overline{0}) < \infty$ is not satisfied. Then, the sums of a convergent series may be different in the sense of our definition and in the sense of M. Stojaković and Z. Stojaković [208]. Nevertheless, there is no need to elaborate on the condition of uniform convergence given in Theorem 6.8.4 for defining a fuzzy number by the series $\sum_k [u_k]_\lambda = [\sum_k u_k^-(\lambda), \sum_k u_k^+(\lambda)]$. Even in finding the sum of two fuzzy numbers we used the sum of λ-level sets instead of Zadeh's Extension Principle. Of course, calculating the sum of a convergent series of fuzzy numbers via Zadeh's Extension Principle is very difficult. So, after calculating the sum of the series of λ-level sets in evaluating the response of the question: *Is the sum a fuzzy number?*, it becomes easier and much more applicable than the usage of the Zadeh's Extension Principle.

If we consider the convergent series $\sum_k u_k$ in the sense of the present section, then

$$\sum_k u_k = u \quad \Leftrightarrow \quad \lim_{n \to \infty} D\left(\sum_{k=0}^{n} u_k, u\right) = 0$$

$$\Leftrightarrow \quad \lim_{n \to \infty} \sup_{\lambda \in [0,1]} d\left(\sum_{k=0}^{n} [u_k]_\lambda, [u]_\lambda\right) = 0,$$

where Z is a Banach space, $u_k \in F(Z)$ for all $k \in \mathbb{N}_0$ and d is the Hausdorff metric. This is equivalent to the fact that $\sum_k [u_k]_\lambda = [u]_\lambda$ uniformly in $\lambda \in [0,1]$.

6.8.2 Determination of Duals of the Classical Sets of Sequences of Fuzzy Numbers

Firstly, we define the alpha-dual, beta-dual and gamma-dual of a set $\mu(F)$ of fuzzy valued sequences which are respectively denoted by $\{\mu(F)\}^\alpha$, $\{\mu(F)\}^\beta$ and $\{\mu(F)\}^\gamma$, as follows:

$$\{\mu(F)\}^\alpha := \{(u_k) \in \omega(F) : (u_k v_k) \in \ell_1(F) \text{ for all } (v_k) \in \mu(F)\},$$

$$\{\mu(F)\}^\beta := \{(u_k) \in \omega(F) : (u_k v_k) \in cs(F) \text{ for all } (v_k) \in \mu(F)\},$$

$$\{\mu(F)\}^\gamma := \{(u_k) \in \omega(F) : (u_k v_k) \in bs(F) \text{ for all } (v_k) \in \mu(F)\}.$$

Now, we may give the results concerning the alpha-, beta- and gamma-duals of the sets $\ell_\infty(F)$, $c(F)$, $c_0(F)$ and $\ell_p(F)$.

We begin with a lemma which yields a useful result on the absolute convergence of the series of fuzzy numbers.

Lemma 6.8.10. [214, Lemma 3.1] *Let m denotes the set of all absolutely convergent series of fuzzy numbers, i.e.,*

$$m := \left\{(u_k) \in \omega(F) : \sum_k |u_k| < \infty\right\}.$$

Then, the set m is identical to the set $\ell_1(F)$.

Proof. Let $(u_k) \in \ell_1(F)$. Then, since $\sum D(u_k, \overline{0}) < \infty$ and $\max\{|u_k^-(\lambda)|,$ $|u_k^+(\lambda)|\} \le D(u_k, \overline{0})$ for $\lambda \in [0,1]$, we can see that

$$\sum_k \max\{|u_k^-(\lambda)|, |u_k^+(\lambda)|\} \le \sum_k D(u_k, \overline{0}) < \infty$$

which leads us to the consequence $(u_k) \in m$. That is, the inclusion $\ell_1(F) \subset m$ holds.

Conversely, let $(u_k) \in m$. Therefore, $\sum_k \max\{|u_k^-(\lambda)|, |u_k^+(\lambda)|\}$ converges for $\lambda \in [0, 1]$. This gives for $\lambda = 0$ that

$$\sum_k D(u_k, \overline{0}) = \sum \max\{|u_k^-(0)|, |u_k^+(0)|\} < \infty$$

which means that $(u_k) \in \ell_1(F)$, as desired. Hence, the inclusion $m \subset \ell_1(F)$ also holds.

This step completes the proof. □

Theorem 6.8.11. [214, Theorem 3.2] *The alpha-dual of the sets $c_0(F)$, $c(F)$ and $\ell_\infty(F)$ of fuzzy valued sequences is the set $\ell_1(F)$.*

Proof. Since the proof is similar for the sets $c_0(F)$ and $c(F)$, we give the proof only for the set $\ell_\infty(F)$.

Let $(u_k) \in \ell_\infty(F)$. Then, there exists a $K > 0$ such that $D(u_k, \overline{0}) \leq K$ for all $k \in \mathbb{N}_0$. Therefore, we derive by using the fact given Part (i) of Lemma 6.8.1 that

$$\sum_k D(u_k v_k, \overline{0}) \leq \sum_k D(u_k, \overline{0}) D(v_k, \overline{0}) \leq K \sum_k D(v_k, \overline{0}).$$

If we take $(v_k) \in \ell_1(F)$, then we have $\sum_k D(u_k v_k, \overline{0}) < \infty$ which gives that

$$\ell_1(F) \subseteq \{\ell_\infty(F)\}^\alpha. \tag{6.8.3}$$

Conversely, let $(v_k) \in \{\ell_\infty(F)\}^\alpha$. If we consider $(u_k) = (\overline{1}) \in \ell_\infty(F)$, then the series $\sum_k D(u_k v_k, \overline{0}) = \sum_k D(v_k, \overline{0})$ converges, that is to say that $(v_k) \in \ell_1(F)$. Therefore, we have

$$\{\ell_\infty(F)\}^\alpha \subseteq \ell_1(F). \tag{6.8.4}$$

Thus, the desired result follows by combining the inclusion relations (6.8.3) and (6.8.4). □

Now, following Sarma [196] we may state the fuzzy analogues of the concept of solidity of a set of sequences of complex terms and nextly give our result concerning the solidity of the sets $\ell_\infty(F)$, $c_0(F)$ and $\ell_p(F)$.

Definition 6.8.12. *A set $\mu(F) \subset \omega(F)$ is said to be solid if $(v_k) \in \mu(F)$ whenever $D(v_k, \overline{0}) \leq D(u_k, \overline{0})$ for all $k \in \mathbb{N}_0$, for some $(u_k) \in \mu(F)$. Therefore, one can conclude that the alpha- and gamma-duals of a set of fuzzy valued sequences are identical if it is solid.*

Theorem 6.8.13. [214, Theorem 3.4] *The sets $\ell_\infty(F)$, $c_0(F)$ and $\ell_p(F)$ are solid.*

Proof. Let $\mu(F)$ denotes anyone of the sets $\ell_\infty(F)$, $c_0(F)$ and $\ell_p(F)$, and suppose that

$$D(v_k, \overline{0}) \leq D(u_k, \overline{0}) \tag{6.8.5}$$

holds for some $(u_k) \in \mu(F)$ and $(v_k) \in \omega(F)$. Therefore, one can easily see by (6.8.5) that

$$\sup_{k \in \mathbb{N}_0} D(v_k, \overline{0}) \leq \sup_{k \in \mathbb{N}_0} D(u_k, \overline{0}) < \infty,$$

$$\lim_{k \to \infty} D(v_k, \overline{0}) \leq \lim_{k \to \infty} D(u_k, \overline{0}) = 0,$$

$$\sum_k [D(v_k, \overline{0})]^p \leq \sum_k [D(u_k, \overline{0})]^p < \infty.$$

The above inequalities yield the desired consequence that $(v_k) \in \mu(F)$. This completes the proof. $\qquad\square$

Theorem 6.8.14. [214, Theorem 3.5] *The beta-dual of the sets $c_0(F)$, $c(F)$ and $\ell_\infty(F)$ of fuzzy valued sequences is the set $\ell_1(F)$.*

Proof. Since the proof is similar for the sets $c_0(F)$ and $c(F)$, we give the proof only for the set $\ell_\infty(F)$.

Let $(u_k) \in \ell_\infty(F)$ and $(v_k) \in \omega(F)$. Then, there exists a $K > 0$ such that $D(u_k, \overline{0}) \leq K$ for all $k \in \mathbb{N}_0$. Since

$$|(u_k v_k)^-(\lambda)| \leq D(u_k v_k, \overline{0}) \leq D(u_k, \overline{0})D(v_k, \overline{0}) \leq KD(v_k, \overline{0})$$
$$|(u_k v_k)^+(\lambda)| \leq D(u_k v_k, \overline{0}) \leq D(u_k, \overline{0})D(v_k, \overline{0}) \leq KD(v_k, \overline{0}),$$

Weierstrass's Test yields that $\sum_k (u_k v_k)^-(\lambda)$ and $\sum_k (u_k v_k)^+(\lambda)$ converge uniformly and hence $\sum_k u_k v_k$ converges whenever $\sum_k D(v_k, \overline{0})$ converges. Therefore, we have

$$\ell_1(F) \subseteq \{\ell_\infty(F)\}^\beta. \tag{6.8.6}$$

Conversely, suppose that $(v_k) \in \{\ell_\infty(F)\}^\beta$. Then, the series $\sum_k u_k v_k$ converges for all $(u_k) \in \ell_\infty(F)$. This also holds for the sequence (u_k) of fuzzy numbers defined by $u_k := \chi_{[-1,1]}$ for all $k \in \mathbb{N}_0$. Then, since $u_k^-(\lambda) = -1$ and $u_k^+(\lambda) = 1$ for all $\lambda \in [0,1]$, the series

$$\sum_k (u_k v_k)^+(\lambda) = \sum_k \max\{u_k^-(\lambda)v_k^-(\lambda), u_k^-(\lambda)v_k^+(\lambda), u_k^+(\lambda)v_k^-(\lambda), u_k^+(\lambda)v_k^+(\lambda)\}$$

$$= \sum_k \max\{-v_k^-(\lambda), -v_k^+(\lambda), v_k^-(\lambda), v_k^+(\lambda)\}$$

$$= \sum_k \max\{|v_k^-(\lambda)|, |v_k^+(\lambda)|\}$$

converges uniformly which gives that $(v_k) \in \ell_1(F)$. Hence,

$$\{\ell_\infty(F)\}^\beta \subseteq \ell_1(F). \tag{6.8.7}$$

Thus, combining the inclusions (6.8.6) and (6.8.7) leads us to the desired result. $\qquad\square$

Theorem 6.8.15. [214, Theorem 3.6] *The alpha- and beta-duals of the set* $\ell_p(F)$ *of fuzzy valued sequences are the set* $\ell_q(F)$, *where* $1 \leq p < \infty$.

Proof. We give the proof only for beta dual. Since the implication "$(u_k) \in \ell_p(F)$ if and only if the series $\sum_k D(u_k, \overline{0})^p$ converges, i.e., $\{D(u_k, \overline{0})\} \in \ell_p$" holds, $\{D(v_k, \overline{0})\} \in \ell_q$ whenever $(v_k) \in \ell_q(F)$. Because of the well-known result $\ell_p^\beta = \ell_q$, the series $\sum D(u_k, \overline{0}) D(v_k, \overline{0})$ converges. Therefore, since

$$|(u_k v_k)^-(\lambda)| \leq D(u_k v_k, \overline{0}) \leq D(u_k, \overline{0}) D(v_k, \overline{0})$$
$$|(u_k v_k)^+(\lambda)| \leq D(u_k v_k, \overline{0}) \leq D(u_k, \overline{0}) D(v_k, \overline{0}),$$

Weierstrass's M test implies that the series $\sum_k (u_k v_k)^-(\lambda)$ and $\sum_k (u_k v_k)^+(\lambda)$ converge uniformly, hence the series $\sum_k u_k v_k$ also converges. Thus, we have

$$\ell_q(F) \subseteq \{\ell_p(F)\}^\beta. \tag{6.8.8}$$

Conversely, let $(v_k) \in \{\ell_p(F)\}^\beta$. Then, the series $\sum_k u_k v_k$ converges for all $(u_k) \in \ell_p(F)$ which yields that the series $\sum_k (u_k v_k)^+(\lambda)$ converges uniformly in $\lambda \in [0, 1]$. Define the sequence $(u_k^{(n)}) \in \ell_p(F)$ by

$$u_k^{(n)} := \begin{cases} \chi_{[-(D(v_k, \overline{0}))^{q-1}, (D(v_k, \overline{0}))^{q-1}]} & , \quad k \leq n, \\ \overline{0} & , \quad \text{otherwise} \end{cases}$$

for all $k, n \in \mathbb{N}_0$. Therefore, $[u_k^{(n)}]_\lambda = [-(D(v_k, \overline{0}))^{q-1}, (D(v_k, \overline{0}))^{q-1}]$ for $k \leq n$. Thus, one can see that

$$\begin{aligned}
\left(u_k^{(n)} v_k\right)^+(\lambda) &= \max\{-[D(v_k, \overline{0})]^{q-1} v_k^-(\lambda), -[D(v_k, \overline{0})]^{q-1} v_k^+(\lambda), [D(v_k, \overline{0})]^{q-1} v_k^-(\lambda), \\
&\qquad [D(v_k, \overline{0})]^{q-1} v_k^+(\lambda)\} \\
&= [D(v_k, \overline{0})]^{q-1} \max\{-v_k^-(\lambda), -v_k^+(\lambda), v_k^-(\lambda), v_k^+(\lambda)\} \\
&= [D(v_k, \overline{0})]^{q-1} \max\{|v_k^-(\lambda)|, |v_k^+(\lambda)|\}
\end{aligned}$$

which gives in the case $\lambda = 0$ that

$$\begin{aligned}
\left(u_k^{(n)} v_k\right)^+(0) &= [D(v_k, \overline{0})]^{q-1} \max\{|v_k^-(0)|, |v_k^+(0)|\} \\
&= [D(v_k, \overline{0})]^{q-1} D(v_k, \overline{0}) = [D(v_k, \overline{0})]^q.
\end{aligned}$$

This leads us to the consequence that $\sum_{k=0}^n [D(v_k, \overline{0})]^q = \sum_k [u_k^{(n)} v_k]^+(0)$ for all $n \in \mathbb{N}_0$ which means that $v = (v_k) \in \ell_q(F)$. Hence, we have

$$\{\ell_p(F)\}^\beta \subseteq \ell_q(F). \tag{6.8.9}$$

The desired result is obtained by combining the inclusions (6.8.8) and (6.8.9). $\square$

By combining Theorem 6.8.13 with Theorem 6.8.14 and Theorem 6.8.15, we have:

Corollary 6.8.16. [214, Corollary 3.7] *The gamma-dual of the sets* $\ell_\infty(F)$, $c_0(F)$ *and* $\ell_p(F)$ *of fuzzy-valued sequences are the sets* $\ell_1(F)$ *and* $\ell_q(F)$, *respectively.*

6.8.3 Matrix Transformations Between Some Sets of Sequences of Fuzzy Numbers

Let $\mu_1(F)$, $\mu_2(F) \subset \omega(F)$ and $A = (a_{nk})$ be an infinite matrix of fuzzy numbers. Then, we say that A defines a matrix mapping from $\mu_1(F)$ into $\mu_2(F)$, and denote it by writing $A : \mu_1(F) \to \mu_2(F)$, if for every sequence $u = (u_k) \in \mu_1(F)$ the sequence $Au = \{(Au)_n\}$, the A-transform of u, exists and is in $\mu_2(F)$; where

$$(Au)_n := \sum_k a_{nk} u_k \quad \text{for each} \ \ n \in \mathbb{N}_0. \tag{6.8.10}$$

By $(\mu_1(F) : \mu_2(F))$, we denote the class of all matrices A such that $A : \mu_1(F) \to \mu_2(F)$. Thus, $A \in (\mu_1(F) : \mu_2(F))$ if and only if the series on the right side of (6.8.10) converges for each $n \in \mathbb{N}_0$ and every $u \in \mu_1(F)$, and we have $Au = \{(Au)_n\}_{n \in \mathbb{N}_0} \in \mu_2(F)$ for all $u \in \mu_1(F)$. A sequence u is said to be A-summable to α if Au converges to α which is called as the A-limit of u. Also, by $(c(F) : c(F); p)$; we mean the class of all regular matrices A such that $A : c(F) \to c(F)$ with A-limit of u equals to limit of u, for all $u \in c(F)$. We denote the nth row of a matrix $A = (a_{nk})$ by A_n for all $n \in \mathbb{N}_0$, i.e., $A_n := (a_{nk})_{k=0}^{\infty}$ for all $n \in \mathbb{N}_0$.

In this subsection, we characterize the classes $(\mu(F) : \ell_\infty(F))$, $(c_0(F) : c(F))$, $(c_0(F) : c_0(F))$, $(c(F) : c(F), p)$, $(\ell_p(F) : c(F))$, $(\ell_p(F) : c_0(F))$ and $(\ell_\infty(F) : c_0(F))$ of infinite matrices of fuzzy numbers, where μ denotes any of the symbols ℓ_∞, c, c_0 or ℓ_p.

Now, we may give the following basic theorem:

Theorem 6.8.17. [214, Basic Theorem] *The following statements hold:*

(i) $A = (a_{nk}) \in (\ell_\infty(F) : \ell_\infty(F))$ *if and only if*

$$M = \sup_{n \in \mathbb{N}_0} \sum_k D(a_{nk}, \overline{0}) < \infty. \tag{6.8.11}$$

(ii) $A = (a_{nk}) \in (c(F) : \ell_\infty(F))$ *if and only if (6.8.11) holds.*

(iii) $A = (a_{nk}) \in (c_0(F) : \ell_\infty(F))$ *if and only if (6.8.11) holds.*

(iv) $A = (a_{nk}) \in (\ell_p(F) : \ell_\infty(F))$ *if and only if*

$$C = \sup_{n \in \mathbb{N}_0} \sum_k [D(a_{nk}, \overline{0})]^q < \infty, \ \text{where} \ p > 1 \ \text{and} \ p^{-1} + q^{-1} = 1. \tag{6.8.12}$$

Proof. Since the proof can also be obtained in the similar way for other parts, to avoid the repetition of the similar statements, we prove only Part (i).

(i) Let $A = (a_{nk}) \in (\ell_\infty(F) : \ell_\infty(F))$ and $u = (u_k) \in \ell_\infty(F)$. Then, the series $\sum_k a_{nk} u_k$ converges for each fixed $n \in \mathbb{N}_0$, since Au exists. Hence, $A_n \in \{\ell_\infty(F)\}^\beta$ for each fixed $n \in \mathbb{N}_0$. Define the sequence $u = (u_k) \in \ell_\infty(F)$

by $u_k := \chi_{[-1,1]}$ for all $k \in \mathbb{N}$. Then, $Au \in \ell_\infty(F)$ which yields for all $n \in \mathbb{N}$ and for all $\lambda \in [0,1]$ that

$$\sum_k (a_{nk}u_k)^+(\lambda) = \sum_k \max\{a_{nk}^-(\lambda)u_k^-(\lambda), a_{nk}^-(\lambda)u_k^+(\lambda), a_{nk}^+(\lambda)u_k^-(\lambda), a_{nk}^+(\lambda)u_k^+(\lambda)\}$$

$$= \sum_k \max\{-a_{nk}^-(\lambda), -a_{nk}^+(\lambda), a_{nk}^-(\lambda), a_{nk}^+(\lambda)\}$$

$$= \sum_k \max\{|a_{nk}^-(\lambda)|, |a_{nk}^+(\lambda)|\}.$$

In the special case $\lambda = 0$, the sequence

$$\left\{\sum_k D(a_{nk}, \bar{0})\right\}_{n \in \mathbb{N}_0} = \left\{\sum_k \max\{|a_{nk}^-(0)|, |a_{nk}^+(0)|\}\right\}_{n \in \mathbb{N}_0}$$

is bounded which means that (6.8.11) holds.

Conversely, suppose that (6.8.11) holds and $u = (u_k) \in \ell_\infty(F)$. Then, since $A_n \in \{\ell_\infty(F)\}^\beta$ for each $n \in \mathbb{N}_0$, Au exists. Therefore, one can observe by using Part (i) of Lemma 6.8.1 together with the condition (6.8.11) that

$$D_\infty(Au, \theta) = \sup_{n \in \mathbb{N}_0} D((Au)_n, \bar{0}) = \sup_{n \in \mathbb{N}_0} D\left(\sum_k a_{nk}u_k, \bar{0}\right)$$

$$\leq \sup_{n \in \mathbb{N}_0} \sum_k D(a_{nk}u_k, \bar{0})$$

$$\leq \sup_{n \in \mathbb{N}_0} \sum_k D(a_{nk}, \bar{0})D(u_k, \bar{0})$$

$$\leq \sup_{k \in \mathbb{N}_0} D(u_k, \bar{0}) \sup_{n \in \mathbb{N}_0} \sum_k D(a_{nk}, \bar{0}) < \infty,$$

i.e., $Au \in \ell_\infty(F)$.

This step concludes the proof.   □

Now, we may give an example of an infinite matrix in the class $(\ell_\infty(F) : \ell_\infty(F))$.

Example 6.8.18. [214, Example 4.1] *Define* $u_k = u_k(t)$ *and* $A = (a_{nk})$ *by*

$$u_k(t) := \begin{cases} 1 + (k+1)t & , \quad -\dfrac{1}{k+1} \leq t \leq 0, \\ 1 - (k+1)t & , \quad 0 \leq t \leq \dfrac{1}{k+1}, \\ 0 & , \quad otherwise \end{cases} \quad and \quad a_{nk} := \begin{cases} u_k & , \quad k = n, \\ \bar{0} & , \quad k \neq n \end{cases}$$

for all $k, n \in \mathbb{N}_0$, *respectively. Then, since*

$$D(a_{nk}, \bar{0}) = \begin{cases} 1/(n+1) & , \quad k = n, \\ 0 & , \quad k \neq n, \end{cases}$$

$\sup_{n \in \mathbb{N}_0} \sum_k D(a_{nk}, \overline{0}) = \sup_{n \in \mathbb{N}_0} \dfrac{1}{n+1} < \infty$. *This means by Part (i) of Theorem 6.8.17 that* $A = (a_{nk}) \in (\ell_\infty(F) : \ell_\infty(F))$.

Theorem 6.8.19. [214, Theorem 4.2] $A = (a_{nk}) \in (c_0(F) : c(F))$ *if and only if (6.8.11) holds and there exists* $(\alpha_k) \in \omega(F)$ *such that*

$$\lim_{n \to \infty} D(a_{nk}, \alpha_k) = 0 \qquad (6.8.13)$$

for each $k \in \mathbb{N}_0$. *If* $A = (a_{nk}) \in (c_0(F) : c(F))$, *then* $(\alpha_k) \in \ell_1(F)$ *and* $\sum_k a_{nk} u_k \to \sum_k \alpha_k u_k$, *as* $n \to \infty$.

Proof. Suppose that (6.8.11) and (6.8.13) hold. Then, there exists an $n_K \in \mathbb{N}_0$ for $K \in \mathbb{N}_0$ and $\varepsilon > 0$ such that $\sum_{k=0}^{K} D(a_{nk}, \alpha_k) < \varepsilon$ for all $n \geq n_K$. Since

$$
\begin{aligned}
\sum_{k=0}^{K} D(\alpha_k, \overline{0}) &= \sum_{k=0}^{K} D(a_{nk} + \alpha_k, a_{nk}) \\
&\leq \sum_{k=0}^{K} D(a_{nk}, \alpha_k) + \sum_{k=0}^{K} D(a_{nk}, \overline{0}) \\
&\leq \varepsilon + M
\end{aligned}
$$

for $n \geq n_K$ by (6.8.13), one can see that $(\alpha_k) \in \ell_1(F)$ and $\sum_k D(\alpha_k, \overline{0}) \leq M$.

Let $u = (u_k) \in c_0(F)$. Then, one can choose a $k_0 \in \mathbb{N}_0$ for $\varepsilon > 0$ such that $D(u_k, \overline{0}) < \varepsilon/[2(2M+1)]$ for all $k \geq k_0$. Additionally, since $a_{nk} \to \alpha_k$, as $n \to \infty$, by (6.8.13), we have $a_{nk} u_k \to \alpha_k u_k$, as $n \to \infty$ for each fixed $k \in \mathbb{N}_0$. That is to say that $D(a_{nk} u_k, \alpha_k u_k) \to 0$, as $n \to \infty$. Hence, there exists an $N = N(k_0) \in \mathbb{N}_0$ such that $\sum_{k=0}^{k_0} D(a_{nk} u_k, \alpha_k u_k) < \varepsilon/2$ for all $n \geq N$. Thus, since

$$
\begin{aligned}
D\left(\sum_k a_{nk} u_k, \sum_k \alpha_k u_k \right) &\leq \sum_k D(a_{nk} u_k, \alpha_k u_k) \\
&= \sum_{k=0}^{k_0} D(a_{nk} u_k, \alpha_k u_k) + \sum_{k=k_0+1}^{\infty} D(a_{nk} u_k, \alpha_k u_k) \\
&\leq \frac{\varepsilon}{2} + \sum_{k=k_0+1}^{\infty} [D(a_{nk} u_k, \overline{0}) + D(\alpha_k u_k, \overline{0})] \\
&\leq \frac{\varepsilon}{2} + \sum_{k=k_0+1}^{\infty} D(a_{nk}, \overline{0}) D(u_k, \overline{0}) + \sum_{k=k_0+1}^{\infty} D(\alpha_k, \overline{0}) D(u_k, \overline{0}) \\
&\leq \frac{\varepsilon}{2} + \frac{\varepsilon}{2(2M+1)} \left[\sum_{k=k_0+1}^{\infty} D(a_{nk}, \overline{0}) + \sum_{k=k_0+1}^{\infty} D(\alpha_k, \overline{0}) \right] \\
&\leq \frac{\varepsilon}{2} + \frac{\varepsilon}{2(2M+1)} (M + M + 1) = \varepsilon
\end{aligned}
$$

for all $n \geq N$, the series $\sum_k a_{nk} u_k$ are convergent for each $n \in \mathbb{N}_0$ and $D\left(\sum_k a_{nk} u_k, \sum_k \alpha_k u_k \right) \to 0$, as $n \to \infty$. This means that $Au \in c(F)$.

Conversely, let $A = (a_{nk}) \in (c_0(F) : c(F))$ and $u = (u_k) \in c_0(F)$. Then, since Au exists and the inclusion $(c_0(F) : c(F)) \subset (c_0(F) : \ell_\infty(F))$ holds, the necessity of (6.8.11) is trivial by Part (iii) of Theorem 6.8.17.

Now, consider the sequence $u^{(n)} = \{u_k^{(n)}\} \in c_0(F)$ defined by

$$u_k^{(n)} := \begin{cases} \overline{1} & , \quad n = k, \\ \overline{0} & , \quad n \neq k \end{cases} \tag{6.8.14}$$

for all $n \in \mathbb{N}_0$. Hence, $Au^{(n)} = (a_{nk})_{n=0}^\infty \in c(F)$ for each fixed $k \in \mathbb{N}_0$, i.e., the condition (6.8.13) is also necessary.

This step completes the proof. $\square$

As an easy consequence of Theorem 6.8.19, we have

Corollary 6.8.20. [214, Corollary 4.3] $A = (a_{nk}) \in (c_0(F) : c_0(F))$ *if and only if (6.8.11) holds and (6.8.13) also holds with* $\alpha_k = \overline{0}$ *for all* $k \in \mathbb{N}_0$.

Now, we can give an example of an infinite matrix belonging to the class $(c_0(F) : c_0(F))$.

Example 6.8.21. [214, Example 4.4] *Let* $u_k(t)$ *be defined by*

$$u_k(t) := \begin{cases} (k+1)t - 1 & , \quad \dfrac{1}{k+1} \leq t \leq \dfrac{2}{k+1}, \\ 3 - (k+1)t & , \quad \dfrac{2}{k+1} \leq t \leq \dfrac{3}{k+1}, \\ 0 & , \quad otherwise \end{cases}$$

for all $k \in \mathbb{N}_0$. *Consider the matrix* $A = (a_{nk})$ *defined by*

$$a_{nk} := \begin{cases} u_n & , \quad k \leq n, \\ \overline{0} & , \quad k > n \end{cases}$$

for all $k, n \in \mathbb{N}_0$. *Then, since* $D(a_{nk}, \overline{0}) = \begin{cases} 3/(n+1) & , \quad k \leq n, \\ 0 & , \quad k > n, \end{cases}$ *we have* $\sup_{n \in \mathbb{N}_0} \sum_k D(a_{nk}, \overline{0}) = \sup_{n \in \mathbb{N}_0} \sum_{k=0}^n 3/(n+1) < \infty$. *Additionally, since* $u_k^-(\lambda) = (\lambda + 1)/(k+1)$, $u_k^+(\lambda) = (3 - \lambda)/(k+1)$ *and* $u_k \to \overline{0}$, *as* $k \to \infty$, $a_{nk} \to \overline{0}$, *as* $n \to \infty$, *for each* $k \in \mathbb{N}_0$. *Hence, since the matrix* A *satisfies the conditions of Corollary 6.8.20,* $A = (a_{nk}) \in (c_0(F) : c_0(F))$.

Example 6.8.22. [214, Example 4.5] *Let us consider the sequence* $\{v_k(t)\}$ *defined by Rojes-Medar and Roman-Flores [191], as follows;*

$$v_k(t) := \begin{cases} 1 - (k+1)t & , \quad 0 \leq t \leq 1/(k+1), \\ 0 & , \quad otherwise \end{cases}$$

for all $k \in \mathbb{N}_0$. *Then, it is trivial that* $v_k \to \overline{0}$, *as* $k \to \infty$, *since* $v_k^-(\lambda) = 0$ *and* $v_k^+(\lambda) = (1 - \lambda)/(k+1)$ *for all* $\lambda \in [0, 1]$. *Therefore, one can see that* $\sum_k (a_{nk}v_k)^-(\lambda) = 0$ *and* $\sum_k (a_{nk}v_k)^+(\lambda) = \sum_{k=0}^n (u_n v_k)^+(\lambda) = (3 - \lambda)$

$(1 - \lambda)/(n + 1) \sum_{k=0}^{n} 1/(k + 1)$ *which yields that* $(3 - \lambda)(1 - \lambda)/(n + 1)$ $\sum_{k=0}^{n} 1/(k + 1) \to 0$, *as* $n \to \infty$. *Thus,* $\sum_{k} a_{nk} v_k \to \overline{0}$, *as* $n \to \infty$. *Let us suppose that* $Av = w$. *Therefore, it is clear that* $w_n^-(\lambda) = 0$ *for all* $\lambda \in [0, 1]$. *Then,* $w_n^+(\lambda) = (3 - \lambda)(1 - \lambda)b_n$ *with* $b_n = 1/(n + 1) \sum_{k=0}^{n} 1/(k + 1)$ *for all* $n \in \mathbb{N}_0$. *Thus, we obtain the A-transform of the sequence* (v_k) *as*

$$(Av)_n(t) := \begin{cases} 2 - \sqrt{\dfrac{t}{b_n} + 1} & , \quad 0 \leq t \leq 3b_n, \\ 0 & , \quad otherwise. \end{cases}$$

Theorem 6.8.23. [214, Theorem 4.6] *Let* $a_{nk} \succeq \overline{0}$ *for all* $n, k \in \mathbb{N}_0$. *Then,* $A = (a_{nk}) \in (c(F) : c(F); p)$ *if and only if (6.8.11) holds, and*

$$\lim_{n \to \infty} a_{nk} = \overline{0} \quad for \ each \ k \in \mathbb{N}_0, \tag{6.8.15}$$

$$\lim_{n \to \infty} \sum_{k} a_{nk} = \overline{1}. \tag{6.8.16}$$

Proof. Suppose that (6.8.11), (6.8.15) and (6.8.16) hold, and $(u_k) \in c(F)$. Then, since Au exists the series $\sum_{k} a_{nk} u_k$ converges for each fixed $n \in \mathbb{N}_0$. Hence, $A_n \in \{c(F)\}^\beta$ for all $n \in \mathbb{N}_0$. It is obvious that (6.8.11) holds if and only if

$$\sup_{n \in \mathbb{N}_0} \sum_{k} \sup_{\lambda \in [0,1]} |a_{nk}^-(\lambda)| < \infty,$$

$$\sup_{n \in \mathbb{N}_0} \sum_{k} \sup_{\lambda \in [0,1]} |a_{nk}^+(\lambda)| < \infty$$

(6.8.15) holds if and only if

$$\lim_{n \to \infty} \sup_{\lambda \in [0,1]} |a_{nk}^-(\lambda)| = 0,$$

$$\lim_{n \to \infty} \sup_{\lambda \in [0,1]} |a_{nk}^+(\lambda)| = 0$$

and (6.8.16) holds if and only if

$$\lim_{n \to \infty} \sup_{\lambda \in [0,1]} \left| \sum_{k} a_{nk}^-(\lambda) - 1 \right| = 0,$$

$$\lim_{n \to \infty} \sup_{\lambda \in [0,1]} \left| \sum_{k} a_{nk}^+(\lambda) - 1 \right| = 0.$$

By taking into account the above conditions we see that the matrices $A^-(\lambda) = (a_{nk}^-(\lambda))_{n,k \in \mathbb{N}_0}$ and $A^+(\lambda) = (a_{nk}^+(\lambda))_{n,k \in \mathbb{N}_0}$ are regular for all $\lambda \in [0, 1]$. Now, suppose that $u_k \to u$, as $k \to \infty$. Then, $u_k^-(\lambda) \to u^-(\lambda)$, as $k \to \infty$ and $u_k^+(\lambda) \to u^+(\lambda)$, as $k \to \infty$, uniformly in λ's. Since the matrices $A^-(\lambda)$ and

$A^+(\lambda)$ are regular, we have uniformly that

$$\lim_{n \to \infty} \sum_k a_{nk}^-(\lambda) u_k^-(\lambda) = u^-(\lambda), \qquad \lim_{n \to \infty} \sum_k a_{nk}^+(\lambda) u_k^-(\lambda) = u^-(\lambda),$$

$$\lim_{n \to \infty} \sum_k a_{nk}^-(\lambda) u_k^+(\lambda) = u^+(\lambda), \qquad \lim_{n \to \infty} \sum_k a_{nk}^+(\lambda) u_k^+(\lambda) = u^+(\lambda),$$

for $\lambda \in [0, 1]$. Indeed since

$$\left| \sum_k a_{nk}^-(\lambda) u_k^-(\lambda) - u^-(\lambda) \right|$$

$$= \left| \sum_k a_{nk}^-(\lambda) u_k^-(\lambda) - u^-(\lambda) \sum_k a_{nk}^-(\lambda) + u^-(\lambda) \sum_k a_{nk}^-(\lambda) - u^-(\lambda) \right|$$

$$\leq \left| \sum_k a_{nk}^-(\lambda) u_k^-(\lambda) - u^-(\lambda) \sum_k a_{nk}^-(\lambda) \right| + \left| u^-(\lambda) \sum_k a_{nk}^-(\lambda) - u^-(\lambda) \right|$$

$$\leq \sum_k |a_{nk}^-(\lambda)| |u_k^-(\lambda) - u^-(\lambda)| + |u^-(\lambda)| \left| \sum_k a_{nk}^-(\lambda) - 1 \right|$$

$$\leq \sum_k \sup_{\lambda \in [0,1]} |a_{nk}^-(\lambda)| \sup_{\lambda \in [0,1]} |u_k^-(\lambda) - u^-(\lambda)| + \sup_{\lambda \in [0,1]} |u^-(\lambda)| \sup_{\lambda \in [0,1]} \left| \sum_k a_{nk}^-(\lambda) - 1 \right|,$$

$\sup_{\lambda \in [0,1]} \left| \sum_k a_{nk}^-(\lambda) u_k^-(\lambda) - u^-(\lambda) \right| \to 0$, as $n \to \infty$. Since $a_{nk} \succeq \overline{0}$ for all $n, k \in \mathbb{N}_0$, $u_k^-(\lambda) \leq u_k^+(\lambda)$ for $\lambda \in [0, 1]$ which implies that

$$a_{nk}^-(\lambda) u_k^-(\lambda) \leq a_{nk}^-(\lambda) u_k^+(\lambda),$$
$$a_{nk}^+(\lambda) u_k^-(\lambda) \leq a_{nk}^+(\lambda) u_k^+(\lambda).$$

Therefore, we have

$$(a_{nk} u_k)^-(\lambda) = \min\{a_{nk}^-(\lambda) u_k^-(\lambda), a_{nk}^-(\lambda) u_k^+(\lambda), a_{nk}^+(\lambda) u_k^-(\lambda), a_{nk}^+(\lambda) u_k^+(\lambda)\}$$
$$= \min\{a_{nk}^-(\lambda) u_k^-(\lambda), a_{nk}^+(\lambda) u_k^-(\lambda)\}$$
$$(a_{nk} u_k)^+(\lambda) = \max\{a_{nk}^-(\lambda) u_k^-(\lambda), a_{nk}^-(\lambda) u_k^+(\lambda), a_{nk}^+(\lambda) u_k^-(\lambda), a_{nk}^+(\lambda) u_k^+(\lambda)\}$$
$$= \max\{a_{nk}^-(\lambda) u_k^+(\lambda), a_{nk}^+(\lambda) u_k^+(\lambda)\}.$$

Consequently,

$$\lim_{n \to \infty} \sum_k (a_{nk} u_k)^-(\lambda) = \lim_{n \to \infty} \sum_k \min\{a_{nk}^-(\lambda) u_k^-(\lambda), a_{nk}^+(\lambda) u_k^-(\lambda)\} = u^-(\lambda)$$

$$\lim_{n \to \infty} \sum_k (a_{nk} u_k)^+(\lambda) = \lim_{n \to \infty} \sum_k \max\{a_{nk}^-(\lambda) u_k^+(\lambda), a_{nk}^+(\lambda) u_k^+(\lambda)\} = u^+(\lambda)$$

uniformly in λ. Hence, $\sum_k a_{nk} u_k \to u$, as $n \to \infty$, as expected.

Conversely, let $A = (a_{nk}) \in (c(F) : c(F); p)$ and $u = (u_k) \in c(F)$. Then, since Au exists and the inclusion $(c(F) : c(F), p) \subset (c(F) : \ell_\infty(F))$ holds, the necessity of (6.8.11) is trivial by Part (ii) of Theorem 6.8.17.

Consider the sequences $u^{(n)} = \{u_k^{(n)}\}$ defined by (6.8.14) and $z = (z_k) = (\bar{1})$ in the set $c(F)$. Then, since $Au^{(n)}$, $Az \in c(F)$, the necessities of (6.8.15) and (6.8.16) are immediate.

This step concludes the proof. $\qquad\square$

We give an example of an infinite matrix in the class $(c(F) : c(F); p)$, below:

Example 6.8.24. [214, Example 4.7] *Define* $u_k(t)$ *by*

$$
u_k(t) := \begin{cases} (k+1)(t-1) & , \quad 1 \le t \le 1 + \dfrac{1}{k+1}, \\ 2 - (k+1)(t-1) & , \quad 1 + \dfrac{1}{k+1} \le t \le 1 + \dfrac{2}{k+1}, \\ 0 & , \quad \text{otherwise} \end{cases}
$$

for all $k \in \mathbb{N}_0$. *Consider the matrix* $A = (a_{nk})$ *defined by*

$$
a_{nk} := \begin{cases} u_k & , \quad k = n, \\ \bar{0} & , \quad k \ne n \end{cases}
$$

for all $k, n \in \mathbb{N}_0$. *Then, since* $D(a_{nk}, \bar{0}) = \begin{cases} 1 + \dfrac{2}{n+1} & , \quad k = n, \\ 0 & , \quad k \ne n, \end{cases}$

$\sup_{n \in \mathbb{N}_0} \sum_k D(a_{nk}, \bar{0}) = \sup_{n \in \mathbb{N}_0} \{1 + 2/(n+1)\} < \infty$. *Additionally, since* $u_k^-(\lambda) = 1 + \lambda/(k+1)$, $u_k^+(\lambda) = 1 + (2-\lambda)/(k+1)$ *and* $u_k \to \bar{1}$, *as* $k \to \infty$, $a_{nk} \to \bar{0}$, *as* $n \to \infty$ *for each* $k \in \mathbb{N}_0$. $\lim_{n\to\infty} \sum_k a_{nk} = \lim_{n\to\infty} u_n = \bar{1}$. *Hence,* A *is a regular matrix since it satisfies the conditions of Theorem 6.8.23.*

Now, we have

Theorem 6.8.25. [214, Theorem 4.8] $A = (a_{nk}) \in (\ell_p(F) : c(F))$ *if and only if (6.8.12) and (6.8.13) hold.*

Proof. Let $A = (a_{nk}) \in (\ell_p(F) : c(F))$ and $u = (u_k) \in \ell_p(F)$. Then, since Au exists and the inclusion $(\ell_p(F) : c(F)) \subset (\ell_p(F) : \ell_\infty(F))$ holds, the necessity of (6.8.12) is trivial by Part (iv) of Theorem 6.8.17.

Consider the sequences $u^{(n)} = \{u_k^{(n)}\}$ in the set $\ell_p(F)$ defined by (6.8.14). Then, since $Au^{(n)} \in c(F)$, the necessity of (6.8.13) is clear.

Conversely, suppose that the conditions (6.8.12) and (6.8.13) hold, and $u = (u_k) \in \ell_p(F) \subset c_0(F)$. Then, since Au exists and $A_n \in \{\ell_p(F)\}^\beta$ for each $n \in \mathbb{N}_0$ by taking into account the validity of (6.8.12) one can see by Hölder's inequality that

$$
\sum_{k=0}^{m} D(\alpha_k u_k, \bar{0}) = \lim_{n \to \infty} \sum_{k=0}^{m} D(a_{nk} u_k, \bar{0})
$$

$$
\le \left\{ \sum_k [D(u_k, \bar{0})]^p \right\}^{1/p} \left\{ \sup_{n \in \mathbb{N}_0} \sum_k [D(a_{nk}, \bar{0})]^q \right\}^{1/q} < \infty
$$

for all $m \in \mathbb{N}_0$ which says us that $(\alpha_k) \in \ell_q(F)$. For simplicity in notation, here and after we write B instead of $\{\sum_k [D(\alpha_k, \bar{0})]^q\}^{1/q}$. Since $u = (u_k) \in \ell_p(F)$, one can choose a $k_0 \in \mathbb{N}_0$ for $\varepsilon > 0$ such that

$$\sum_{k=k_0+1}^{\infty} [D(u_k, \bar{0})]^p < \frac{\varepsilon^p}{2^p [C^{1/q} + B]^p}$$

for each fixed $k \geq k_0$. Additionally, since $a_{nk} \to \alpha_k$, as $n \to \infty$ by (6.8.13), we have $a_{nk} u_k \to \alpha_k u_k$, as $n \to \infty$ for each fixed $k \in \mathbb{N}_0$. That is to say that $D(a_{nk} u_k, \alpha_k u_k) \to 0$, as $n \to \infty$. Hence, there exists an $N = N(k_0) \in \mathbb{N}_0$ such that $\sum_{k=0}^{k_0} D(a_{nk} u_k, \alpha_k u_k) < \varepsilon/2$ for all $n \geq N$. Thus, since

$$
\begin{aligned}
D\left(\sum_k a_{nk} u_k, \sum_k \alpha_k u_k\right) &\leq \sum_k D(a_{nk} u_k, \alpha_k u_k) \\
&= \sum_{k=0}^{k_0} D(a_{nk} u_k, \alpha_k u_k) + \sum_{k=k_0+1}^{\infty} D(a_{nk} u_k, \alpha_k u_k) \\
&\leq \frac{\varepsilon}{2} + \sum_{k=k_0+1}^{\infty} [D(a_{nk} u_k, \bar{0}) + D(\alpha_k u_k, \bar{0})] \\
&\leq \frac{\varepsilon}{2} + \sum_{k=k_0+1}^{\infty} D(a_{nk}, \bar{0}) D(u_k, \bar{0}) + \sum_{k=k_0+1}^{\infty} D(\alpha_k, \bar{0}) D(u_k, \bar{0}) \\
&\leq \frac{\varepsilon}{2} + \left\{\sum_{k=k_0+1}^{\infty} [D(u_k, \bar{0})]^p\right\}^{1/p} \left\{\sum_{k=k_0+1}^{\infty} [D(a_{nk}, \bar{0})]^q\right\}^{1/q} + \\
&\quad + \left\{\sum_{k=k_0+1}^{\infty} [D(u_k, \bar{0})]^p\right\}^{1/p} \left\{\sum_{k=k_0+1}^{\infty} [D(\alpha_k, \bar{0})]^q\right\}^{1/q} \\
&\leq \frac{\varepsilon}{2} + \frac{\varepsilon}{2(C^{1/q} + B)}(C^{1/q} + B) = \varepsilon
\end{aligned}
$$

for all $n \geq N$, the series $\sum_k a_{nk} u_k$ are convergent for each $n \in \mathbb{N}_0$ and $D(\sum_k a_{nk} u_k, \sum_k \alpha_k u_k) \to 0$, as $n \to \infty$. This means that $Au \in c(F)$. This step terminates the proof.   □

Finally, we have

Corollary 6.8.26. [214, Corollary 4.9] $A = (a_{nk}) \in (\ell_p(F) : c_0(F))$ *if and only if (6.8.12) holds and (6.8.13) also holds with* $\alpha_k = 0$ *for all* $k \in \mathbb{N}_0$.

Theorem 6.8.27. [214, Theorem 4.10] $A = (a_{nk}) \in (\ell_\infty(F) : c_0(F))$ *if and only if*

$$\lim_{n \to \infty} \sum_k D(a_{nk}, \bar{0}) = 0. \tag{6.8.17}$$

Proof. Let $A = (a_{nk}) \in (\ell_\infty(F) : c_0(F))$ and $u = (u_k) \in \ell_\infty(F)$. Then, the series $\sum_k a_{nk} u_k$ converges for each fixed $n \in \mathbb{N}_0$, since Au exists. Hence, $A_n \in \{\ell_\infty(F)\}^\beta$ for all $n \in \mathbb{N}_0$. Define the sequence $u = (u_k) \in \ell_\infty(F)$ by

$u_k := \chi_{[-1,1]}$ for all $k \in \mathbb{N}_0$. Then, $Au \in c_0(F)$ which yields for all $n \in \mathbb{N}_0$ and for all $\lambda \in [0,1]$ that

$$\sum_k (a_{nk}u_k)^+(\lambda) = \sum_k \max\{a_{nk}^-(\lambda)u_k^-(\lambda), a_{nk}^-(\lambda)u_k^+(\lambda), a_{nk}^+(\lambda)u_k^-(\lambda), a_{nk}^+(\lambda)u_k^+(\lambda)\}$$

$$= \sum_k \max\{-a_{nk}^-(\lambda), -a_{nk}^+(\lambda), a_{nk}^-(\lambda), a_{nk}^+(\lambda)\}$$

$$= \sum_k \max\{|a_{nk}^-(\lambda)|, |a_{nk}^+(\lambda)|\}.$$

In the special case $\lambda = 0$, we have

$$\lim_{n\to\infty} \sum_k D(a_{nk}, \overline{0}) = \lim_{n\to\infty} \sum_k \max\{|a_{nk}^-(0)|, |a_{nk}^+(0)|\} = 0.$$

Conversely, suppose that (6.8.17) holds and $u = (u_k) \in \ell_\infty(F)$. Then, since $A_n \in \{\ell_\infty(F)\}^\beta$ for each $n \in \mathbb{N}_0$, Au exists. Therefore, one can observe by using Part (ii) of Lemma 6.8.1 together with the condition (6.8.17) that

$$\lim_{n\to\infty} D[(Au)_n, \overline{0}] = \lim_{n\to\infty} D\left(\sum_k a_{nk}u_k, \overline{0}\right)$$

$$\leq \lim_{n\to\infty} \sum_k D(a_{nk}u_k, \overline{0})$$

$$\leq \lim_{n\to\infty} \sum_k D(a_{nk}, \overline{0})D(u_k, \overline{0})$$

$$\leq \sup_{k\in\mathbb{N}_0} D(u_k, \overline{0}) \lim_{n\to\infty} \sum_k D(a_{nk}, \overline{0}) = 0$$

which means that $Au \in c_0(F)$, as desired.

This step completes the proof. $\square$

6.9 On Some Sets of Fuzzy-Valued Sequences with the Level Sets

Following Kadak and Başar [106], in this section, we introduce the sets of bounded, convergent and null series and the set of fuzzy valued sequences of bounded variation with the level sets. We investigate the relationships between these sets and their classical forms. Furthermore, we study some of their properties like completeness, duality and present some illustrative examples related to these sets. Finally, we obtain the alpha-, beta- and gamma-duals of the sets with respect to the level sets.

We define the sets $bs(F)$, $cs(F)$, $cs_0(F)$ and $bv(F)$ consisting of the sets of all bounded, convergent, null series and the set of bounded variation fuzzy valued sequences, respectively, that is

$$bs(F) := \left\{ u = (u_k) \in \omega(F) : \sup_{n \in \mathbb{N}_0} D \left(\sum_{k=0}^{n} u_k, \overline{0} \right) < \infty \right\},$$

$$cs(F) := \left\{ u = (u_k) \in \omega(F) : \exists l \in L(\mathbb{R}) \ni \lim_{n \to \infty} D \left(\sum_{k=0}^{n} u_k, l \right) = 0 \right\},$$

$$cs_0(F) := \left\{ u = (u_k) \in \omega(F) : \lim_{n \to \infty} D \left(\sum_{k=0}^{n} u_k, \overline{0} \right) = 0 \right\},$$

$$bv(F) := \left\{ u = (u_k) \in \omega(F) : \sum_{k} D \left[(\triangle u)_k, \overline{0} \right] < \infty \right\},$$

where $\triangle$ denotes the forward difference operator, that is, $(\triangle u)_k = u_k - u_{k+1}$ for all $k \in \mathbb{N}_0$. Define D_∞ by means of the Hausdorff metric d on the space $X(F)$ by

$$D_\infty : X(F) \times X(F) \longrightarrow \mathbb{R} \qquad\qquad (6.9.1)$$

$$(u, v) \longrightarrow D_\infty(u, v) := \sup_{n \in \mathbb{N}_0} \sum_{k=0}^{n} D(u_k, v_k)$$

$$= \sup_{n \in \mathbb{N}_0} \sum_{k=0}^{n} \sup_{\lambda \in [0,1]} d([u_k]_\lambda, [v_k]_\lambda);$$

where $u = (u_k)$, $v = (v_k) \in X(F)$ and here and after X denotes any of the sets bs, cs or cs_0. Then, we can show that $bs(F)$, $cs(F)$ and $cs_0(F)$ are complete metric spaces with the metric D_∞. $bv(F)$ is complete metric space with the metric $D_\triangle$ defined by

$$D_\triangle(u, v) := \sum_{k} D \left[(\triangle u)_k, (\triangle v)_k \right] = \sum_{k} \sup_{\lambda \in [0,1]} \{ d([(\triangle u)_k]_\lambda, [(\triangle v)_k]_\lambda) \},$$

where $u = (u_k)$, $v = (v_k)$ are the elements of the set $bv(F)$.

Theorem 6.9.1. (cf. [45]) *The following statements hold:*

(i) $\overline{0}$ *is neutral element with respect to* $+$, *i.e.,* $u + \overline{0} = \overline{0} + u = u$ *for all* $u \in L(\mathbb{R})$.

(ii) *With respect to* $\overline{0}$, *non of* $u \neq \overline{r}, r \in \mathbb{R}$ *has opposite in* $L(\mathbb{R})$.

(iii) *For any* $\alpha, \beta \in \mathbb{R}$ *with* $\alpha, \beta \geq 0$ *or* $\alpha, \beta \leq 0$, *and any* $u \in L(\mathbb{R})$, *we have* $(\alpha + \beta)u = \alpha u + \beta u$.
For any $\alpha, \beta \in \mathbb{R}$, *the above property does not hold.*

(iv) *For any* $\alpha \in \mathbb{R}$ *and any* $u, v \in L(\mathbb{R})$, *we have* $\alpha(u + v) = \alpha u + \alpha v$.

(v) *For any* $\alpha, \beta \in \mathbb{R}$ *and any* $u \in L(\mathbb{R})$, *we have* $\alpha(\beta u) = (\alpha \beta)u$.

6.9.1 Completeness of the Sets of Bounded, Convergent and Null Series of Fuzzy Numbers with the Level Sets

In this subsection, we emphasize on the completeness of the metric spaces $bs(F)$, $cs(F)$, $cs_0(F)$ and $bv(F)$.

Proposition 6.9.2. [106, Proposition 3.1] *Define D_∞ on the space $X(F)$ by (6.9.1). Then, $(X(F), D_\infty)$ is a metric space.*

Proof. Let $u = (u_k)$, $v = (v_k) \in X(F)$.

(M1) It is immediate for all $\lambda \in [0,1]$ that

$$D_\infty(u,v) = 0 \quad \Leftrightarrow \quad \sup_{n \in \mathbb{N}_0} \sum_{k=0}^{n} \sup_{\lambda \in [0,1]} \left\{ d([u_k]_\lambda, [v_k]_\lambda) \right\} = 0$$

$$\Leftrightarrow \quad \sup_{n \in \mathbb{N}_0} \sup_{\lambda \in [0,1]} \max \left\{ \left| \sum_{k=0}^{n} (u_k)_\lambda^- - \sum_{k=0}^{n} (v_k)_\lambda^- \right|, \right.$$
$$\left. \left| \sum_{k=0}^{n} (u_k)_\lambda^+ - \sum_{k=0}^{n} (v_k)_\lambda^+ \right| \right\} = 0$$

$$\Leftrightarrow \quad \sum_{k=0}^{n} (u_k)_\lambda^- = \sum_{k=0}^{n} (v_k)_\lambda^- \quad \text{and} \quad \sum_{k=0}^{n} (u_k)_\lambda^+ = \sum_{k=0}^{n} (v_k)_\lambda^+$$

$$\Leftrightarrow \quad \sum_{k=0}^{n} [u_k]_\lambda = \sum_{k=0}^{n} [v_k]_\lambda \Leftrightarrow u_k = v_k \quad \text{for all} \quad k \in \mathbb{N}_0 \Leftrightarrow u = v.$$

(M2) One can easily see that

$$D_\infty(u,v) = \sup_{n \in \mathbb{N}_0} \sum_{k=0}^{n} D(u_k, v_k) = \sup_{n \in \mathbb{N}_0} \sum_{k=0}^{n} D(v_k, u_k) = D_\infty(v, u).$$

(M3) Let $u = (u_k)$, $v = (v_k)$, $w = (w_k) \in X(F)$ and by taking into account the triangle inequality and the condition

$$\max\{a+c, b+d\} \le \max\{a,b\} + \max\{c,d\} \quad \text{for all} \quad a,b,c,d > 0, \quad (6.9.2)$$

we observe that

$$D_\infty(u,w) = \sup_{n \in \mathbb{N}_0} \sum_{k=0}^{n} \sup_{\lambda \in [0,1]} \left\{ d([u_k]_\lambda, [w_k]_\lambda) \right\}$$

$$\le \sup_{n \in \mathbb{N}_0} \sup_{\lambda \in [0,1]} \max \left\{ \left| \sum_{k=0}^{n} (u_k)_\lambda^- - \sum_{k=0}^{n} (w_k)_\lambda^- \right|, \left| \sum_{k=0}^{n} (u_k)_\lambda^+ - \sum_{k=0}^{n} (w_k)_\lambda^+ \right| \right\}$$

$$\le \sup_{n \in \mathbb{N}_0} \sup_{\lambda \in [0,1]} \max \left\{ \left| \sum_{k=0}^{n} (u_k)_\lambda^- - \sum_{k=0}^{n} (v_k)_\lambda^- \right|, \left| \sum_{k=0}^{n} (u_k)_\lambda^+ - \sum_{k=0}^{n} (v_k)_\lambda^+ \right| \right\}$$

$$+ \sup_{n \in \mathbb{N}_0} \sup_{\lambda \in [0,1]} \max \left\{ \left| \sum_{k=0}^{n} (v_k)_\lambda^- - \sum_{k=0}^{n} (w_k)_\lambda^- \right|, \left| \sum_{k=0}^{n} (v_k)_\lambda^+ - \sum_{k=0}^{n} (w_k)_\lambda^+ \right| \right\}$$

$$= D_\infty(u,v) + D_\infty(v,w),$$

where

$$a = \left| \sum_{k=0}^{n} (u_k)_{\lambda}^{-} - \sum_{k=0}^{n} (v_k)_{\lambda}^{-} \right|, \quad b = \left| \sum_{k=0}^{n} (u_k)_{\lambda}^{+} - \sum_{k=0}^{n} (v_k)_{\lambda}^{+} \right|,$$

$$c = \left| \sum_{k=0}^{n} (v_k)_{\lambda}^{-} - \sum_{k=0}^{n} (w_k)_{\lambda}^{-} \right|, \quad d = \left| \sum_{k=0}^{n} (v_k)_{\lambda}^{+} - \sum_{k=0}^{n} (w_k)_{\lambda}^{+} \right|$$

for all $\lambda \in [0,1]$.

Since (M1)-(M3) are satisfied, $(X(F), D_{\infty})$ is a metric space. $\qquad\square$

By the following examples, we calculate the distance function for the spaces $bs(F)$ and $cs(F)$ with respect to the level sets.

Example 6.9.3. [106, Example 3.1] *Consider the sequences $u = (u_k)$ and $v = (v_k)$ defined by the triangular fuzzy numbers as*

$$u_k(t) := \begin{cases} k(k+1)t - 1 & , \quad \dfrac{1}{k(k+1)} \le t \le \dfrac{2}{k(k+1)}, \\ 3 - k(k+1)t & , \quad \dfrac{2}{k(k+1)} < t \le \dfrac{3}{k(k+1)}, \\ 0 & , \quad otherwise, \end{cases}$$

$$v_k(t) := \begin{cases} (k+1)^2 t - 1 & , \quad \dfrac{1}{(k+1)^2} \le t \le \dfrac{2}{(k+1)^2}, \\ 3 - (k+1)^2 t & , \quad \dfrac{2}{(k+1)^2} < t \le \dfrac{3}{(k+1)^2}, \\ 0 & , \quad otherwise \end{cases}$$

for all $k \in \mathbb{N}_0$. It is trivial that $u_k^{-}(\lambda) = (\lambda+1)/[k(k+1)]$ and $u_k^{+}(\lambda) = (3-\lambda)/[k(k+1)]$ for all $\lambda \in [0,1]$. Therefore, we see that $\sum_k (u_k)_{\lambda}^{-} = \lambda + 1$ and $\sum_k (u_k)_{\lambda}^{+} = 3 - \lambda$. Then, it is conclude that $(u_k) \in bs(F)$. Similarly, $v_k^{-}(\lambda) = (\lambda+1)/(k+1)^2$ and $v_k^{+}(\lambda) = (3-\lambda)/(k+1)^2$ for all $\lambda \in [0,1]$. It is clear that $\sum_k (v_k)_{\lambda}^{-} = [(\lambda+1)\pi^2]/6$ and $\sum_k (v_k)_{\lambda}^{+} = [(3-\lambda)\pi^2]/6$. Then, $(v_k) \in bs(F)$. Now, we can calculate the distance between the sequences $u = (u_k)$ and $v = (v_k)$ in $bs(F)$ that

$$\begin{aligned} D_{\infty}(u,v) &= \sup_{n \in \mathbb{N}_0} \sum_{k=0}^{n} \sup_{\lambda \in [0,1]} d([u_k]_{\lambda}, [v_k]_{\lambda}) \\ &= \sup_{\lambda \in [0,1]} \max \left\{ \sup_{n \in \mathbb{N}_0} \left| \sum_{k=0}^{n} (u_k)_{\lambda}^{-} - \sum_{k=0}^{n} (v_k)_{\lambda}^{-} \right|, \sup_{n \in \mathbb{N}_0} \left| \sum_{k=0}^{n} (u_k)_{\lambda}^{+} - \sum_{k=0}^{n} (v_k)_{\lambda}^{+} \right| \right\} \\ &= \sup_{\lambda \in [0,1]} \max \left\{ \left| \lambda + 1 - \frac{(\lambda+1)\pi^2}{6} \right|, \left| 3 - \lambda - \frac{(3-\lambda)\pi^2}{6} \right| \right\} \\ &= \sup_{\lambda \in [0,1]} \left| (3-\lambda)(1 - \frac{\pi^2}{6}) \right| \cong \frac{3}{2}. \end{aligned}$$

Example 6.9.4. [106, Example 3.2] *Consider the sequences $u = (u_k)$ and $v = (v_k)$ defined by the trapezoidal fuzzy numbers as*

$$
u_k(t) := \begin{cases}
2^k t - 1 & , \quad \dfrac{1}{2^k} \le t \le \dfrac{2}{2^k}, \\[2mm]
1 & , \quad \dfrac{2}{2^k} < t \le \dfrac{4}{2^k}, \\[2mm]
2 - 2^{k-2} t & , \quad \dfrac{4}{2^k} < t \le \dfrac{8}{2^k}, \\[2mm]
0 & , \quad otherwise,
\end{cases}
$$

$$
v_k(t) := \begin{cases}
6k(k+1)t - 3 & , \quad \dfrac{1}{2k(k+1)} \le t \le \dfrac{2}{3k(k+1)}, \\[2mm]
1 & , \quad \dfrac{2}{3k(k+1)} < t \le \dfrac{3}{4k(k+1)}, \\[2mm]
4 - 4k(k+1)t & , \quad \dfrac{3}{4k(k+1)} < t \le \dfrac{1}{k(k+1)}, \\[2mm]
0 & , \quad otherwise
\end{cases}
$$

for all $k \in \mathbb{N}_0$. It is obvious that $u_k^-(\lambda) = (\lambda+1)/2^k$ and $u_k^+(\lambda) = 4(2-\lambda)/2^k$ for all $\lambda \in [0,1]$. Then, $\sum_k (u_k)_\lambda^- = 2(\lambda+1)$ and $\sum_k (u_k)_\lambda^+ = 8(2-\lambda)$. Similarly, $v_k^-(\lambda) = (\lambda+3)/[6k(k+1)]$ and $v_k^+(\lambda) = (4-\lambda)/[4k(k+1)]$ for all $\lambda \in [0,1]$. Then, $\sum_k (v_k)_\lambda^- = (\lambda+3)/6$ and $\sum_k (v_k)_\lambda^+ = (4-\lambda)/4$. Now, we can calculate the distance between the sequences $u = (u_k)$ and $v = (v_k)$ in $cs(F)$ that

$$
D_\infty(u,v) = \sup_{n \in \mathbb{N}_0} \sup_{\lambda \in [0,1]} d\left(\sum_{k=0}^n [u_k]_\lambda, \sum_{k=0}^n [v_k]_\lambda \right)
$$

$$
= \sup_{\lambda \in [0,1]} \max \left\{ \sup_{n \in \mathbb{N}_0} \left| \sum_{k=0}^n (u_k)_\lambda^- - \sum_{k=0}^n (v_k)_\lambda^- \right|, \sup_{n \in \mathbb{N}_0} \left| \sum_{k=0}^n (u_k)_\lambda^+ - \sum_{k=0}^n (v_k)_\lambda^+ \right| \right\}
$$

$$
= \sup_{\lambda \in [0,1]} \max \left\{ \left| \frac{11\lambda + 9}{6} \right|, \left| \frac{60 - 31\lambda}{4} \right| \right\} = \sup_{\lambda \in [0,1]} \left| \frac{60 - 31\lambda}{4} \right| = 15.
$$

Theorem 6.9.5. [106, Theorem 3.2] *The space $(X(F), D_\infty)$ is complete.*

Proof. Since the proof is similar for the spaces $cs(F)$ and $cs_0(F)$, we prove the theorem only for the space $bs(F)$. Let (x_m) be any Cauchy sequence in the space $bs(F)$, where $x_m = \left\{ x_0^{(m)}, x_1^{(m)}, x_2^{(m)}, \dots \right\}$. Then, for every $\epsilon > 0$, there exists $n_0 \in \mathbb{N}_0$ such that for all $m, r > n_0$,

$$
D_\infty(x_m, x_r) = \sup_{n \in \mathbb{N}_0} D\left[\sum_{k=0}^n x_k^{(m)}, \sum_{k=0}^n x_k^{(r)} \right] < \epsilon.
$$

A fortiori, for every fixed $k \in \mathbb{N}_0$ and for $m, r > n_0$

$$
D\left[\sum_{k=0}^n x_k^{(m)}, \sum_{k=0}^n x_k^{(r)} \right] < \epsilon. \tag{6.9.3}
$$

Hence, for every fixed $k \in \mathbb{N}_0$, by taking into account the completeness of the space $(L(\mathbb{R}), D)$, the sequence $\left\{x_k^{(m)}\right\}$ is a Cauchy sequence and so, is convergent. Now, we suppose that $x_k^{(m)} \to x_k$, as $m \to \infty$, and $x = (x_0, x_1, x_2, \ldots)$. We have to show that

$$\lim_{m \to \infty} D_\infty(x_m, x) = 0 \quad \text{and} \quad x \in bs(F).$$

Letting $r \to \infty$ in (6.9.3), we get for all $m, n \in \mathbb{N}_0$ with $m > n_0$ that

$$D\left[\sum_{k=0}^{n} x_k^{(m)}, \sum_{k=0}^{n} x_k\right] < \epsilon. \tag{6.9.4}$$

Since the sequence $\left\{x_k^{(m)}\right\}$ is in $bs(F)$, there exists a non-negative number M such that $D\left[\sum_{k=0}^{n} x_k^{(m)}, \overline{0}\right] < M$ for all $k \in \mathbb{N}_0$, where

$$M = \max\left\{\sup_{n \in \mathbb{N}_0} \sup_{\lambda \in [0,1]} \left|\sum_{k=0}^{n}(x_k^{(m)})_\lambda^-\right|, \sup_{n \in \mathbb{N}_0} \sup_{\lambda \in [0,1]} \left|\sum_{k=0}^{n}(x_k^{(m)})_\lambda^+\right|\right\}.$$

Thus, (6.9.4) gives together with the triangle inequality for $m > n_0$ that

$$D\left(\sum_{k=0}^{n} x_k, \overline{0}\right) \leq D\left[\sum_{k=0}^{n} x_k, \sum_{k=0}^{n} x_k^{(m)}\right] + D\left[\sum_{k=0}^{n} x_k^{(m)}, \overline{0}\right] \leq \epsilon + M. \tag{6.9.5}$$

It is clear that (6.9.5) holds for every fixed $k \in \mathbb{N}_0$ whose right-hand side does not involve k. Hence, $x \in bs(F)$. Also from (6.9.4), we obtain for $m > n_0$ that

$$D_\infty(x_m, x) = \sup_{n \in \mathbb{N}_0} D\left[\sum_{k=0}^{n} x_k^{(m)}, \sum_{k=0}^{n} x_k\right] \leq \epsilon.$$

This shows that $D_\infty(x_m, x) \to 0$, as $m \to \infty$. Therefore, the space $(bs(F), D_\infty)$ is complete. $\square$

Proposition 6.9.6. [106, Proposition 3.3] $(bv(F), D_\triangle)$ *is a metric space.*

Proof. Since the metric axioms $(M1)$ and $(M2)$ are easily satisfied, we omit details.

Let $u = (u_k)$, $v = (v_k)$, $w = (w_k) \in bv(F)$ and by taking into account the triangle inequality with the condition (6.9.2), we see that

$$
\begin{aligned}
D_\triangle(u, w) &\leq \sup_{\lambda \in [0,1]} \max\left\{\left|\sum_k (\triangle u_k)_\lambda^- - \sum_k (\triangle w_k)_\lambda^-\right|, \left|\sum_k (\triangle u_k)_\lambda^+ - \sum_k (\triangle w_k)_\lambda^+\right|\right\} \\
&\leq \sup_{\lambda \in [0,1]} \max\left\{\left|\sum_k (\triangle u_k)_\lambda^- - \sum_k (\triangle v_k)_\lambda^-\right|, \left|\sum_k (\triangle u_k)_\lambda^+ - \sum_k (\triangle v_k)_\lambda^+\right|\right\} \\
&\quad + \sup_{\lambda \in [0,1]} \max\left\{\left|\sum_k (\triangle v_k)_\lambda^- - \sum_k (\triangle w_k)_\lambda^-\right|, \left|\sum_k (\triangle v_k)_\lambda^+ - \sum_k (\triangle w_k)_\lambda^+\right|\right\} \\
&= D_\triangle(u, v) + D_\triangle(v, w),
\end{aligned}
$$

where

$$a = \left| \sum_k (\triangle u_k)^-_\lambda - \sum_k (\triangle v_k)^-_\lambda \right|, \quad b = \left| \sum_k (\triangle u_k)^+_\lambda - \sum_k (\triangle v_k)^+_\lambda \right|,$$

$$c = \left| \sum_k (\triangle v_k)^-_\lambda - \sum_k (\triangle w_k)^-_\lambda \right|, \quad d = \left| \sum_k (\triangle v_k)^+_\lambda - \sum_k (\triangle w_k)^+_\lambda \right|.$$

Hence, triangle inequality holds. Since the metric axioms (M1)-(M3) are satisfied, $(bv(F), D_\triangle)$ is a metric space. $\qquad \square$

Example 6.9.7. [106, Example 3.3] *Consider the membership functions $u_k(t)$ and $v_k(t)$ defined by the triangular fuzzy numbers as*

$$u_k(t) := \begin{cases} kt - 1 &, \quad \dfrac{1}{k} \le t \le \dfrac{2}{k}, \\ 3 - kt &, \quad \dfrac{2}{k} < t \le \dfrac{3}{k}, \\ 0 &, \quad otherwise, \end{cases}$$

$$v_k(t) := \begin{cases} \dfrac{12^k t - 3^k}{4^k - 3^k} &, \quad \dfrac{1}{4^k} \le t \le \dfrac{1}{3^k}, \\ \dfrac{3^k - 6^k t}{3^k - 2^k} &, \quad \dfrac{1}{3^k} < t \le \dfrac{1}{2^k}, \\ 0 &, \quad otherwise \end{cases}$$

for all $k \in \mathbb{N}_0$. Then, $u_k^-(\lambda) = (\lambda+1)/k$ and $u_k^+(\lambda) = (3-\lambda)/k$. Additionally, since

$$[(\triangle u)_k]_\lambda = [(u_k)^-_\lambda - (u_{k+1})^-_\lambda, (u_k)^+_\lambda - (u_{k+1})^+_\lambda] = \left[\frac{\lambda + 1}{k(k+1)}, \frac{3 - \lambda}{k(k+1)} \right],$$

$(u_k) \in bv(F)$. Similarly, $v_k^-(\lambda) = 3^{-k}\lambda - (\lambda - 1)4^{-k}$ and $v_k^+(\lambda) = 3^{-k}\lambda + (1-\lambda)2^{-k}$ then,

$$[(\triangle v)_k]_\lambda = [(v_k)^-_\lambda - (v_{k+1})^-_\lambda, (v_k)^+_\lambda - (v_{k+1})^+_\lambda] = \left[\frac{2\lambda}{3^{k+1}} - \frac{3(\lambda - 1)}{4^{k+1}}, \frac{2\lambda}{3^{k+1}} + \frac{1 - \lambda}{2^{k+1}} \right]$$

and $(v_k) \in bv(F)$. Now, we can calculate that

$$D(u,v) = \sum_k D[(\triangle u)_k, (\triangle v)_k] = \sup_{\lambda \in [0,1]} \sum_k d\{[(\triangle u)_k]_\lambda, [(\triangle v)_k]_\lambda\}$$

$$= \sup_{\lambda \in [0,1]} \max \left\{ \left| \sum_k [(\triangle u)_k]^-_\lambda - \sum_k [(\triangle v)_k]^-_\lambda \right|, \left| \sum_k [(\triangle u)_k]^+_\lambda - \sum_k [(\triangle v)_k]^+_\lambda \right| \right\}$$

$$= \sup_{\lambda \in [0,1]} \max \left\{ \left| \sum_k \frac{\lambda + 1}{k(k+1)} - \sum_k \left[\frac{2\lambda}{3^{k+1}} - \frac{3(\lambda - 1)}{4^{k+1}} \right] \right|, \right.$$

$$\left. \left| \sum_k \frac{3 - \lambda}{k(k+1)} - \sum_k \left(\frac{2\lambda}{3^{k+1}} + \frac{1 - \lambda}{2^{k+1}} \right) \right| \right\}$$

$$= \sup_{\lambda \in [0,1]} \max \{|\lambda|, |2 - \lambda|\}$$

$$= \sup_{\lambda \in [0,1]} \{2 - \lambda\} = 2.$$

Theorem 6.9.8. [106, Theorem 3.4] $(bv(F), D_\triangle)$ *is a complete metric space.*

Proof. Let (u_m) be any Cauchy sequence in the space $bv(F)$, where $u_m = \left\{u_0^{(m)}, u_1^{(m)}, u_2^{(m)}, \ldots\right\}$ for all $m \in \mathbb{N}_0$. Then, for every $\varepsilon > 0$, there exists $N(\varepsilon) \in \mathbb{N}_0$ such that for all $m, r > N(\varepsilon)$,

$$D_\triangle(u_m, u_r) = \sum_k D\left[(\triangle u)_k^{(m)}, (\triangle u)_k^{(r)}\right] < \varepsilon.$$

A fortiori, for every fixed $k \in \mathbb{N}_0$ and for $m, r > N(\varepsilon)$

$$D\left[(\triangle u)_k^{(m)}, (\triangle u)_k^{(r)}\right] < \varepsilon. \tag{6.9.6}$$

Hence, for every fixed $k \in \mathbb{N}_0$, by taking into account the completeness of the space $(L(\mathbb{R}), D)$, the sequence $\{(\triangle u)_k^{(m)}\}$ is a Cauchy sequence and so, it converges. Now, we suppose that $(\triangle u_k)^{(m)} \to (\triangle u)_k$, as $m \to \infty$, $\triangle u = \{(\triangle u)_0, (\triangle u)_1, \ldots, (\triangle u)_k, \ldots\}$. We have to show that $D_\triangle(u_m, u) \to 0$, as $m \to \infty$, and $u = (u_k) \in bv(F)$.

We have from (6.9.6) for each $j \in \mathbb{N}_0$ and $m, r > N(\varepsilon)$ that

$$\sum_{k=0}^{j} D\left[(\triangle u)_k^{(m)}, (\triangle u)_k^{(r)}\right] \leq D_\triangle(u_m, u_r) < \varepsilon. \tag{6.9.7}$$

Take any $m > N(\varepsilon)$. By letting firstly $r \to \infty$ and next $j \to \infty$ in (6.9.7), we obtain $D_\triangle(u_m, u) \leq \varepsilon$. Since the sequence (u_m) is in $bv(F)$, there exists a non-negative number M such that

$$D_\triangle(u_m, \overline{0}) < M = \sup_{\lambda \in [0,1]} \max\left\{\left|\sum_k \left((\triangle u)_k^{(m)}\right)_\lambda^-\right|, \left|\sum_k \left((\triangle u)_k^{(m)}\right)_\lambda^+\right|\right\}$$

for all $m \in \mathbb{N}_0$. By using (6.9.7) and Minkowski's inequality we see that

$$\sum_{k=0}^{j} D[(\triangle u)_k, \overline{0}] \leq D_\triangle(u_m, u) + D_\triangle(u_m, \overline{0}) \leq \varepsilon + M$$

for each $j \in \mathbb{N}_0$, which implies that $u \in bv(F)$. Since $D_\triangle(u_m, u) \leq \epsilon$ for all $m > N(\varepsilon)$, it follows that $D_\triangle(u_m, u) \to 0$, as $m \to \infty$. Since (u_m) is an arbitrary Cauchy sequence, the space $(bv(F), D_\triangle)$ is complete. This step completes the proof. $\square$

Theorem 6.9.9. [213, Theorem 7] *Define the relation D_p on the space $bv_p(F)$ by*

$$D_p : bv_p(F) \times bv_p(F) \longrightarrow \mathbb{R}^+$$
$$(u, v) \longmapsto D_p(u, v) = \left\{\sum_k D\left[(\triangle u)_k, (\triangle v)_k\right]^p\right\}^{1/p},$$

where $1 \leq p < \infty$, $u = (u_k), v = (v_k) \in bv_p(F)$ and the difference sequence $\triangle u$ is defined by the backward difference matrix $\triangle$ as $\triangle u = \{(\triangle u)_k\}_{k \in \mathbb{N}_0} = (u_k - u_{k-1})_{k \in \mathbb{N}_0}$ with $u_{-1} = 0$. Then, $(bv_p(F), D_p)$ is a complete metric space.

6.9.2 The Duals of the Sets of Sequences of Fuzzy Numbers with the Level Sets

We define the classes $bv_p(F)$ and $bv_\infty(F)$ consisting of the p-bounded variation and bounded difference fuzzy valued sequences, i.e.,

$$bv_p(F) := \left\{ u = (u_k) \in \omega(F) : \sum_{k=0}^{\infty} \left\{ D\left((\Delta u)_k, \bar{0}\right) \right\}^p < \infty \right\},$$

$$bv_\infty(F) := \left\{ u = (u_k) \in \omega(F) : \sup_{k \in \mathbb{N}_0} D\left((\Delta u)_k, \bar{0}\right) < \infty \right\},$$

where the difference sequence $\Delta u = \{(\Delta u)_k\}_{k \in \mathbb{N}_0}$ is defined as in Theorem 6.9.9 and $1 \leq p < \infty$. Additionally, the space $bv_0(F)$ is the intersection of the spaces $bv(F)$ and $c_0(F)$.

Now, we give the alpha-, beta- and gamma-duals of the sets $bs(F)$, $cs(F)$, $cs_0(F)$ and $bv(F)$.

Theorem 6.9.10. [106, Theorem 4.7] *The alpha-dual of the sets $cs(F)$, $bs(F)$, $bv_1(F)$ and $bv_0(F)$ is the set $\ell_1(F)$.*

Proof. We prove the case $\{cs(F)\}^\alpha = \ell_1(F)$, and the rest can be obtained, similarly.

Let $u = (u_k) \in \ell_1(F)$. Then, $\sum_k D(u_k, \bar{0})$ converges. Therefore, we derive by using the fact given in Part (i) of Lemma 6.3.16 that

$$\sum_k D(u_k v_k, \bar{0}) \leq \sum_k D(u_k, \bar{0}) D(v_k, \bar{0}) \leq M \sum_k D(u_k, \bar{0}).$$

If we take $v = (v_k) \in cs_0(F) \subset cs(F)$, then we have $\sum_k D(u_k v_k, \bar{0}) < \infty$ which gives that $\ell_1(F) \subseteq \{cs(F)\}^\alpha$.

Conversely, suppose that $u = (u_k) \in \{cs(F)\}^\alpha$ and $v = (v_k) \in cs_0(F)$. Then, the series $\sum_k D(v_k, \bar{0})$ converges. Since

$$|(u_k v_k)_\lambda^-| \leq \sum_k D(u_k v_k, \bar{0}) \leq \sum_k D(u_k, \bar{0}) D(v_k, \bar{0})$$

$$|(u_k v_k)_\lambda^+| \leq \sum_k D(u_k v_k, \bar{0}) \leq \sum_k D(u_k, \bar{0}) D(v_k, \bar{0}),$$

Weierstrass's M Test yields that $\sum_k (u_k v_k)_\lambda^-$ and $\sum_k (u_k v_k)_\lambda^+$ converge uniformly and hence $\sum_k u_k v_k$ converges whenever $\sum_k D(u_k, \bar{0})$ converges. Therefore, we have $\{cs(F)\}^\alpha \subseteq \ell_1(F)$.

This step concludes the proof. $\qquad\square$

Theorem 6.9.11. [106, Theorem 4.8] *The following statements hold:*

(i) $\{cs(F)\}^\beta = bv_1(F)$.

(ii) $\{bv(F)\}^\beta = cs(F)$.

(iii) $\{bv_0(F)\}^\beta = bs(F)$.

(iv) $\{bs(F)\}^\beta = bv_0(F)$.

Proof. Since the other parts can be similarly proved, we consider only Part (i).

(i) Let $u = (u_k) \in \{cs(F)\}^\beta$ and $w = (w_k) \in c_0(F)$. Define the sequence $v = (v_k) \in cs(F)$ by $v_k = w_k - w_{k+1}$ for all $k \in \mathbb{N}_0$. Therefore, $\sum_k u_k v_k$ converges, but

$$\sum_{k=0}^{n} (w_k - w_{k+1}) u_k = \sum_{k=0}^{n-1} w_k (u_k - u_{k-1}) - w_{n+1} u_n \tag{6.9.8}$$

and the inclusion $\ell_1(F) \subset cs(F)$ yields that $(u_k) \in \{cs(F)\}^\beta \subset \{\ell_1(F)\}^\beta = \ell_\infty(F)$. Then, we derive by passing to limit in (6.9.8), as $n \to \infty$, that

$$\sum_k (w_k - w_{k+1}) u_k = \sum_k w_k (u_k - u_{k-1}).$$

Hence, $(u_k - u_{k-1}) \in \{c_0(F)\}^\beta = \{c_0(F)\}^\alpha = \ell_1(F)$, i.e., $u \in bv_1(F)$. Therefore, $\{cs(F)\}^\beta \subseteq bv_1(F)$.

Conversely, suppose that $u = (u_k) \in bv_1(F)$. Then, $(u_k - u_{k-1}) \in \ell_1(F)$. Further, if $v = (v_k) \in cs(F)$, the sequence (w_n) defined by $w_n = \sum_{k=0}^{n} v_k$ for all $n \in \mathbb{N}_0$, is an element of the space $c(F)$. Since $\{c(F)\}^\alpha = \ell_1(F)$, the series $\sum_k w_k (u_k - u_{k+1})$ is absolutely convergent. Also, we have

$$\left| \sum_{k=m}^{n} (w_k - w_{k-1}) u_k \right| \le \left| \sum_{k=m}^{n-1} w_k (u_k - u_{k+1}) \right| + |w_n u_n - w_{m-1} u_m|. \tag{6.9.9}$$

Since $(w_n) \in c(F)$ and $(u_k) \in bv_1(F) \subset c(F)$, the right-hand side of inequality (6.9.9) converges to zero, as $m, n \to \infty$. Hence, the series $\sum_k (w_k - w_{k-1}) u_k$ or $\sum_k u_k v_k$ converges and so, $bv_1(F) \subseteq \{cs(F)\}^\beta$. Thus, $\{cs(F)\}^\beta = bv_1(F)$. $\square$

Lemma 6.9.12. [214, Corollary 3.7] *The gamma-dual of the sets* $\ell_\infty(F)$, $c_0(F)$ *and* $\ell_p(F)$ *of sequences of fuzzy numbers is the set* $\ell_1(F)$, *where* $0 < p < 1$.

Theorem 6.9.13. [106, Theorem 4.8] *The following statements hold:*

(i) $\{cs(F)\}^\gamma = \{bs(F)\}^\gamma = bv_1(F)$.

(ii) $\{bv_0(F)\}^\gamma = \{bv_1(F)\}^\gamma = bs(F)$.

Proof. We prove only Part (i) for $\{cs(F)\}^\gamma$, since Part (ii) can be proved in a similar way.

By Theorem 6.9.11, we have $bv_1(F) \subseteq \{cs(F)\}^\beta$ and since $\{cs(F)\}^\beta \subset \{cs(F)\}^\gamma$, so $bv_1(F) \subset \{cs(F)\}^\gamma$. We need to show that $\{cs(F)\}^\gamma \subset bv_1(F)$. Let $u = (u_n) \in \{cs(F)\}^\gamma$ and $v = (v_n) \in c_0(F)$. Then, for the sequence

$(w_n) \in cs(F)$ defined by $w_n = v_n - v_{n+1}$ for all $n \in \mathbb{N}_0$, we can find a constant $K > 0$ such that $\left|\sum_{k=0}^{n} u_k w_k\right| \leq K$ for all $n \in \mathbb{N}_0$. Since $(v_n) \in c_0(F)$ and $(u_n) \in \{cs(F)\}^\gamma \subset \ell_\infty(F)$, there exists a constant $M > 0$ such that $|u_n v_n| \leq M$ for all $n \in \mathbb{N}_0$. Therefore,

$$\left|\sum_{k=0}^{n}(u_k - u_{k-1})v_k\right| \leq \left|\sum_{k=0}^{n+1} u_k(v_k - v_{k+1})\right| + |v_{n+2}u_{n+1}| \leq K + M.$$

Hence, $(u_k - u_{k-1}) \in \{c_0(F)\}^\gamma = \{c_0(F)\}^\alpha = \ell_1(F)$ from Lemma 6.9.12, i.e., $(u_n) \in bv_1(F)$. Therefore, since the inclusion $\{cs(F)\}^\gamma \subset bv_1(F)$ holds, we conclude that $\{cs(F)\}^\gamma = bv_1(F)$, as desired. □

Now, we give an example related to the alpha, beta and gamma duals of the set $cs(F)$.

Example 6.9.14. *Consider the triangular fuzzy numbers with the membership functions u_k and v_k defined by*

$$u_k(t) := \begin{cases} k(k+1)t - 1 & , \quad \dfrac{1}{k(k+1)} \leq t \leq \dfrac{2}{k(k+1)}, \\ 3 - k(k+1)t & , \quad \dfrac{2}{k(k+1)} < t \leq \dfrac{3}{k(k+1)}, \\ 0 & , \quad otherwise, \end{cases}$$

$$v_k(t) := \begin{cases} (2k-1)(2k+1)t - 1 & , \quad \dfrac{1}{(2k-1)(2k+1)} \leq t \leq \dfrac{2}{(2k-1)(2k+1)}, \\ 8 - 2t(2k-1)(2k+1) & , \quad \dfrac{2}{(2k-1)(2k+1)} < t \leq \dfrac{4}{(2k-1)(2k+1)}, \\ 0 & , \quad otherwise \end{cases}$$

for all $k \in \mathbb{N}_0$. Since $u_k^-(\lambda) = (\lambda+1)/[k(k+1)]$ and $u_k^+(\lambda) = (3-\lambda)/[k(k+1)]$, $(u_k) \in \ell_1(F)$. Similarly, since $v_k^-(\lambda) = (\lambda+1)/[(2k-1)(2k+1)]$ and $v_k^+(\lambda) = (8-\lambda)/[2(2k-1)(2k+1)]$, $\sum_k v_k = [(\lambda+1)/2, (8-\lambda)/4]$, i.e., $(v_k) \in cs(F)$. Therefore, a straightforward calculation yields that

$$\sum_k (u_k v_k)_\lambda^- = \sum_k \min\left\{(u_k)_\lambda^-(v_k)_\lambda^-, (u_k)_\lambda^-(v_k)_\lambda^+, (u_k)_\lambda^+(v_k)_\lambda^-, (u_k)_\lambda^+(v_k)_\lambda^+\right\}$$

$$= \sum_k \min\left\{\frac{(\lambda+1)^2}{k(k+1)(2k-1)(2k+1)}, \frac{(\lambda+1)(8-\lambda)}{2k(k+1)(2k-1)(2k+1)}, \right.$$

$$\left. \frac{(3-\lambda)(\lambda+1)}{2k(k+1)(2k-1)(2k+1)}, \frac{(3-\lambda)(8-\lambda)}{2k(k+1)(2k-1)(2k+1)}\right\}$$

$$= \sum_k \frac{(\lambda+1)^2}{k(k+1)(4k^2-1)}. \tag{6.9.10}$$

Then, we see for the sequences $(u_k) = \{(\lambda+1)^2/[k(k+1)(2k-1)(2k+1)]\}$ and $(v_k) = \{(\lambda+1)^2/[(2k-1)(2k+1)]\}$ of fuzzy numbers that $u_k \preceq v_k$ for all $k \in \mathbb{N}_0$. Since the series $\sum_k v_k$ converges, then the series (6.9.10) *also converges by using the Comparison Test in Part (i) of Lemma 6.3.16. Similarly,*

$$\sum_k (u_k v_k)_\lambda^+ = \sum_k \frac{(3-\lambda)(8-\lambda)}{2k(k+1)(4k^2-1)}. \tag{6.9.11}$$

Taking the sequences

$$(u_k) = \left\{ \frac{(3-\lambda)(8-\lambda)}{2k(k+1)(2k-1)(2k+1)} \right\} \quad \text{and} \quad (v_k) = \left\{ \frac{(3-\lambda)(8-\lambda)}{(2k-1)(2k+1)} \right\}$$

of fuzzy numbers, we have $u_k \preceq v_k$ for all $k \in \mathbb{N}_0$. Since the series $\sum_k v_k$ converges, then the series (6.9.11) also converges. Therefore,

$$
\begin{aligned}
\sum_k D(u_k v_k, \overline{0}) &= \sup_{\lambda \in [0,1]} \max \left\{ \left| \sum_k (u_k v_k)_\lambda^- \right|, \left| \sum_k (u_k v_k)_\lambda^+ \right| \right\} \\
&= \sup_{\lambda \in [0,1]} \max \left\{ \left| \sum_k \frac{(\lambda+1)^2}{k(k+1)(4k^2-1)} \right|, \left| \sum_k \frac{(3-\lambda)(8-\lambda)}{2k(k+1)(4k^2-1)} \right| \right\} < \infty.
\end{aligned}
$$

Hence, $(u_k v_k) \in \ell_1(F) = \{cs(F)\}^\alpha$. Additionally, by using $\sum_{k=0}^n D(u_k v_k, l) \to 0$, as $n \to \infty$, for some $l \in L(\mathbb{R})$, one can see that $(u_k v_k) \in cs(F)$. It is obvious that $(u_k) \in bv_1(F) = \{cs(F)\}^\beta = \{cs(F)\}^\gamma$. Indeed,

$$
\begin{aligned}
\sum_k D\left[\Delta(u_k), \overline{0}\right] &= \sup_{\lambda \in [0,1]} \sum_k d\left([u_k - u_{k-1}]_\lambda, \overline{0}\right) \\
&= \sup_{\lambda \in [0,1]} \max \left\{ \left| \sum_k (u_k)_\lambda^- - (u_{k-1})_\lambda^- \right|, \left| \sum_k (u_k)_\lambda^+ - (u_{k-1})_\lambda^+ \right| \right\} \\
&= \sup_{\lambda \in [0,1]} \max \left\{ \left| \sum_k \left[\frac{\lambda+1}{k(k+1)} - \frac{\lambda+1}{k(k-1)} \right] \right|, \left| \sum_k \left(\frac{3-\lambda}{k(k+1)} - \frac{3-\lambda}{k(k-1)} \right) \right| \right\} \\
&= \sup_{\lambda \in [0,1]} \max \left\{ \sum_k \left[\frac{2(\lambda+1)}{k(k+1)(k-1)} \right], \sum_k \left[\frac{2(3-\lambda)}{k(k+1)(k-1)} \right] \right\} < \infty.
\end{aligned}
$$

Therefore, $(u_k) \in bv_1(F)$.

6.10   Certain Sequence Spaces of Fuzzy Numbers Defined by a Modulus Function

In the present section, following Talo and Başar [215], we essentially deal with the metric spaces $\ell_\infty(F, f)$, $c(F, f)$, $c_0(F, f)$ and $\ell_p(F, f, s)$ of fuzzy valued sequences defined by a modulus function f which are the generalization of the metric spaces $\ell_\infty(F)$, $c(F)$, $c_0(F)$ and $\ell_p(F)$ of fuzzy valued sequences. Additionally, we state and prove some inclusion theorems related to those spaces. Finally, we establish that the spaces $\ell_\infty(F, f)$, $c_0(F, f)$ and $\ell_p(F, f, s)$ are solid as a consequence of the fact that the sets $\ell_\infty(F)$, $c_0(F)$ and $\ell_p(F)$ are solid.

The notion of modulus function was introduced by Nakano [173], as follows;

Definition 6.10.1. [215, Definition 1.6] *A function* $f : [0, \infty) \to [0, \infty)$ *is called a modulus if the following conditions hold:*

(a) $f(x) = 0$ *if and only if* $x = 0$.

(b) $f(x + y) \leq f(x) + f(y)$ *for all* $x, y \geq 0$.

(c) f *is increasing.*

(d) f *is continuous from the right at* 0.

Hence, f is continuous on the interval $[0, \infty)$.

Zadeh introduced the concept of fuzzy sets and define the fuzzy set operations, in his significant article [236]. Subsequently several authors discussed various aspects of the theory and applications of fuzzy sets such as fuzzy topological spaces, similarity relations and fuzzy orderings, fuzzy measures of fuzzy events, fuzzy mathematical programming. Especially, El Naschie [175] studied the E infinity theory which has very important applications in quantum particle physics.

6.10.1 The Spaces $\ell_\infty(F, f)$, $c(F, f)$, $c_0(F, f)$ and $\ell_p(F, f, s)$ of Sequences of Fuzzy Numbers Defined by a Modulus Function

We extend the classical metric spaces $\ell_\infty(F)$, $c(F)$, $c_0(F)$ and $\ell_p(F)$ of fuzzy valued sequences to the metric spaces $\ell_\infty(F, f)$, $c(F, f)$, $c_0(F, f)$ and $\ell_p(F, f, s)$ of fuzzy valued sequences, via a modulus function f. Throughout this subsection, we suppose that $1 \leq p < \infty$.

Let f be a modulus function. Following Talo and Başar [215], we introduce the sets $\ell_\infty(F, f)$, $c(F, f)$, $c_0(F, f)$ and $\ell_p(F, f, s)$ of fuzzy valued sequences defined by a modulus function f by

$$\ell_\infty(F, f) \quad := \quad \left\{ x = (x_k) \in \omega(F) : \sup_{k \in \mathbb{N}_0} f[\overline{d}(x_k, \overline{0})] < \infty \right\},$$

$$c(F, f) \quad := \quad \left\{ x = (x_k) \in \omega(F) : \exists l \in L(\mathbb{R}) \text{ such that } \lim_{k \to \infty} f[\overline{d}(x_k, l)] = 0 \right\},$$

$$c_0(F, f) \quad := \quad \left\{ x = (x_k) \in \omega(F) : \lim_{k \to \infty} f[\overline{d}(x_k, \overline{0})] = 0 \right\},$$

$$\ell_p(F, f, s) \quad := \quad \left\{ x = (x_k) \in \omega(F) : \sum_{k \in \mathbb{N}_1} \frac{\{f[\overline{d}(x_k, \overline{0})]\}^p}{k^s} < \infty \right\}, \ (s \geq 0).$$

Now, we may begin with the following theorem which is essential in this subsection:

Theorem 6.10.2. [215, Theorem 2.1] *The sets* $\ell_\infty(F, f)$, $c(F, f)$, $c_0(F, f)$ *and* $\ell_p(F, f, s)$ *of fuzzy valued sequences defined by a modulus function* f *are closed under the coordinatewise addition and scalar multiplication.*

Proof. Since it is not hard to show that the sets $\ell_\infty(F, f)$, $c(F, f)$, $c_0(F, f)$ and $\ell_p(F, f, s)$ are closed with respect to the coordinatewise addition and scalar multiplication, we omit the detail. □

Theorem 6.10.3. [215, Theorem 2.2] *The sets* $\ell_\infty(F, f)$, $c(F, f)$, $c_0(F, f)$ *and* $\ell_p(F, f, s)$ *of fuzzy valued sequences defined by a modulus function* f *are complete metric spaces with respect to the metrics* $\overline{D}_\infty$ *and* $\overline{D}_p$ *defined by*

$$\overline{D}_\infty(x, y) := \sup_{k \in \mathbb{N}_0} f[\overline{d}(x_k, y_k)],$$

$$\overline{D}_p(x, y) := \left\{ \sum_{k \in \mathbb{N}_1} \frac{\{f[\overline{d}(x_k, y_k)]\}^p}{k^s} \right\}^{1/p},$$

respectively; where $x = (x_k)$, $y = (y_k)$ *are the elements of the sets* $\ell_\infty(F, f)$, $c(F, f)$, $c_0(F, f)$ *and* $\ell_p(F, f, s)$.

Proof. Since the proof is analogue for the spaces $\ell_\infty(F, f)$, $c(F, f)$ and $\ell_p(F, f, s)$, we consider only the space $c_0(F, f)$. One can easily establish that $\overline{D}_\infty$ defines a metric on $c_0(F, f)$ which is a routine verification, so we leave it to the reader.

It remains to prove the completeness of the space $c_0(F, f)$. Let (x^i) be any Cauchy sequence in the space $c_0(F, f)$, where $x^i := \left\{ x_0^{(i)}, x_1^{(i)}, x_2^{(i)}, \ldots \right\}$. Then, for a given $\varepsilon > 0$ there exists a positive integer $n_0(\varepsilon)$ such that

$$\overline{D}_\infty(x^i, x^j) = \sup_{k \in \mathbb{N}_0} f\left[\overline{d}\left(x_k^{(i)}, x_k^{(j)} \right) \right] < \varepsilon \tag{6.10.1}$$

for all $i, j \geq n_0(\varepsilon)$. We obtain for each fixed $k \in \mathbb{N}_0$ from (6.10.1) that

$$f\left[\overline{d}\left(x_k^{(i)}, x_k^{(j)} \right) \right] < \varepsilon \tag{6.10.2}$$

for every $i, j \geq n_0(\varepsilon)$. (6.10.2) means that

$$\lim_{i,j \to \infty} f\left[\overline{d}\left(x_k^{(i)}, x_k^{(j)} \right) \right] = 0. \tag{6.10.3}$$

Since f is continuous, we see from (6.10.3) that

$$f\left[\lim_{i,j \to \infty} \overline{d}\left(x_k^{(i)}, x_k^{(j)} \right) \right] = 0. \tag{6.10.4}$$

Therefore, since f is a modulus function one can derive by (6.10.4) that

$$\lim_{i,j \to \infty} \overline{d}\left(x_k^{(i)}, x_k^{(j)} \right) = 0, \tag{6.10.5}$$

which means that $\left\{ x_k^{(i)} \right\}$ is a Cauchy sequence in $L(\mathbb{R})$ for every fixed $k \in \mathbb{N}_0$. Since $L(\mathbb{R})$ is complete, it converges, say $x_k^{(i)} \to x_k$, as $i \to \infty$. Using these infinitely many limits, we define the sequence $x = (x_0, x_1, x_2, \ldots)$. Then, by letting $j \to \infty$ and taking supremum over $k \in \mathbb{N}_0$ in (6.10.2), we obtain $\overline{D}_\infty(x^i, x) \leq \varepsilon/2$. Since $x^i = \{x_k^{(i)}\} \in c_0(F, f)$ for each $i \in \mathbb{N}_0$, there exists $k_0(\varepsilon) \in \mathbb{N}_0$ such that $f\left[\overline{d}\left(x_k^{(i)}, \overline{0}\right)\right] \leq \dfrac{\varepsilon}{2}$ for every $k \geq k_0(\varepsilon)$ and for each fixed $i \in \mathbb{N}_0$. Therefore, since

$$f\left[\overline{d}(x_k, \overline{0})\right] < f\left[\overline{d}\left(x_k^{(i)}, x_k\right)\right] + f\left[\overline{d}\left(x_k^{(i)}, \overline{0}\right)\right]$$

holds by triangle inequality for all $i, k \in \mathbb{N}_0$ for fixed $i \geq n_0(\varepsilon)$, we have

$$f\left[\overline{d}(x_k, \overline{0})\right] \leq \varepsilon \quad \text{for all} \quad k \geq k_0(\varepsilon).$$

This shows that $x \in c_0(F, f)$. Since (x^i) was an arbitrary Cauchy sequence, the space $c_0(F, f)$ is complete.

This step concludes the proof. ☐

Theorem 6.10.4. [215, Theorem 2.3] *Let $\lambda(F, f)$ denotes anyone of the spaces $\ell_\infty(F, f)$, $c(F, f)$, $c_0(F, f)$ and f_1, f_2 be two modulus functions. Then, the following inclusion relations hold:*

(a) $\lambda(F, f_1) \cap \lambda(F, f_2) \subseteq \lambda(F, f_1 + f_2)$.

(b) $\lambda(F, f_1) \subseteq \lambda(F, f_2 \circ f_1)$.

(c) If $f_1(t) \leq f_2(t)$ for all $t \in [0, \infty)$, $\lambda(F, f_2) \subseteq \lambda(F, f_1)$.

Proof. (a) Let $x = (x_k) \in \lambda(F, f_1) \cap \lambda(F, f_2)$. Since

$$(f_1 + f_2)[\overline{d}(x_k, \overline{0})] = f_1[\overline{d}(x_k, \overline{0})] + f_2[\overline{d}(x_k, \overline{0})], \tag{6.10.6}$$

one can see by passing to limit, as $k \to \infty$, and taking supremum over $k \in \mathbb{N}_0$ from (6.10.6) that $x \in \lambda(F, f_1 + f_2)$, where $\lambda \in \{\ell_\infty, c, c_0\}$.

(b) Let $x = (x_k) \in \lambda(F, f_1)$. Since f_2 is continuous, there exists an $\eta > 0$ such that $f_2(\eta) = \varepsilon$ for all $\varepsilon > 0$. Since $x = (x_k) \in \lambda(F, f_1)$ there exists an $k_0 \in \mathbb{N}_0$ such that $f_1[\overline{d}(x_k, \overline{0})] < \eta$ for all $k \geq k_0$. Therefore, one can derive by applying f_2 that $f_2\{f_1[\overline{d}(x_k, \overline{0})]\} < f_2(\eta) = \varepsilon$, i.e., $x \in \lambda(F, f_2 \circ f_1)$.

(c) Since $f_1(t) \leq f_2(t)$ for all $t \in [0, \infty)$, we have $f_1[\overline{d}(x_k, \overline{0})] \leq f_2[\overline{d}(x_k, \overline{0})]$. This leads us to the consequence that $x \in \lambda(F, f_2)$ which implies that $x \in \lambda(F, f_1)$, as expected. ☐

Lemma 6.10.5. (cf. [145]) *Let f_1 and f_2 be two modulus functions, and $0 < \delta < 1$. If $f_1(t) > \delta$ for $t \in [0, \infty)$, then*

$$(f_2 \circ f_1)(t) \leq \frac{2f_2(1)}{\delta} f_1(t)$$

holds for all $t \in [0, \infty)$.

Theorem 6.10.6. [215, Theorem 2.5] *Let f_1 and f_2 are two modulus functions. Then, the following inclusion relations hold:*

(a) $\ell_p(F, f_1, s) \cap \ell_p(F, f_2, s) \subseteq \ell_p(F, f_1 + f_2, s)$.

(b) If $s > 1$, then $\ell_p(F, f_1, s) \subseteq \ell_p(F, f_2 \circ f_1, s)$.

(c) If $\limsup_{t \to \infty}[f_1(t)/f_2(t)] < \infty$, $\ell_p(F, f_2, s) \subseteq \ell_p(F, f_1, s)$.

(d) If $s_1 \leq s_2$, $\ell_p(F, f_1, s_1) \subseteq \ell_p(F, f_1, s_2)$.

Proof. (a) $\{(f_1 + f_2)[\overline{d}(x_k, \overline{0})]\}^p \leq 2^{p-1}\left\{[f_1[\overline{d}(x_k, \overline{0})]]^p + [f_2[\overline{d}(x_k, \overline{0})]]^p\right\}$ holds, which yields us by taking summation over $k \in \mathbb{N}_0$ that $x \in \ell_p(F, f_1 + f_2, s)$, as desired.

(b) Since f_2 is continuous from the right at 0, there is δ with $0 < \delta < 1$ such that $f_2(t) < \varepsilon$ for all $\varepsilon > 0$ whenever $0 \leq t \leq \delta$. Define the sets N_1 and N_2 by

$$
\begin{aligned}
N_1 &:= \{k \in \mathbb{N}_1 : f_1[\overline{d}(x_k, \overline{0})] \leq \delta\}, \\
N_2 &:= \{k \in \mathbb{N}_1 : f_1[\overline{d}(x_k, \overline{0})] > \delta\}.
\end{aligned}
$$

Then, we obtain from Lemma 6.2.2 for $f_1[\overline{d}(x_k, \overline{0})] > \delta$ that

$$(f_2 \circ f_1)[\overline{d}(x_k, \overline{0})] \leq \frac{2f_2(1)}{\delta} f_1[\overline{d}(x_k, \overline{0})].$$

Therefore, we derive for $x = (x_k) \in \ell_p(F, f_1, s)$ with $s > 1$ that

$$
\begin{aligned}
&\sum_{k \in \mathbb{N}_1} \frac{\{(f_2 \circ f_1)[\overline{d}(x_k, \overline{0})]\}^p}{k^s} \\
&= \sum_{k \in N_1} \frac{\{(f_2 \circ f_1)[\overline{d}(x_k, \overline{0})]\}^p}{k^s} + \sum_{k \in N_2} \frac{\{(f_2 \circ f_1)[\overline{d}(x_k, \overline{0})]\}^p}{k^s} \\
&\leq \sum_{k \in N_1} \frac{\varepsilon^p}{k^s} + \sum_{k \in N_2} \frac{\left\{\dfrac{2f_2(1)}{\delta} f_1[\overline{d}(x_k, \overline{0})]\right\}^p}{k^s} \\
&= \varepsilon^p \sum_{k \in N_1} \frac{1}{k^s} + \left[\frac{2f_2(1)}{\delta}\right]^p \sum_{k \in N_2} \frac{\{f_1[\overline{d}(x_k, \overline{0})]\}^p}{k^s} < \infty.
\end{aligned}
$$

Hence, $x = (x_k) \in \ell_p(F, f_2 \circ f_1, s)$.

(c) Suppose that $\limsup_{t \to \infty}[f_1(t)/f_2(t)] < \infty$. Then, there is a number $M > 0$ such that $[f_1(t)/f_2(t)] \leq M$ for all $t \in [0, \infty)$. Since $\overline{d}(x_k, \overline{0}) \geq 0$ for all $k \in \mathbb{N}_1$ and for all $x = (x_k) \in \ell_p(F, f_2, s)$, we have $f_1[\overline{d}(x_k, \overline{0})] \leq M f_2[\overline{d}(x_k, \overline{0})]$, which leads to

$$
\begin{aligned}
\sum_{k \in \mathbb{N}_1} \frac{\{f_1[\overline{d}(x_k, \overline{0})]\}^p}{k^s} &\leq \sum_{k \in \mathbb{N}_1} \frac{\{M f_2[\overline{d}(x_k, \overline{0})]\}^p}{k^s} \\
&= M^p \sum_{k \in \mathbb{N}_1} \frac{\{f_2[\overline{d}(x_k, \overline{0})]\}^p}{k^s} < \infty.
\end{aligned}
$$

Thus $x = (x_k) \in \ell_p(F, f_1, s)$, as desired.

(d) Let $s_1 \leq s_2$. Since $0 < k^{-1} \leq 1$ for all $k \in \mathbb{N}_1$, it is immediate that $k^{-s_2} \leq k^{-s_1}$. Then, one can see that

$$\sum_{k \in \mathbb{N}_1} \frac{1}{k^{s_2}} \{f_1[\bar{d}(x_k, \bar{0})]\}^p \leq \sum_{k \in \mathbb{N}_1} \frac{1}{k^{s_1}} \{f_1[\bar{d}(x_k, \bar{0})]\}^p < \infty$$

holds for all $x = (x_k) \in \ell_p(F, f_1, s_1)$. This means that $x = (x_k) \in \ell_p(F, f_1, s_2)$ which completes the proof. $\square$

Corollary 6.10.7. [215, Corollary 2.6] *Define the sets $\ell_p(F, s)$ and $\ell_p(F, f)$ by*

$$\ell_p(F, s) := \left\{ x = (x_k) \in \omega(F) : \sum_{k \in \mathbb{N}_1} \frac{1}{k^s} [\bar{d}(x_k, \bar{0})]^p < \infty \right\}; \ (s \geq 0),$$

$$\ell_p(F, f) := \left\{ x = (x_k) \in \omega(F) : \sum_k \{f[\bar{d}(x_k, \bar{0})]\}^p < \infty \right\}.$$

Then, the following statements hold:

(a) If $s > 1$, $\ell_p(F, s) \subseteq \ell_p(F, f, s)$.

(b) $\ell_p(F, f) \subseteq \ell_p(F, f, s)$.

Proof. (a) This follows from Part (b) of Theorem 6.10.6 with $f_1(t) = t$ and $f_2 = f$.

(b) This is immediate from Part (d) of Theorem 6.10.6 with $s_1 = 0$, $s_2 = s$ and $f_1 = f$. $\square$

Now, we can give the next theorem.

Theorem 6.10.8. [215, Theorem 2.7] *Let $s > 1$. Then, the following relations hold:*

(a) $\ell_\infty(F) \subseteq \ell_p(F, f, s)$.

(b) If f is bounded, then $\ell_p(F, f, s) = \omega(F)$.

Proof. (a) Let $x = (x_k) \in \ell_\infty(F)$. Then, there is a number $M > 0$ such that $\bar{d}(x_k, \bar{0}) \leq M$ for all $k \in \mathbb{N}_1$. Since f is continuous and increasing, there is a number $N > 0$ such that $f[\bar{d}(x_k, \bar{0})] \leq f(M) \leq N$. Therefore, we get for $s > 1$ that

$$\sum_{k \in \mathbb{N}_1} \frac{\{f[\bar{d}(x_k, \bar{0})]\}^p}{k^s} \leq N^p \sum_{k \in \mathbb{N}_1} \frac{1}{k^s} < \infty.$$

Hence, $x = (x_k) \in \ell_p(F, f, s)$.

(b) Suppose that f is bounded. Then, one can find a number $N > 0$ such that $f(t) \leq N$ for all $t \in [0, \infty)$. Thus, for $x = (x_k) \in \omega(F)$ we have

$$\sum_{k \in \mathbb{N}_1} \frac{\{f[\overline{d}(x_k, \overline{0})]\}^p}{k^s} \leq N^p \sum_{k \in \mathbb{N}_1} \frac{1}{k^s} < \infty$$

which says that the inclusion $\omega(F) \subseteq \ell_p(F, f, s)$ holds. Since the inclusion $\ell_p(F, f, s) \subseteq \omega(F)$ also holds we conclude that $\omega(F) = \ell_p(F, f, s)$, as desired. $\square$

The final theorem of this section is on the solidity of the spaces $\ell_\infty(F, f)$, $c_0(F, f)$ and $\ell_p(F, f, s)$.

Theorem 6.10.9. [215, Theorem 2.9] *The spaces $\ell_\infty(F, f)$, $c_0(F, f)$ and $\ell_p(F, f, s)$ are solid.*

Proof. This is immediate by Theorem 6.8.13, since the modulus function f is increasing. $\square$

Conclusion

In this chapter, alternating and binomial series of fuzzy numbers with level sets have been introduced. This offers a new tool for the description and analysis of different series of fuzzy numbers with level sets. We have studied the fuzzy analogues of the beginning results related to the power series of real or complex numbers. Quite recently, Kadak and Başar [105] have investigated the power series of fuzzy numbers with real or fuzzy coefficients and related results. Furthermore, it is quite meaningful to extend the results to Fourier series and to the geometric series of fuzzy numbers [107]. We note that following Kadak and Başar [104, 105], we study on the convergence of the series of fuzzy numbers by using the concept of the λ-level set and give the fuzzy analogues of certain convergence tests and results concerning the series of non-negative or arbitrary real terms.

Although some beginning results concerning the series of fuzzy numbers and power series of fuzzy numbers with real or fuzzy coefficients are derived by Kadak and Başar [104], as a natural continuation of the subsection, one can study the alternating series of power series and obtain the Fourier expansion of a fuzzy-valued function for adding some new results to the fuzzy calculus.

By using the sum of the series of λ-level sets, Talo and Başar [213] have recently determined the alpha-, beta- and gamma-duals of the classical sets $\ell_\infty(F)$, $c(F)$, $c_0(F)$ and $\ell_p(F)$ consisting of the bounded, convergent, null and absolutely p-summable fuzzy valued sequences and given the necessary and sufficient conditions on an infinite matrix of fuzzy numbers transforming one of the classical sets to the another one together with certain basic results related to the convergence and absolute convergence of series of fuzzy numbers.

Bibliography

[1] A. M. Akhmedov, F. Başar, *On the fine spectrum of the Cesàro operator in c_0*, Math. J. Ibaraki Univ. **36** (2004), 25–32.

[2] A. M. Akhmedov, F. Başar, *On the spectra of the difference operator Δ over the sequence space ℓ_p*, Demonstratio Math. **39** (3) (2006), 585–595.

[3] A. M. Akhmedov, F. Başar, *On the fine spectra of the difference operator Δ over the sequence space bv_p, $(1 \leq p < \infty)$*, Acta Math. Sin. Eng. Ser. **23** (10) (2007), 1757–1768.

[4] A. M. Akhmedov, S.R. El-Shabrawy, *On the fine spectrum of the operator $\Delta_{a,b}$ over the sequence space c*, Comput. Math. Appl. **61** (10) (2011), 2994–3002.

[5] A. Aizpuru, R. Armario, F. J. García-Pacheco, F. J. Pérez-Fernández, *Vector-valued almost convergence and classical properties in normed spaces*, Proc. Indian Acad. Sci. (Math. Sci.) **124** (1) (2014), 93–108.

[6] B. Altay, *On the space of p-summable difference sequences of order m, $(1 \leq p < \infty)$*, Stud. Sci. Math. Hungar. **43** (4) (2006), 387–402.

[7] B. Altay, F. Başar, *On the paranormed Riesz sequence spaces of non-absolute type*, Southeast Asian Bull. Math. **26** (5) (2002), 701–715.

[8] B. Altay, F. Başar, *On the fine spectrum of the difference operator Δ on c_0 and c*, Inform. Sci. **168** (2004), 217–224.

[9] B. Altay, F. Başar, *On the fine spectrum of the generalized difference operator $B(r, s)$ over the sequence spaces c_0 and c*, Int. J. Math. Math. Sci. **18** (2005), 3005–3013.

[10] B. Altay, F. Başar, *Some paranormed Riesz sequence spaces of non-absolute type*, Southeast Asian Bull. Math. **30** (5) (2006), 591–608.

[11] B. Altay, F. Başar, *Some paranormed sequence spaces of non-absolute type derived by weighted mean*, J. Math. Anal. Appl. **319** (2) (2006), 494–508.

[12] B. Altay, F. Başar, *Generalization of the sequence space $\ell(p)$ derived by weighted mean*, J. Math. Anal. Appl. **330** (1) (2007), 174–185.

271

[13] B. Altay, F. Başar, *The fine spectrum and the matrix domain of the difference operator* Δ *on the sequence space* ℓ_p, $(0 < p < 1)$, Commun. Math. Anal. **2** (2) (2007), 1–11.

[14] B. Altay, F. Başar, E. Malkowsky, *Matrix transformations on some sequence spaces related to strong Cesàro summability and boundedness*, Appl. Math. Comput. **211** (2) (2009), 255–264.

[15] B. Altay, M. Karakuş, *On the spectrum and fine spectrum of the Zweier matrix as an operator on some sequence spaces*, Thai J. Math. **3** (2005), 153–162.

[16] Y. Altın, M. Et, R. Çolak, *Lacunary statistical and lacunary strongly convergence of generalized difference sequences of fuzzy numbers*, Comput. Math. Appl. **52** (2006), 1011–1020.

[17] Y. Altın, M. Mursaleen, H. Altınok, *Statistical summability* $(C, 1)$-*for sequences of fuzzy real numbers and a Tauberian theorem*, J. Intell. Fuzzy Syst. **21** (2010), 379–384.

[18] M. Altun, *On the fine spectra of triangular Toeplitz operators*, Appl. Math. Comput. **217** (20) (2011), 8044–8051.

[19] M. Altun, V. Karakaya, *Fine spectra of lacunary matrices*, Commun. Math. Anal. **7** (2009), 1–10.

[20] G. A. Anastassiou, *Fuzzy Mathematics: Approximation Theory*, Studies in Fuzziness and Soft Computing, Vol: 251. Springer-Verlag, Berlin, 2010.

[21] J. Appell, E. Pascale, A. Vignoli, *Nonlinear Spectral Theory*, Walter de Gruyter, Berlin · New York, 2004.

[22] Ç. Asma, R. Çolak, *On the Köthe-Toeplitz duals of some generalized sets of difference sequences*, Demonstratio Math. **33** (2000), 797–803.

[23] C. Aydın, F. Başar, *Some generalizations of the sequence space* a_p^r, Iran. J. Sci. Technol. Trans. A, Sci. **30** (2006), No. A2, 175–190.

[24] S. Aytar, *Statistical limit points of sequences of fuzzy numbers*, Inform. Sci. **165** (2004), 129–138.

[25] S. Aytar, S. Pehlivan, *Statistical cluster and extreme limit points of sequences of fuzzy numbers*, Inform. Sci. **177** (16) (2007), 3290–3296.

[26] S. Aytar, M. Mammadov, S. Pehlivan, *Statistical limit inferior and limit superior for sequences of fuzzy numbers*, Fuzzy Sets Syst. **157** (7) (2006), 976–985.

[27] P. Baliarsingh, *Some new difference sequence spaces of fractional order and their dual spaces*, Appl. Math. Comput. **219** (2013), 9737–9742.

[28] S. Banach, *Théorie des Operations Lineaires*, Warszawa, 1932.

[29] J. Banaś, M. Mursaleen, *Sequence Spaces and Measures of Noncompactness with Applications to Differential and Integral Equations*, Springer, 2014.

[30] F. Başar, *Infinite matrices and almost boundedness*, Boll. Unione Mat. Ital. Sez. A (7) 6 (1992), 395–402.

[31] F. Başar, *Dual summability methods with a new approach*, Modern Methods in Analysis and its Applications, (Ed. M. Mursaleen), **2010**, Anamaya Publishers, New Delhi, pp. 56–67.

[32] F. Başar, *Summability Theory and its Applications*, Bentham Science Publishers, e-books, Monograph, İstanbul, 2012.

[33] F. Başar, *Linear Algebra*, Sürat Üniversite Yayınları, 3rd ed. vii+460 pages, İstanbul, 2012, ISBN: 9-786055-301071, (in Turkish).

[34] F. Başar, B. Altay, *Matrix mappings on the space bs(p) and its α, β and γ-duals*, Aligarh Bull. Math. **21** (1) (2001), 79–91.

[35] F. Başar, B. Altay, *On the space of sequences of p-bounded variation and related matrix mappings*, Ukrain. Mat. Zh. **55** (1) (2003), 108–118; reprinted in Ukrainian Math. J. **55** (1) (2003), 136–147.

[36] F. Başar, B. Altay, M. Mursaleen, *Some generalizations of the space bv_p of p-bounded variation sequences*, Nonlinear Anal. **68** (2) (2008), 273–287.

[37] F. Başar, N. Durna, M. Yıldırım, *Subdivisions of the spectra for difference operator over certain sequence spaces*, Malays. J. Math. Sci. **6** (S) (2012), 151–165.

[38] F. Başar, A. Karaisa, *On the fine spectrum of the generalized difference operator defined by a double sequential band matrix over the sequence space ℓ_p, $(1 < p < \infty)$*, Hacet. J. Math. Stat. **44** (6) (2015), 1315–1332.

[39] F. Başar, M. Kirişçi, *Almost convergence and generalized difference matrix*, Comput. Math. Appl. **61** (3) (2011), 602–611.

[40] F. Başar, İ. Solak, *Almost-coercive matrix transformations*, Rend. Mat. Appl. (7) **11** (2) (1991), 249–256.

[41] M. Başarır, *On some new sequence spaces and related matrix transformations*, Indian J. Pure Appl. Math. **26** (10) (1995), 1003–1010.

[42] M. Başarır, *On the generalized Riesz B-difference sequence spaces*, Filomat **24** (4) (2010), 35–52.

[43] M. Başarır, M. Kayıkçı, *On the generalized B^m-Riesz sequence space and property*, J. Inequal. Appl. **2009** Article ID 385029, 18 pages.

[44] M. Başarır, M. Öztürk, *On the Riesz difference sequence space*, Rend. Circ. Mat. Palermo (2) **57** (2008), 377–389.

[45] B. Bede, S. G. Gal, *Almost periodic fuzzy-number-valued functions*, Fuzzy Sets Syst. **147** (2004), 385–403.

[46] Ç. A. Bektas, *On some new generalized sequence spaces*, J. Math. Anal. Appl. **277** (2003), 681–688.

[47] T. D. Benavides, *Set-contractions and ball-contractions in some classes of spaces*, J. Math. Anal. Appl. **136** (1) (1988), 131–140.

[48] G. Bennett, N. J. Kalton, *Consistency theorems for almost convergence*, Trans. Amer. Math. Soc. **198** (1974), 23–43.

[49] R. Bernatz, *Fourier Series and Numerical Methods For Partial Differential Equations*, Luther College, Published by John Wiley & Sons Inc., Hoboken · New Jersey, 2010.

[50] H. Bilgiç, H. Furkan, *On the fine spectrum of the operator $B(r, s, t)$ over the sequence spaces ℓ_1 and bv*, Math. Comput. Modelling **45** (2007), 883–891.

[51] H. Bilgiç, H. Furkan, *On the fine spectrum of the generalized difference operator $B(r, s)$ over the sequence spaces ℓ_p and bv_p $(1 < p < \infty)$*, Nonlinear Anal. **68** (3) (2008), 499–506.

[52] J. Boos, *Classical and Modern Methods in Summability*, Oxford University Press, New York, 2000.

[53] P. J. Cartlidge, Weighted mean matrices as operators on ℓ^p, Ph.D. Dissertation, Indiana University, 1978.

[54] F. P. Cass, B. E. Rhoades, *Mercerian theorems via spectral theory*, Pacific J. Math. **73** (1977), 63–71.

[55] Y. Chalco-Cano, H. Román-Flores, *On new solutions of fuzzy differential equations*, Chaos Solitons Fractals **38** (2008), 112–119.

[56] S. T. Chen, *Geometry of Orlicz Spaces*, Dissertations Math. **CCCLVI**, Warszawa, 1996.

[57] B. Choudhary, S. Nanda, *Functional Analysis with Applications*, John Wiley & Sons Inc. New York, 1989.

[58] J. Christopher, *The asymptotic density of some k-dimensional sets*, Amer. Math. Monthly **63** (1956), 399–401.

[59] J. A. Clarkson, *Uniformly convex spaces*, Trans. Amer. Math. Soc. **40** (1936), 396–414.

[60] R. G. Cooke, *Infinite Matrices and Sequence Spaces*, Dover Publication Inc., New York, 1955.

[61] R. Çolak, M. Et, *On some generalized difference sequence spaces and related matrix transformations*, Hokkaido Math. J. **26** (3) (1997), 483–492.

[62] R. Çolak, M. Et, E. Malkowsky, *Some Topics of Sequence Spaces*, Lecture Notes in Mathematics, Fırat Univ. Elâzığ, Turkey, 2004, pp. 1–63, Fırat Univ. Press, 2004, ISBN: 975-394-038-6.

[63] J. S. Connor, *R-type summability methods, Cauchy criteria, p-sets, and statistical convergence*, Proc. Amer. Math. Soc. **115** (1992), 319–327.

[64] C. Coşkun, *The spectra and fine spectra for p-Cesàro operators*, Turkish J. Math. **21** (1997), 207–212.

[65] Y. Cui, H. Hudzik, *On the Banach-Saks and weak Banach-Saks properties of some Banach sequence spaces*, Acta Sci. Math. (Szeged) **65** (1999), 179–187.

[66] Y. Cui, H. Hudzik, *Some geometric properties related to fixed point theory in Cesàro spaces*, Collect. Math. **50** (3) (1999), 277–288.

[67] Y. Cui, H. Hudzik, *Packing constant for Cesàro sequence spaces*, Nonlinear Anal. **47** (4) (2001), 2695–2702.

[68] Y. A. Cui, C. Meng, *Banach-Saks property and property (β) in Cesàro sequence spaces*, Southeast Asian Bull. Math. **24** (2000), 201–210.

[69] L. Crone, *A characterization of matrix mappings on ℓ_2*, Math. Z. **123** (1971), 315–317.

[70] P. Diamond, P. Kloeden, *Metric spaces of fuzzy sets*, Fuzzy Sets Syst. **35** (1990), 241–249.

[71] J. Diestel, *Sequence and Series in Banach Spaces*, in: Graduate Texts in Math. **92**, Springer-Verlag, 1984.

[72] J. P. Duran, *Infinite matrices and almost convergence*, Math. Z. **128**, (1972), 75–83.

[73] J. P. Duran, *Almost convergence, summability and ergodicity*, Canad. J. Math. **26** (1974), 372–387.

[74] N. Durna, M. Yildirim, *Subdivision of the spectra for factorable matrices on c and ℓ^p*, Math. Commun. **16** (2) (2011), 519–530.

[75] E. Dündar, F. Başar, *On the fine spectrum of the upper triangle double band matrix* Δ^+ *on the sequence space* c_0, Math. Commun. **18** (2013), 337–348.

[76] C. Eizen, G. Laush, *Infinite matrices and almost convergence*, Math. Japon. **14** (1969), 137–143.

[77] P. Enflo, *A counter example to the approximation property*, Acta Math. **130** (1973), 309–317.

[78] M. Et, *On some difference sequence spaces*, Turkish J. Math. **17** (1993), 18–24.

[79] M. Et, *On some generalized Cesàro difference sequence spaces*, İstanbul Üniv. Fen Fak. Mat. Derg. **55-56** (1996-1997), 221–229.

[80] M. Et, M. Başarır, *On some new generalized difference sequence spaces*, Period. Math. Hung. **35** (3) (1997), 169–175.

[81] J. Fang, H. Huang, *On the level convergence of sequence of fuzzy numbers*, Fuzzy Sets Syst. **147** (2004), 417–435.

[82] A. Farés, B. de Malafosse, *Spectra of the operator of the first difference in* s_α, s_α^0, $s_\alpha^{(c)}$ *and* $l_p(\alpha)$ $(1 \le p < \infty)$ *and application to matrix transformations*, Demonstratio Math. **41** (3) (2008), 661–676.

[83] H. Fast, *Sur la convergence statistique*, Colloq. Math. **2** (1951), 241–244.

[84] J. A. Fridy, M. K. Khan, *Statistical extensions of some classical Tauberian theorems*, Proc. Amer. Math. Soc. **128** (2000), 2347–2355.

[85] H. Furkan, H. Bilgiç, B. Altay, *On the fine spectrum of the operator* $B(r, s, t)$ *over* c_0 *and* c, Comput. Math. Appl. **53** (2007), 989–998.

[86] H. Furkan, H. Bilgiç, F. Başar, *On the fine spectrum of the operator* $B(r, s, t)$ *over the sequence spaces* ℓ_p *and* bv_p, $(1 < p < \infty)$, Comput. Math. Appl. **60** (7) (2010), 2141–2152.

[87] H. Furkan, H. Bilgiç, K. Kayaduman, *On the fine spectrum of the generalized difference operator* $B(r, s)$ *over the sequence spaces* ℓ_1 *and* bv, Hokkaido Math. J. **35** (2006), 897–908.

[88] J. Garcia-Falset, *Stability and fixed points for nonexpansive mappings*, Houston J. Math. **20** (1994), 495–505.

[89] J. Garcia-Falset, *The fixed point property in Banach spaces with NUS-property*, J. Math. Anal. Appl. **215** (2) (1997), 532–542.

[90] R. Goetschel, W. Voxman, *Elementary fuzzy calculus*, Fuzzy Sets Syst. **18** (1986), 31–43.

[91] S. Goldberg, *Unbounded Linear Operators*, Dover Publications Inc. New York, 1985.

[92] Z. Gong, C. Wu, *Bounded variation, absolute continuity and absolute integrability for fuzzy-number-valued functions*, Fuzzy Sets Syst. **129** (2002), 83–94.

[93] M. Gonzàlez, *The fine spectrum of the Cesàro operator in ℓ_p ($1 < p < \infty$)*, Arch. Math. **44** (1985), 355–358.

[94] V. I. Gurariĭ, *On moduli of convexity and flattering of Banach spaces*, Soviet Math. Dokl. **161** (5) (1965), 1003–1006.

[95] V. I. Gurariĭ, *On differential properties of the convexity moduli of Banach spaces*, Math. Issled. **2** (1969), 141–148.

[96] N. I. Gurariĭ, Y. U. Sozonov, *Normed spaces that do not have distortion of the unit sphere*, Mat. Zametki **7** (1970), 307–310, (Russian).

[97] G. H. Hardy, *Divergent Series*, Oxford, 1956.

[98] H. Hudzik, V. Karakaya, M. Mursaleen, N. Şimsek, *Banach-Saks type and Gurariĭ modulus of convexity of some Banach sequence spaces*, Abstract Appl. Anal. **2014**, Article ID 427382, 9 pages.

[99] M. Hukuhara, *Intégration des applications mesurables dont la valeur est un compact convex*, Funkcial. Ekvac **10** (1967), 205–229.

[100] A. Jakimovski, D. C. Russell, *Matrix mappings between BK-spaces*, Bull. London Math. Soc. **4** (3) (1972), 345–353.

[101] R. C. James, *Characterizations of reflexivity*, Studia Math. **23** (1964), 205–216.

[102] P. D. Johnson Jr., R. N. Mohapatra, *On inequalities related to sequence spaces ces[p, q]*, in: General Inequalities 4 (4th Oberwolfach Conf. 1983), Inter. Series Numer. Math. 71, Birkhäuser Verlag, Besel, 1984, pp. 191–201.

[103] U. Kadak, *On the sets of fuzzy-valued function with the level sets*, J. Fuzzy Set Valued Anal. **2013** (2013), Article ID jfsva-00171, 13 pages, doi: 10.5899/2013/jfsva-00171.

[104] U. Kadak, F. Başar, *Power series of fuzzy numbers*, AIP Conference Proceedings **1309** (2010), 538–550.

[105] U. Kadak, F. Başar, *Power series of fuzzy numbers with reel or fuzzy coefficients*, Filomat **25** (3) (2012), 519–528.

[106] U. Kadak, F. Başar, *On some sets of fuzzy-valued sequences with the level sets*, Contemp. Anal. Appl. Math. **1** (2) (2013), 70–90.

[107] U. Kadak, F. Başar, *On Fourier series of fuzzy-valued functions*, The Scientific World Journal, **2014**, Article ID 782652, 13 pages, 2014.

[108] U. Kadak, F. Başar, *Alternating and binomial series of fuzzy numbers with the level sets*, Contemp. Anal. Appl. Math. **3** (2) (2015), 310–328.

[109] S. Kakutani, *Weak convergence in uniformly convex spaces*, Tohoku Math. J. **45** (1938), 188–193.

[110] O. Kaleva, S. Seikkala, *On fuzzy metric spaces*, Fuzzy Sets Syst. **12** (1984), 215–229.

[111] P. K. Kamthan, M. Gupta, *Sequence Spaces and Series*, Marcel Dekker Inc. New York and Basel, 1981.

[112] E. E. Kara, *Some topological and geometrical properties of new Banach sequence spaces*, J. Inequal. Appl. **2013**, 2013: 38.

[113] E. E. Kara, M. Öztürk, M. Başarır, *Some topological and geometric properties of generalized Euler sequence spaces*, Math. Slovaca **60** (3) (2010), 385–398.

[114] A. Karaisa, *Fine spectra of upper triangular double-band matrices over the sequence space ℓ_p, $(1 < p < \infty)$*, Discrete Dyn. Nature Soc. **2012** Article ID 381069 (2012).

[115] A. Karaisa, F. Başar, *Fine spectra of upper triangular triple-band matrix over the sequence space ℓ_p where $(0 < p < 1)$*, AIP Conference Proceedings **1470** (2012), 134–137.

[116] A. Karaisa, F. Başar, *Fine spectra of upper triangular triple-band matrices over the sequence space ℓ_p, $(0 < p < \infty)$*, Abstr. Appl. Anal. **2013** Article ID 342682, 10 pages, 2013. doi:10.1155/2013/342682.

[117] V. Karakaya, *Some geometric properties of sequence spaces involving lacunary sequence*, J. Inequal. Appl. (2007), Article ID 81028, doi:10.1155/2007/81028.

[118] V. Karakaya, M. Altun, *Fine spectra of upper triangular double-band matrices*, J. Comput. Appl. Math. **234** (2010), 1387–1394.

[119] V. Karakaya, F. Altun, *On some geometric properties of a new paranormed sequence space*, J. Funct. Spaces Appl. **2014**, Article ID 685382, 8 pages.

[120] V. Karakaya, M. Dzh. Manafov, N. Şimşek, *On the fine spectrum of the second order difference operator over the sequence spaces ℓ_p and bv_p*, Math. Comput. Modelling **55** (3) (2012), 426–436.

[121] K. Kayaduman, H. Furkan, *The fine spectra of the difference operator* Δ *over the sequence spaces* ℓ_1 *and* bv, Int. Math. Forum **1** (24) (2006), 1153–1160.

[122] F. M. Khan, M. F. Rahman, *Matrix transformations on Cesàro sequence spaces of nonabsolute type*, J. Analysis **4** (1996), 97–101.

[123] Y. K. Kim, B. M. Ghil, *Integrals of fuzzy-number-valued functions*, Fuzzy Sets Syst. **86** (1997), 213–222.

[124] J. P. King, *Almost summable sequences*, Proc. Amer. Math. Soc. **17** (1966), 1219–1225.

[125] J. P. King, *The Lototsky transform and Bernstein polynomials*, Canad. J. Math. **18** (1966), 89–91.

[126] J. P. King, J. J. Swetits, *Positive linear operators and summability*, J. Australian Math. Soc. **11** (1970), 281–290.

[127] M. Kirişçi, F. Başar, *Some new sequence spaces derived by the domain of generalized difference matrix*, Comput. Math. Appl. **60** (5) (2010), 1299–1309.

[128] H. Knaust, *Orlicz sequence spaces of Banach-Saks type*, Arch. Math. **59** (1992), 562–565.

[129] K. Knopp, *Infinite Sequences and Series*, Dover Publications Inc. New York, 1956.

[130] K. Knopp, *Theory and Application of Infinite Series*, Dover Publications Inc. New York, 1990.

[131] J. Korevaar, *Tauberian Theorems: A Century of Development*, Springer, New York, 2014.

[132] P. P. Korovkin, *Linear Operators and Approximation Theory*, Hindustan Publ. Co., Delhi, 1960.

[133] T. Koshy, *Fibonacci and Lucas Numbers with Applications*, Wiley, 2001.

[134] E. Kreyszig, *Introductory Functional Analysis with Applications*, John Wiley & Sons Inc. New York · Chichester · Brisbane · Toronto, 1978.

[135] D. N. Kutzarova, $k-\beta$ *and* $k-NUC$ *Banach spaces*, C. R. Acad. Bulgare Sci. **43** (1990), no. 3, 13–15.

[136] J. S. Kwon, *On statistical and p-Cesàro convergence of fuzzy numbers*, Korean J. Comput. Appl. Math. **7** (2000), 195–203.

[137] C. G. Lascarides, I. J. Maddox, *Matrix transformations between some classes of sequences*, Proc. Camb. Phil. Soc. **68** (1970), 99–104.

[138] H. Li, C. Wu, *The integral of a fuzzy mapping over a directed line*, Fuzzy Sets Syst. **158** (2007), 2317–2338.

[139] B. V. Limaye, *Functional Analysis*, New Age International (P) Ltd., Publishers, New Delhi, Second Edition 1996, Reprint 2008.

[140] G. G. Lorentz, *A contribution to the theory of divergent sequences*, Acta Math. **80** (1948), 167–190.

[141] G. G. Lorentz, K. Zeller, *Summation of sequences and summations of series*, Proc. Amer. Math. Soc. **15** (1964), 743–746.

[142] I. J. Maddox, *Spaces of strongly summable sequences*, Quart. J. Math. Oxford (2) **18** (1967), 345–355.

[143] I. J. Maddox, *Paranormed sequence spaces generated by infinite matrices*, Proc. Comb. Phil. Soc. **64** (1968), 335–340.

[144] I. J. Maddox, *On strong almost convergence*, Math. Proc. Camb. Phil. Soc. **85** (1979), 345–350.

[145] I. J. Maddox, *Sequence spaces defined by a modulus*, Math. Proc. Camb. Phil. Soc. **100** (1986), 161–166.

[146] I. J. Maddox, *A Tauberian theorem for ordered spaces*, BUSEFAL, **28** (1986), 28–37.

[147] I. J. Maddox, *Elements of Functional Analysis*, The University Press, 2nd ed., Cambridge, 1988.

[148] B. de Malafosse, *Properties of some sets of sequences and application to the spaces of bounded difference sequences of order μ*, Hokkaido Math. J. **31** (2002), 283–299.

[149] B. de Malafosse, E. Malkowsky, *Sets of difference sequences of order m*, Acta Sci. Math. (Szeged) **70** (3–4) (2004), 659–682.

[150] E. Malkowsky, *Klassen von Matrix abbildungen in paranormierten FK–Raumen*, Analysis **7** (1987), 275–292.

[151] E. Malkowsky, *Modern functional analysis in the theory of sequence spaces and matrix transformations*, Jordan J. Math. Stat. **1** (1) (2008), 1–29.

[152] E. Malkowsky, *Compact matrix operators between some BK spaces*, (Ed. M. Mursaleen), Modern Methods in Analysis and its Applications, **2010**, Anamaya Publishers, New Delhi, pp. 86–120.

[153] E. Malkowsky, M. Mursaleen, *Matrix transformations between FK-spaces and the sequence spaces $m(\phi)$ and $n(\phi)$*, J. Math. Anal. Appl. **196** (1995), no. 2, 659–665.

[154] E. Malkowsky, V. Rakočević, *An introduction into the theory of sequence spaces and measures of noncompactness*, Zbornik Radova, Mat. Institut SANU (Beograd) **9** (17) (2000), 143–234.

[155] E. Malkowsky, V. Rakočević, *On matrix domains of triangles*, Appl. Math. Comput. **189** (2) (2007), 1146–1163.

[156] E. Malkowsky, V. Rakočević, S. Zivkovic, *Matrix transformations between the sequence space bv_p and certain BK spaces*, Bull. Acad. Serbe Sci. Arts Math. **123** (27) (2002), 33–46.

[157] M. Mares, *Weak arithmetics of fuzzy numbers*, Fuzzy Sets Syst. **91** (1997), 143–153.

[158] M. Matloka, *Sequences of fuzzy numbers*, BUSEFAL **28** (1986), 28–37.

[159] M. Matloka, *Fuzzy mappings-sequences and series*, Institute of Economical Cybernetics, Department of Mathematics **146/150**, 60–967, Poland.

[160] M. Matloka, *Sequence of fuzzy numbers*, Analysis **9** (1989), 297–302.

[161] F. M. Mears, *The inverse Nörlund mean*, Annals Math. **44** (3) (1943), 401–410.

[162] S. A. Mohiuddine, *An application of almost convergence in approximation theorems*, Appl. Math. Lett. **24** (2011), 1856–1860.

[163] S. A. Mohiuddine, A. Alotaibi, M. Mursaleen, *Statistical summability $(C,1)$ and a Korovkin type approximation theorem*, J. Inequal. Appl. **2012**, 2012:172, doi:10.1186/1029-242X-2012-172.

[164] F. Móricz, *Ordinary convergence follows from statistical summability $(C,1)$ in the case of slowly decreasing or oscillating sequences*, Colloq. Math. **99** (2004), 207–219.

[165] M. Mursaleen, *Generalized spaces of difference sequences*, J. Math. Anal. Appl. **203** (3) (1996), 738–745.

[166] M. Mursaleen, *On some geometric properties of a sequence space related to ℓ_p*, Bull. Australian Math. Soc. **67** (2003), no. 2, 343–347.

[167] M. Mursaleen, *Some matrix transformations on sequence spaces of invariant means*, Hacettepe J. Math. Stat. **38** (3) (2009), 259-264.

[168] M. Mursaleen, *Applied Summability Methods*, Springer Briefs, 2014.

[169] M. Mursaleen, F. Başar, B. Altay, *On the Euler sequence spaces which include the spaces ℓ_p and ℓ_∞ II*, Nonlinear Anal. **65** (3) (2006), 707–717.

[170] M. Mursaleen, M. Başarır, On some new sequence spaces of fuzzy numbers, Indian J. Pure Appl. Math. **34** (9) (2003), 1351–1357.

[171] M. Mursaleen, A. K. Noman, *On generalized means and some related sequence spaces*, Comput. Math. Appl. **61** (2011), 988–999.

[172] H. Nakano, *Modulared sequence spaces*, Proc. Japan Acad. **27** (2) (1951), 508–512.

[173] H. Nakano, *Concave modulars*, J. Math. Soc. Japan **5** (1953), 29–49.

[174] S. Nanda, On sequences of fuzzy numbers, Fuzzy Sets Syst. **33** (1989), 123–126.

[175] M. S. El Naschie, *A review of $\varepsilon^{(\infty)}$ theory and the mass spectrum of high energy particle physics*, Chaos, Solitons & Fractals **19** (1) (2004), 209–236.

[176] P. N. Natarajan, Criterion for regular matrices in non-archimedean fields, J. Ramanujan Math. Soc. **6** (1991), 185-195.

[177] P. N. Ng, P. Y. Lee, *Cesàro sequence spaces of non-absolute type*, Comment. Math. Prace Mat. **20** (2) (1978), 429–433.

[178] F. Nuray, E. Savaş, *Statistical convergence of sequences of fuzzy numbers*, Math. Slovaca **45** (3) (1995), 269–273.

[179] J. I. Okutoyi, *On the spectrum of C_1 as an operator on bv_0*, J. Austral. Math. Soc. Ser. A. **48** (1990), 79–86.

[180] J. I. Okutoyi, *On the spectrum of C_1 as an operator on bv*, Commun. Fac. Sci. Univ. Ank. Ser. A_1. **41** (1992), 197–207.

[181] Z. Opial, *Weak convergence of the sequence of successive approximations for non expensive mappings*, Bull. Amer. Math. Soc. **73** (1967), 591–597.

[182] C. Orhan, E. Öztürk, *On f-regular dual summability methods*, Bull. Inst. Math. Acad. Sinica **14** (1) (1986), 99–104.

[183] E. Öztürk, *On strongly regular dual summability methods*, Comm. Fac. Sci. Univ. Ank. Ser. A_1 Math. Statist. **32** (1983), 1–5. Octobre 1989.

[184] A. Peyerimhoff, *Über ein Lemma von Herrn Chow*, J. London Math. Soc. **32** (1957), 33–36.

[185] A. Peyerimhoff, *Lectures on Summability*, Lecture Notes in Mathematics, Springer-Verlag, Berlin, 1969.

[186] H. Polat, F. Başar, *Some Euler spaces of difference sequences of order m*, Acta Math. Sci. Ser. B Engl. Ed. **27B** (2) (2007), 254–266.

[187] S. Prus, *Banach spaces with uniform Opial property*, Nonlinear Anal. **8** (1992), 697–704.

[188] M. L. Puri, D. A. Ralescu, *Differentials for fuzzy functions*, J. Math. Anal. Appl. **91** (1983), 552–558.

[189] J.B. Reade, *On the spectrum of the Cesaro operator*, Bull. Lond. Math. Soc. **17** (1985), 263–267.

[190] B. E. Rhoades, *The fine spectra for weighted mean operators*, Pacific J. Math. **104** (1) (1983), 219–230.

[191] M. Rojes-Medar, H. Roman-Flores, *On the equivalence of convergences of fuzzy sets*, Fuzzy Sets Syst. **80** (1996), 217–224.

[192] W. Sanhan, S. Suantai, *Some geometric properties of Cesàro sequence space*, Kyungpook Math. J. **43** (2003), 191–197.

[193] W. L. C. Sargent, *Some sequence spaces related to the l^p spaces*, J. London Math. Soc. **35** (1960), 161–171.

[194] W. L. Sargent, *On sectionally bounded BK spaces*, Math. Z. **83** (1964), 57–66.

[195] W. L. Sargent, *On compact matrix transformations between sectionally bounded BK-spaces*, J. London Math. Soc. **41** (1) (1966), 79–87.

[196] B. Sarma, *On a class of sequences of fuzzy numbers defined by modulus function*, Internat. J. Sci. Technol. **2** (1) (2007), 25–28.

[197] S. Simons, *The sequence spaces $\ell(p_v)$ and $m(p_v)$*, Proc. London Math. Soc. (3), **15** (1965), 422–436.

[198] P. Schaefer, *Almost convergent and almost summable sequences*, Proc. Amer. Math. Soc. **20** (1) (1969), 51-54.

[199] P. Schaefer, *Infinite matrices and invariant means*, Proc. Amer. Math. Soc. **36** (1972), 104–110.

[200] E. Savaş, *On statistical convergent sequences of fuzzy numbers*, Inform. Sci. **137** (2001), 277–282.

[201] R. Schmidt, *Uber divergente Folgen und Mittelbildungen*, Math. Z. **22** (1925), 89–152.

[202] J. S. Shiue, *On the Cesàro sequence spaces*, Tamkang J. Math. **1** (1970), 143–150.

[203] B. Sims, *A class of spaces with weak normal structure*, Bull. Austral. Math. Soc. **50** (1994), 523–528.

[204] P. D. Srivastava, S. Kumar, *Fine spectrum of the generalized difference operator Δ_ν on sequence space ℓ_1*, Thai J. Math. **8** (2) (2010), 7–19.

[205] P. D. Srivastava, S. Kumar, *Fine spectrum of the generalized difference operator* Δ_{uv} *on sequence space* l_1, Appl. Math. Comput. **218** (11) (2012), 6407–6414.

[206] L. Stefanini, *A Generalization of Hukuhara Difference For Interval and Fuzzy Arithmetic*, D. Dubois, M. A. Lubiano, H. Prade, M. A. Gil, P. Grzegorzewski, O. Hryniewicz (Eds.), Soft Methods for Handling Variability and Imprecision, Series on Advances in Soft Computing, vol. 48, Springer, Berlin, 2008.

[207] L. Stefanini, B. Bede, *Generalized Hukuhara differentiability of interval-valued functions and interval differential equations*, Nonlinear Anal. **71** (2009), 1311–1328.

[208] M. Stojaković, Z. Stojaković, *Addition and series of fuzzy sets*, Fuzzy Sets Syst. **83**(1996), 341–346.

[209] M. Stojakovic, Z. Stojakovic, *Series of fuzzy sets*, Fuzzy Sets Syst. **160** (21) (2009), 3115–3127.

[210] P. V. Subrahmanyam, *Cesàro summability of fuzzy real numbers*, J. Anal. **7** (1999), 159–168.

[211] M. Şengönül, F. Başar, *Some new Cesàro sequence spaces of non-absolute type which include the spaces* c_0 *and* c, Soochow J. Math. **31** (1) (2005), 107–119.

[212] N. Şimşek, V. Karakaya, *Structure and some geometric properties of generalized Cesaro sequence space*, Int. J. Contemp. Math. Sci. **3** (8) (2008), 389–399.

[213] Ö. Talo, F. Başar, *On the space* $bv_p(F)$ *of sequences of p-bounded variation of fuzzy numbers*, Acta Math. Sin. Eng. Ser. **24** (7) (2008), 1205–1212.

[214] Ö. Talo, F. Başar, *Determination of the duals of classical sets of sequences of fuzzy numbers and related matrix transformations*, Comput. Math. Appl. **58** (2009), 717–733.

[215] Ö. Talo, F. Başar, *Certain spaces of sequences of fuzzy numbers defined by a modulus function*, Demonstratio Math. **43** (1) (2010), 139–149.

[216] Ö. Talo, F. Başar, *Quasilinearity of the classical sets of sequences of the fuzzy numbers and some applications*, Taiwanese J. Math. **14** (5) (2010), 1799–1819.

[217] Ö. Talo, F. Başar, *On the slowly decreasing sequences of fuzzy numbers*, Abstr. Appl. Anal. **2013**, Article ID 891986, 7 pages, 2013.

[218] Ö. Talo, C. Çakan, *On the Cesàro convergence of sequences of fuzzy numbers*, Appl. Math. Lett. **25** (4) (2012), 676–681.

[219] Ö. Talo, U. Kadak, F. Başar, *On series of fuzzy numbers*, Contemp. Anal. Appl. Math. **4** (1) (2016), 132–155.

[220] B. S. Thomson, J. B. Bruckner, A. M. Bruckner, *Elementary Real Analysis*, Second edition, Prentice Hall (Pearson), 2008.

[221] W. F. Trench, *Introduction to Real Analysis*, Free Edition 1, March 2009.

[222] B. C. Tripathy, M. Sen, *On a new class of sequences related to the space l^p*, Tamkang J. Math. **33** (2) (2002), 167–171.

[223] C. S. Wang, *On Nörlund sequence space*, Tamkang J. Math. **9** (1978), 269–274.

[224] J. R. L. Webb, W. Zhao, *On connections between set and ball measures of noncompactness*, Bull. London Math. Soc. **22** (1990), 471–477.

[225] L. Wen, *A nowhere differentiable continuous function constructed by infinite products*, Amer. Math. Monthly **109** (4) (2002), 378–380.

[226] R. B. Wenger, *The fine spectra of Hölder summability operators*, Indian J. Pure Appl. Math. **6** (1975), 695–712.

[227] A. Wilansky, *Functional Analysis*, Blaisdell, New York, 1964.

[228] A. Wilansky, *Summability through Functional Analysis*, North-Holland Mathematics Studies **85**, Amsterdam · New York · Oxford, 1984.

[229] J. Wu, *Lecture Notes of Advanced Calculus II*, Department of Mathematics of National University of Singapore.

[230] C. -x. Wu, C. Wu, *Some notes on the supremum and infimum of the set of fuzzy numbers*, Fuzzy Sets Syst. **103** (1999), 183-187.

[231] M. Yeşilkayagil, F. Başar, *On the fine spectrum of the operator defined by a lambda matrix over the sequence space c_0 and c*, AIP Conference Proceedings **1470** (2012), 199–202.

[232] M. Yeşilkayagil, F. Başar, *On the fine spectrum of the operator defined by a lambda matrix over the sequence spaces of null and convergent sequences*, Abstr. Appl. Anal. **2013**, Article ID 687393, 13 pages, 2013. doi: 10.1155/2013/687393.

[233] M. Yeşilkayagil, F. Başar, *On the paranormed Nörlund sequence space of non-absolute type*, Abstr. Appl. Anal. **2014**, Article ID 858704, 9 pages, 2014. doi:10.1155/2014/858704.

[234] M. Yeşilkayagil, F. Başar, *A survey for the spectrum of triangles over sequence spaces*, Numer. Funct. Anal. Optim., (in press).

[235] M. Yıldırım, *On the spectrum and fine spectrum of the compact Rhally operators*, Indian J. Pure Appl. Math. **27** (8) (1996), 779–784.

[236] L. A. Zadeh, *Fuzzy sets*, Information and Control **8** (1965), 338–353.

[237] C. Zanco, A. Zucchi, *Moduli of rotundity and smoothness for convex bodies*, Boll. Unione Mat. Ital. Sez. B (7) 7 (1993), 833–855.

[238] K. Zeller, *Allgemeine Eigenschaften von Limitierungsverfahren*, Math. Z. **53** (1951), 463–487.

[239] K. Zeller, W. Beekmann, *Theorie der Limiierungsverfahren (2. Aufl.)*, Springer, Berlin, 1970.

Index

Printed in the United States
by Baker & Taylor Publisher Services